Genes and Viruses in Multiple Sclerosis

GENES AND VIRUSES IN MULTIPLE SCLEROSIS

EDITED BY

O.R. HOMMES
European Charcot Foundation, Heiweg 97, 6533 PA, Nijmegen, The Netherlands

M. CLANET
Fédération de Neurologie, CHU Toulouse Purpan, 31059 Toulouse Cedex, France

H. WEKERLE
*Max-Planck-Institut für Neurobiologie, Am Klopferspitz 18a, D-82152,
Planegg-Martinsried, Germany*

2001
ELSEVIER
Amsterdam – Lausanne – New York – Oxford – Shannon – Singapore – Tokyo

ELSEVIER SCIENCE B.V.
Sara Burgerhartstraat 25
P.O. Box 211, 1000 AE Amsterdam, The Netherlands

© 2001 Elsevier Science B.V. All rights reserved.

This work is protected under copyright by Elsevier Science, and the following terms and conditions apply to its use:

Photocopying
Single photocopies of single chapters may be made for personal use as allowed by national copyright laws. Permission of the Publisher and payment of a fee is required for all other photocopying, including multiple or systematic copying, copying for advertising or promotional purposes, resale, and all forms of document delivery. Special rates are available for educational institutions that wish to make photocopies for non-profit educational classroom use.

Permissions may be sought directly from Elsevier Science Rights & Permissions Department, PO Box 800, Oxford OX5 1DX, UK; phone: (+44) 1865 843830, fax: (+44) 1865 853333, e-mail: permissions@elsevier.co.uk. You may also contact Rights & Permissions directly through Elsevier's home page (http://www.elsevier.nl), selecting first 'Customer Support', then 'General Information', then 'Permissions Query Form'.

In the USA, users may clear permissions and make payments through the Copyright Clearance Center, Inc., 222 Rosewood Drive, Danvers, MA 01923, USA; phone: (978) 7508400, fax: (978) 7504744, and in the UK through the Copyright Licensing Agency Rapid Clearance Service (CLARCS), 90 Tottenham Court Road, London W1P 0LP, UK; phone: (+44) 171 631 5555; fax: (+44) 171 631 5500. Other countries may have a local reprographic rights agency for payments.

Derivative Works
Tables of contents may be reproduced for internal circulation, but permission of Elsevier Science is required for external resale or distribution of such material.
Permission of the Publisher is required for all other derivative works, including compilations and translations.

Electronic Storage or Usage
Permission of the Publisher is required to store or use electronically any material contained in this work, including any chapter or part of a chapter.

Except as outlined above, no part of this work may be reproduced, stored in a retrieval system or transmitted in any form or by any means, electronic, mechanical, photocopying, recording or otherwise, without prior written permission of the Publisher.
Address permissions requests to: Elsevier Science Rights & Permissions Department, at the mail, fax and e-mail addresses noted above.

Notice
No responsibility is assumed by the Publisher for any injury and/or damage to persons or property as a matter of products liability, negligence or otherwise, or from any use or operation of any methods, products, instructions or ideas contained in the material herein. Because of rapid advances in the medical sciences, in particular, independent verification of diagnoses and drug dosages should be made.

First edition 2001

Library of Congress Cataloging in Publication Data
A catalog record from the Library of Congress has been applied for.

ISBN 0 444 50694 2

∞ The paper used in this publication meets the requirements of ANSI/NISO Z39.48-1992 (Permanence of Paper).
Printed in The Netherlands.

Introduction

It is the expressed goal of the European Charcot Foundation to support and stimulate information on new developments in Multiple Sclerosis (MS).

The European Charcot Foundation Symposium on "Genes and Viruses" attempted to collect the essential information up till 1999 on this subject.

In a way this Symposium was necessary, because a year before in France questions on the safety of hepatitis-B vaccination were raised. It was suggested that in some cases onset of MS was related to a preceding vaccination against hepatitis-B.

The epidemiological investigations so far did not produce evidence that such a relation existed [1–3]. But many researchers felt that in general some viral vectors may be active in the pathogenesis of MS.

That genetic susceptibility might play an important role in the way the immunological system handles viral infection, or in the way brain cells (oligodendrocytes, asctrocytes, neurons) react to viral invasion, has in recent years been extensively studied.

Experimental evidence from animals makes a viral oligodendrocytopathy a certainty. And this in itself opens the perspective of such a disease in human beings. New insights in MS-pathology indicate that in a number of MS-patients an oligodendrocytopathy is paramount. Proof that this may be virus-induced is absent so far.

In this volume several viruses are discussed, of which a relation to MS and its activity seems likely, although not proven.

Viruses may infect humans and remain present in the brain for many years. How these neurotropic viruses survive and how they become active in producing inflammatory reactions are unanswered questions.

The authors have, on our request, updated their contribution, so that this volume is a reflection of our knowledge per 2001.

Prof. Otto R. Hommes
Chairman
European Charcot Foundation

1. Confavreux C, et al. The New England J. of Med. 2001;344:319–326.
2. Ascherio A, et al. The New England J. of Med. 2001;344:327–332.
3. Editorial: Gellin BG, The New England J. of Med. 2001; 372–373.

List of contributors

O. Andersen
Sahlgrenska University Hospital
Göteborg
Sweden

N. Arbour
INRS-Institut Armand-Frappier
Laval (Québec)
Canada

S. Aubagnac
Institut Pasteur
Paris Cedex
France

J.D. Bannan
Vanderbilt School of Medicine
Nashville, Tennessee
USA

L. Behrens
Institut Pasteur
Paris Cedex
France

A. Bergami
San Raffaele Scientific Institute
Milan
Italy

F. Bihl
Montreal General Hospital
Montreal (Québec)
Canada

L. De Bolle
Katholieke Universiteit Leuven
Leuven
Belgium

A. Boucher
INRS-Institut Armand-Frappier
Laval (Québec)
Canada

M. Brahic
Institut Pasteur
Paris Cedex
France

E. Brambilla
San Raffaele Scientific Institute
Milan
Italy

D. Brassat
Fédération de Neurologie
Toulouse
France

J.H. Brewer
St. Luke's Hospital
Kansas City, Missouri
USA

J.-F. Bureau
Institut Pasteur
Paris Cedex
France

N. Canal
San Raffaele Scientific Institute
Milan
Italy

D. Caputo
University of Milan
Milan
Italy

D.R. Carrigan
Institute for Viral Pathogenesis
Milwaukee, Wisconsin
USA

R. Cavarretta
University of Milan
Milan
Italy

T. Christensen
University of Aarhus
Aarhus C
Denmark

M. Clanet
Fédération de Neurologie
Toulouse
France

E. de Clerq
Katholieke Universiteit Leuven
Leuven
Belgium

F. Clerget-Darpoux
INSERM U 535
Le Kremlin Bicêtre
France

G. Comi
San Raffaele Scientific Institute
Milan
Italy

A. Compston
University of Cambridge
Cambridge
United Kingdom

E. Corcuff
Hôpital de la Salpêtriere
Paris
France

R. Day
Sherbrooke University
Sherbrooke (Québec)
Canada

S. Della Bella
University of Milan
Milan
Italy

F. Denis
INRS-Institut
Laval (Québec)
Canada

P. Dissing Sørensen
Biotechnological Institute
Hørsholm
Denmark

P. Duclos
World Health Organization
Geneva
Switzerland

P. Duquette
Hôpital Notre-Dame
Montréal (Québec)
Canada

G. Edan
University Hospital
Rennes
France

J. Edwards
Sherbrooke University
Sherbrooke (Québec)
Canada

P. Ferrante
University of Milan
Milan
Italy

M. Filippi
Ospedale San Raffaele
Milan
Italy

D. Franciotta
University of Pavia
Pavia
Italy

R. Furlan
San Raffaele Scientific Institute
Milan
Italy

S. Gofette
Université Catholique de Louvain
Brussels
Belgium

O. Gout
Hôpital de la Salpêtriere
Paris
France

L.M.E. Grimaldi
San Raffaele Scientific Institute
Milan
Italy

S. Haahr
Aarhus University
Aarhus
Denmark

H. Hall
London School of Hygiene &
Tropical Medicine
London
UK

H.J. Hansen
Aarhus University Hospital
Aarhus C
Denmark

J. Hillert
Karolinska Institutet
Huddinge
Sweden

P. Höllsberg
Aarhus University
Aarhus
Denmark

O.R. Hommes
European Charcot Foundation
Nijmegen
The Netherlands

G. Iannucci
Ospedale San Raffaele
Milan
Italy

A. Iglesias
Max-Planck-Institut für Neurobiologie
Martinsried
Germany

J. De Keyser
Academisch Ziekenhuis Groningen
Groningen
The Netherlands

K.K. Knox
Institute for Viral Pathogenesis
Milwaukee, Wisconsin
USA

N. Kruse
Julius-Maximilians-University
Würzberg
Germany

R. Liblau
Hôpital de la Salpêtriere
Paris
France

Tobias Litzenburger
Max-Planck-Institut für Neurobiologie
Martinsried
Germany

G. Locatelli
San Raffaele Institute
Milan
Italy

P. Lusso
San Raffaele Institute
Milan
Italy

O. Lyon-Caen
Hôpital de la Salpêtriere
Paris
France

S. Malnati
San Raffaele Institute
Milan
Italy

R. Mancuso
University of Milan
Milan
Italy

P. Marconi
San Raffaele Scientific Institute
Milan
Italy

G. Martino
San Raffaele Scientific Institute
Milan
Italy

D.B. McGavern
Mayo Medical and Graduate School
Rochester, Minnesota
USA

P. Melzi
University of Milan
Milan
Italy

V. ter Meulen
Julius-Maximilians-University
Würzberg
Germany

W.M. Mitchell
Vanderbilt School of Medicine
Nashville, Tennessee
USA

A. Møller-Larsen
Aarhus University Hospital
Aarhus
Denmark

P. Monteyne
SmithKline Beecham Biologicals
Rixensart
Belgium

N. Moriabadi
Julius-Maximilians-University
Würzberg
Germany

L. Naesens
Katholieke Universiteit Leuven
Leuven
Belgium

J. Newcombe
University College London
London
United Kingdom

S. Niewisk
Julius-Maximilians-University
Würzberg
Germany

E. Pagani
University of Milan
Milan
Italy

P.L. Poliani
San Raffaele Scientific Institute
Milan
Italy

G. Ramsaransing
Academisch Ziekenhuis Groningen
Groningen
The Netherlands

J. Reboul
Hôpital de la Salpêtrière
Paris
France

P. Rieckmann
Julius-Maximilians-University
Würzberg
Germany

M. Rodriguez
Mayo Clinic
Rochester, Minnesota
USA

F. Ruffini
San Raffaele Scientific Institute
Milan
Italy

S. Sathornsumetee
Mayo Medical and Graduate School
Rochester, Minnesota
USA

F.L. Sciacca
National Neurologic Institute
'C. Besta'
Milan
Italy

W.A. Sibley
University of Arizona
Tucson, Arizona
USA

C.J.M. Sindic
Université Catholique de Louvain
Brussels
Belgium

S. Sriram
Vanderbilt School of Medicine
Nashville, Tennessee
USA

C.W. Stratton
Vanderbilt School of Medicine
Nashville, Tennessee
USA

J.L. Strominger
Harvard University
Cambridge, Massachusetts
USA

A. Svejgaard
Rigshospitalet
Copenhagen
Denmark

P.J. Talbot
INRS-Institut
Laval (Québec)
Canada

A. Tourbah
Hôpital de la Salpêtriere
Paris
France

K.V. Toyka
Julius-Maximilians-University
Würzberg
Germany

M. Tremblay
INRS-Institut Armand-Frappier
Laval (Québec)
Canada

B.M.J. Uitdehaag
Academic Hospital 'Vrije Universiteit'
Amsterdam
The Netherlands

D.R. Ure
Mayo Medical and Graduate School
Rochester, Minnesota
USA

P. Van Damme
University of Antwerp
Antwerp
Belgium

S. Vigneau
Institut Pasteur
Paris Cedex
France

H. Wekerle
Max-Planck-Institut für Neurobiologie
Martinsried
Germany

S. Yao
Vanderbilt School of Medicine
Nashville, Tennessee
USA

C. Zwanikken
Academisch Ziekenhuis Groningen
Groningen
The Netherlands

Contents

Introduction, *O.R. Hommes*		*v*
List of Contributors		*vii*
1.	Pathogenesis of MS – Tracking the elusive role of B Lymphocytes, *H. Wekerle, A. Iglesias, T. Litzenburger*	*1*
2.	Genetic susceptibility to multiple sclerosis, *A. Compston*	*9*
3.	HLA controlled susceptibility to multiple sclerosis – a survey, *A. Svejgaard*	*25*
4.	Nordic contributions in multiple sclerosis genetics, *J. Hillert*	*37*
5.	Genomic screening in multiple sclerosis, *F. Clerget-Darpoux*	*41*
6.	Cytokine gene polymorphisms in multiple sclerosis, *B.M.J. Uitdehaag*	*49*
7.	Genetic control of Theiler's virus infection, *M. Brahic, S. Aubagnac, L. Behrens, F. Bihl, J.-F. Bureau, S. Vigneau*	*57*
8.	MHC proteins and human diseases: a tale of recognition in two immune systems, *J.L. Strominger*	*61*
9.	Central nervous system delivery of therapeutic genes using viral vectors as an alternative strategy in autoimmune demyelinating diseases, *R. Furlan, P.L. Poliani, P. Marconi, E. Brambilla, F. Ruffini, A. Bergami, G. Comi, G. Martino*	*71*
10.	The contribution of axonal injury to neurologic dysfunction in Theiler's virus-induced inflammatory demyelinating disease, *D.R. Ure, D.B. McGavern, S. Sathornsumetee, M. Rodriguez*	*79*
11.	The effect of virus-like infections on the course of multiple sclerosis, *W.A. Sibley*	*89*
12.	A clinical neuroradiological and virological longitudinal study to verify the role of virus in inducing relapses in multiple sclerosis patients; preliminary results, *P. Ferrante, R. Mancuso, R. Cavarretta, E. Pagani, M. Filippi, G. Comi, G. Iannucci, P. Melzi, S. Della Bella, D. Caputo*	*97*
13.	Viral and bacterial specificities of oligoclonal IgG bands in multiple sclerosis, *S. Gofette, C.J.M. Sindic*	*109*
14.	Hepatitis B and multiple sclerosis. Hepatitis B vaccination in France, *M. Clanet, D. Brassat*	*115*

15. Hepatitis B vaccination and central nervous system demyelination: an immunological approach,
    E. Corcuff, J. Reboul, A. Tourbah, G. Edan, O. Gout, O. Lyon-Caen, R. Liblau   121
16. Is there a causal link between hepatitis B vaccination and multiple sclerosis?
    P. Monteyne   131
17. Public health aspects of hepatitis B vaccination and multiple sclerosis,
    Ph. Duclos, H. Hall, P. Van Damme   135
18. Influenza vaccination in multiple sclerosis,
    J. De Keyser, G. Ramsaransing, C. Zwanikken   143
19. Infection, immunization and immunomodulation – what's the difference?
    P. Rieckmann, N. Moriabadi, S. Niewiesk, N. Kruse, K.V. Toyka, V. ter Meulen   147
20. Detection of human herpesvirus 6 by quantitative real-time PCR in serum and cerebrospinal fluid of patients with multiple sclerosis,
    G. Locatelli, M.S. Malnati, D. Franciotta, R. Furlan, G. Comi, G. Martino, P. Lusso   153
21. The ability of candidate viruses to explain epidemiological findings in multiple sclerosis,
    S. Haahr, P. Höllsberg   163
22. Active human herpesvirus six viremia in patients with multiple sclerosis,
    K.K. Knox, J.H. Brewer, D.R. Carrigan   185
23. Retroviruses in multiple sclerosis,
    T. Christensen, P. Dissing Sørensen, H.J. Hansen, A. Møller-Larsen   195
24. The role of neuroinvasive human coronaviruses in autoimmune processes associated with multiple sclerosis,
    A. Boucher, M. Tremblay, N. Arbour, J. Edwards, R. Day, J. Newcombe, P. Duquette, F. Denis, P.J. Talbot   209
25. Association of multiple sclerosis with Chlamydia pneumoniae: demonstration of the 16S rRNA gene and immunoreactivity of CSF cationic antibodies against C. pneumoniae antigens,
    S. Sriram, C.W. Stratton, S. Yao, J.D. Bannan, W.M. Mitchell   221
26. A rationale for antiviral therapy in multiple sclerosis,
    O. Andersen   231
27. Antiviral activity of antiherpetic drugs in lymphoblast cells infected with human herpes virus 6,
    L. Naesens, L. De Bolle, E. De Clercq   241
28. Il-1 Receptor antagonist: a possible circulating and genetic marker of interferon-beta1b therapeutic effectiveness in MS,
    F.L. Sciacca, L.M.E. Grimaldi, G. Comi, N. Canal   251

Subject index   257

Pathogenesis of multiple sclerosis – tracking the elusive role of B lymphocytes

Antonio Iglesias, Tobias Litzenburger and Hartmut Wekerle*

Department of Neuroimmunology, Max-Planck-Institut für Neurobiologie, D-82152 Martinsried, Germany

Introduction

Current concepts of MS pathogenesis stress the central role of myelin autoreactive T cells controlling the initiation, localization and state of activity of the individual MS lesions [1]. Such myelin specific T cell clones are present in the repertoires of MS patients, and also in healthy persons. Although the pathogenic contribution of autoreactive T cells in MS patients has not been formally proven, studies in experimental animal models – rodents [2] as well as primates [3;4] – have unequivocally proven their pathogenic potential. At the same time the studies revealed that an autoaggressive potential only unfolds following pathological activation, most probable in the context of anti-microbial responses, which in MS as in other autoimmune diseases include molecular mimicry, superantigens, pro-inflammatory milieus [5]. Once activated, the T cells cross BBB, enter the CNS and start an interaction which escalates till a full inflammatory response is formed. Inflammatory infiltrates around small blood vessels, along with parenchymal infiltrates, BBB perforation, and glial activation are the result of such a scenario [6].

This concept explains the *inflammatory* changes in the MS lesion, but not *demyelination*. This is strikingly illustrated by experimental autoimmune encephalitis (EAE), a most popular model of the inflammatory phase of the MS plaque, not of demyelination. It should be noted that in EAE of Lewis rats, the CNS lesions are purely inflammatory, with no confluent demyelinating process demonstrable [7].

Which autoimmune mechanisms would then be responsible for large-scale myelin destruction? In the following we shall discuss the role of B lymphocytes – the most probable culprits in MS related demyelination.

B cell participation in the pathogenesis MS

Multiple lines of evidence indicate a crucial role of B cells in the pathogenesis of myelin autoimmune diseases. In fact, it appears that B cells operate at diverse points of the pathogenesis. As antigen presenting partners of T cells, B cells may take up myelin autoantigen, process it and present the recognizable peptide fragments in the context of MHC determinants to autoimmune T lymphocytes. First

---

*Correspondence address: Max-Planck-Institut für Neurobiologie, Am Klopferspitz 18a, D-82152 Martinsried, Germany. E-mail: hwekerle@neuro.mpg.de

indication of such an antigen presenting role came from EAE experiments testing the induction of EAE in rodents which had been depleted of recirculating B cells. It has been known for more almost 20 years that rats lacking B cells are resistant to attempts to induce EAE by conventional means [8]. Similar findings were obtained in mouse EAE [9].

More recent experiments relying on the use of transgenic mice created a complex picture. Mutant mice lacking the CD40 ligand (CD154), a co-stimulatory molecule involved in interactions between T and B cells, failed to mount an encephalitogenic T cell response upon immunization with myelin protein, an aberration that could be corrected by transfer of fully functional APC [10]. This may point towards a role of B cells in antigen presentation, though other APC types must be considered, too [11]. B cell deprived transgenic mice showed slightly reduced susceptibility to EAE after immunization with MBP peptides [12], but were unchanged in response to MOG peptides [13]. Recent work suggests that B cell involvement in the initial phase of EAE depends on the nature of the immunizing autoantigen. While B-cell deficient mice resisted to EAE induction by immunization against MOG protein, they were susceptible to MOG peptide [14].

A second, more spectacular function of B cells is primary demyelination, destruction of myelin sheaths and myelin forming oligodendrocytes. B cells and their immunoglobulin products have been implied in MS for many years. Morphological studies have demonstrated B cells and plasma cells in perivascular infiltrates as well as in the parenchyma proper [15,16], although their nature and function, however have remained enigmatic to date. CNS infiltrating B cells are most probably responsible for the secretion of the oligoclonal immunoglobulins, which are commonly found in the cerebrospinal fluid (CSF) of MS patients, but also in other inflammatory brain disorders [17].

It has been difficult to assign any autoimmune specificity to the MS related B cells and immunoglobulins. Interestingly, however, a recent analysis of immunoglobulin gene rearrangements of CSF B cells provides first evidence of the selective expansion of individual B cell clones, possible result of an (auto-)antigen driven process [18]. Concordant results have been reported in a study of plaque infiltrating B cells [19].

Demyelinating B cells and autoantibodies have been identified with certainty in EAE models. Linington et al. were the first to create EAE with large confluent demyelinating areas by co-transferring MBP specific T cell lines (which, by themselves would mediate inflammation, but no demyelination) plus monoclonal antibodies specific for MOG [20]. Similar MS plaque-like lesions were recently produced in primates. The lesions contained myelin debris lined with MOG specific immunoglobulin, structures which were also seen in plaque material from brains of MS patients [21,22].

Thus, recent investigations redirect our attention to the B cell as a key factor in the demyelinating pathogenesis of MS, and they point to MOG as a potential autoantigenic target. In the following we will describe a new experimental model which is of particular promise to study the elusive function of MOG specific B cells in vivo and in vitro.

Autoimmunity and tolerance of myelin specific B lymphocytes in gene-targeted, Ig-transgenic mice

As pointed out above, in EAE, the disease caused by immunisation with different encephalitogenic myelin antigens varies with respect to clinical severity and pathology. The Myelin Oligodendrocyte Glycoprotein (MOG) is a minor component, representing 0.05% of the total protein content of myelin, and mostly concentrated in the outer most sheath covering axons [23]. However, disease induced with immunized MOG is characterized by severe clinical course, strong primary demyelination, irreversible neurological deficits and frequent fatal outcome, normally not seen in active immunization with other myelin antigens like MBP or PLP. Antibodies directed against MOG, but not against other myelin components, have a strong pathogenic potential, as they increase disease severity and participate in demyelination in experimental animals and probably also in humans [24,25]. In addition, MOG is not expressed outside the CNS [23,26]. In contrast, MBP and PLP are detectable in peripheral immune organs. Expression of MBP in thymic cells has been claimed to suffice for activation of MBP-specific T cells in the absence of added MBP protein [27]. Thymic expression of MBP and PLP is presumably responsible for shaping the immune repertoire in the murine [28–31], and possibly also in the human [32] immune system.

Therefore, the tolerance status to a myelin antigen sequestered in the CNS (like MOG), as compared with peripherally accessible antigens (like MBP and PLP) might explain the enhanced pathology induced in MOG-EAE [26,33] as well as the predominance of MOG-specific clones in MS [34,35]. Recent findings of MOG binding autoantibodies co-localised with myelin lesions in MS brains have additionally focused attention on the role of B cells in CNS pathology [36]. What is the tolerance status for such CNS-secluded autoantigens like MOG? Are increased levels of MOG-specific autoantibodies pathogenic in experimental animals?

In an effort to understand B cell tolerance and -autoimmunity against MOG, we generated mice whose germline was manipulated to insert the functionally rearranged V region of the heavy chain of the anti-MOG antibody 8.18C5 [37] into its natural genomic location via gene-targeting. These "knock-in" mice (named TH) thus express the targeted, transgenic H chain in most of their B lymphocytes. In addition, the combination of the transgenic H chain with endogenous L chains generates MOG specificity in a high proportion of B cells (30%) of the knock-in TH mice [38]. Although TH mice contain high titers of MOG-specific autoantibodies and large numbers of autoreactive, MOG specific, B lymphocytes, they do not develop spontaneous EAE under normal circumstances. In spite of their pathogenic potential, MOG-specific B cells arise and mature in the bone marrow without tolerogenic restriction whatsoever [39]. But when challenged with CNS antigens (also different from MOG) or encephalitic T cells (e.g. PLP-specific), the TH knock-in mice develop an exacerbated and accelerated form of EAE [38]. The mechanisms involved are not fully elucidated, but it seems likely that the MOG-specific Abs exert their pathogenic potential after encephalitic T cells provoked an inflammatory cascade and opened the BBB. The target antigen on the surface of myelin is then amenable for binding by incoming MOG-specific Abs and complement, thus resulting in widespread demyelination. The subsequent engagement of MOG-specific T cell clones would then be facilitated by the release of MOG and the high frequency of MOG-specific B cells. Indeed, when stimulated with splenocytes from TH mice, T cells specific for MOG peptides are 10 fold better stimulated by MOG protein than by the cognate MOG peptide (our unpublished results). The cumulative effect of pathogenic MOG-specific Abs infiltrating the CNS and of enhanced antigen presentation by MOG-specific B lymphocytes in peripheral immune organs can explain both acceleration and exacerbation of EAE typically found in the TH knock-in mice. The role of B cells as APC has been recently documented also for other autoimmune diseases [40,41]. Likewise, the importance of B cells in EAE has been challenged in studies using the B cell deficient μMT knock-out mice [13,42,43]. However, all these studies were performed in EAE induced with encephalitogenic peptides from MOG (35–55) or MBP (Ac1-11). In contrast, a recent publication has compared the susceptibility of μMT mice to EAE disease induced either with the MOG(35–55) peptide or with recombinant protein, and shown that B deficient μMT mice are susceptible to MOG(35–55) peptide EAE but resistant to EAE induced with recombinant MOG protein [14]. Therefore, at least for MOG-induced EAE, autoreactive B cells are important pathogenic components of the disease. Such autoreactive, pathogenic B cell clones specific for antigens that, like MOG, are secluded behind the blood-brain-barrier cannot be actively tolerized. Hence, such crucial autoantigens, once their pathogenic role has been defined, represent interesting targets for tolerization in immune therapeutic strategies.

A widespread strategy for immune therapy concerns DNA vaccination with relevant antigens. DNA vaccination has proven useful and safe in a large number of cases involving viral or bacterial antigens and in cases of allergens. DNA vaccination has also been tried to protect against autoimmune disease. In mice the immunization with DNA encoding epitopes of MBP has led to protection from a subsequent, encephalitogenic, challenge with MBP. Whereas vaccination with DNA encoding microbial antigens primes the immune system to generate a rapid and stronger response upon secondary challenge, the protection achieved with allergens or autoantigens relates to the production of a rather suppressive response. In contrast, vaccination of experimental mice with DNA encoding the extracellular part (aa 1–120) of MOG protein generated a MOG specific Ab response, that, upon secondary challenge (both with MOG and unrelated myelin antigens), provoked exacerbated EAE [44]. Therefore, vaccination with DNA encoding self-antigens cannot be predicted to generate a protective, suppressive, response. In particular if such autoantigens are sequestered, they can be dealt with like foreign antigens by the immune system and provoke strong (auto)immune responses. Nonetheless, the generation of therapeutic immune strategies to protect from MOG dependent autoimmune attacks remains an important and challenging goal, in light of the crucial role played by MOG in EAE and, apparently, also multiple sclerosis [45,46].

Detailed knowledge of the origin and development of MOG-specific lymphocytes will thus be of pivotal importance in the therapy of MS. We are currently studying the shaping of the MOG-specific repertoire in transgenic mice, which express MOG outside the CNS. Furthermore, knock-out mice deficient for MOG are far more resistant to EAE than their normal counterparts (A. Dautigny, personal communication).

To study the generation of B cells expressing the same autoreactive Ig receptor as the original anti-MOG specific hybridoma 8.18C5, we have also produced double transgenic animals by crossing knock-in TH mice with transgenic mice bearing the corresponding L chain.

While B cells in knock-in animals expressing a MOG specific H chain develop a pathogenic potential without tolerogenic restriction, we found that transgenic B cells expressing both the knock-in H chain and its MOG-specific L chain partner are subjected to tolerance induction processes. In such double transgenic animals the transgenic L chain is silenced in a process of receptor editing that results in its substitution by endogenous L chains, both in fetal liver and adult bone marrow [39]. However, receptor editing affected the L chain exclusively, and the intact knock-in H chain associates with endogenous L chains to generate a MOG specific repertoire indistinguishable from that of H chain only knock-in mice. In addition, the tolerogenic silencing of the transgenic L chain in double transgenic animals is independent of MOG, as it is also observed in MOG-deficient mice [39]. Thus MOG itself cannot be the tolerogenic principle behind the observed L chain editing process. Clearly, the combination of the MOG specific, transgenic H and L chain is cross-reactive with a second, peripherally expressed, antigen distinct from MOG. Notably, the unidentified, cross-reacting, self-antigen induces receptor editing of the transgenic L chain in immature B cells of double transgenic mice without eliminating MOG specificity. These experiments demonstrated that an Ig receptor cross-reacting with two different self-antigens can be triggered by both, but tolerance can be finely tuned for only one of them, while leaving intact the highly pathogenic potential for the second.

References

1. Noseworthy JH. Progress in determining the causes and treatment of multiple sclerosis. N 1999;399:A40–A47.
2. Schluesener HJ, Wekerle H. Autoaggressive T lymphocyte lines recognizing the encephalitogenic region of myelin basic protein: In vitro selection from unprimed rat T lymphocyte populations. J Immunol 1985;135:3128–3133.
3. Genain CP, Lee-Parritz D, Nguyen M-H, Massacesi L, Joshi N, Ferrante R et al. In healthy primates, circulating autoreactive T cells mediate autoimmune disease. J Clin Invest 1994;94:1339–1345.
4. Meinl E, Hoch RM, Dornmair K, De Waal Malefijt R, Bontrop RE, Jonker M et al. Encephalitogenic potential of myelin basic protein-specific T cells isolated from normal rhesus macaques. Am J Pathol 1997;150:445–453.

5. Wekerle H. The viral triggering of autoimmune disease (News & Views). Nature Med 1998;4(7):770–771.
6. Lassmann H, Wekerle H. Experimental models of multiple sclerosis. In: Compston A, Ebers G, Lassmann H, Matthews B, Wekerle H, editors. McAlpine's Multiple Sclerosis. London: Churchill Livingston, 1998: 409–434.
7. Wekerle H, Kojima K, Lannes-Vieira J, Lassmann H, Linington C. Animal models. Ann Neurol 1994;36:S47–S53.
8. Gausas J, Paterson PY, Day ED, Dal Canto MC. Intact B-cell activity is essential for complete expression of experimental allergic encephalomyelitis in Lewis rats. Cell Immunol 1982;72:360–366.
9. Myers KJ, Sprent J, Dougherty JP, Ron Y. Synergy between encephalitogenic T cells and myelin basic protein-specific antibodies in the induction of experimental autoimmune encephalomyelitis. J Neuroimmunol 1992;41:1–8.
10. Grewal IS, Flavell RA. A central role of CD40 ligand in the regulation of $CD4^+$ T cell responses. Immunol Today 1996;17:410–414.
11. Gerritse K, Laman JD, Noelle RJ, Aruffo A, Ledbetter JA, Boersma WJA et al. CD40-CD40 ligand interactions in experimental allergic encephalomyelitis and multiple sclerosis. Proc Natl Acad Sci USA 1996;93:2499–2504.
12. Wolf SD, Dittel BN, Hardardottir F, Janeway CA. Experimental autoimmune encephalomyelitis induction in genetically B cell-deficient mice. J Exp Med 1996;184:2271–2278.
13. Hjelmström P, Juedes AE, Fjell J, Ruddle NH. B cell deficient mice develop experimental allergic encephalomyelitis with demyelination after myelin oligodendrocyte glycoprotein immunization. J Immunol 1998;161:4480–4483.
14. Lyons J-A, San M, Happ MP, Cross AH. B cells are critical to induction of experimental allergic encephalomyelitis by protein but not by a short encephalitogenic peptide. Eur J Immunol 1999;29(11):3432–3439.
15. Esiri MM. Multiple sclerosis: A quantitative and qualitative study of immunoglobulin-containing cells in the central nervous system. Neuropathol Appl Neurobiol 1980;6:9–21.
16. Prineas JW, Wright RG. Macrophages, lymphocytes and plasma cells in the perivascular compartment in chronic multiple sclerosis. Lab Invest 1978;38:409–421.
17. Thompson EJ. Cerebrospinal fluid. J Neurol Neurosurg Psych 1995;59:349–357.
18. Qin Y, Duquette P, Zhang Y, Poole R, Antel JP. Clonal expansion and somatic hypermutation of V_H genes of B cells from cerebrospinal fluid in multiple sclerosis. J Clin Invest 1998;102(5):1045–1050.
19. Owens GP, Kraus H, Burgoon MP, Smith-Jensen T, Devlin ME, Gilden DH. Restricted use of V_H4 germline segments in an acute multiple sclerosis brain. Ann Neurol 1998;43(2):236–243.
20. Linington C, Bradl M, Lassmann H, Brunner C, Vass K. Augmentation of demyelination in rat acute allergic encephalomyelitis by circulating mouse monoclonal antibodies directed against a myelin/oligodendrocyte glycoprotein. Am J Pathol 1988;130:443–454.
21. Genain CP, Cannella B, Hauser SL, Raine CS. Identification of autoantibodies associated with myelin damage in multiple sclerosis. Nature Med 1999;5(2):170–175.
22. Raine CS, Cannella B, Hauser SJ, Genain CP. Demyelination in primate autoimmune encephalomyelitis and acute multiple sclerosis lesions: A case for antigen specific antibody mediation. Ann Neurol 1999;46:144–160.
23. Bernard CCA, Johns TG, Slavin A, Ichikawa M, Ewing C, Liu J et al. Myelin oligodendrocyte glycoprotein: A novel candidate autoantigen in multiple sclerosis. J Mol Med 1997;75:77–88.
24. Genain CP, Nguyen M-H, Letvin NL, Pearl R, Davis RL, Adelmann M et al. Antibody facilitation of multiple sclerosis-like lesions in a nonhuman primate. J Clin Invest 1995;96:2966–2974.
25. Storch MK, Stefferl A, Brehm U, Weissert R, Wallström E, Kerschensteiner M et al. Autoimmunity to myelin oligodendrocyte glycoprotein in rats mimics the spectrum of multiple sclerosis pathology. Brain Pathol 1998;8:681–694.
26. Gardinier MV, Amiguet P, Linington C, Matthieu J-M. Myelin/oligodendrocyte glycoprotein is

a unique member of the immunoglobulin superfamily. J Neurosci Res 1992;33:177–187.
27. Fritz RB, Zhao M-L. Thymic expression of myelin basic protein (MBP). Activation of MBP-specific T cells by thymic cells in the absence of exogenous MBP. J Immunol 1996;157:5249–5253.
28. Anderson AC, Nicholson LB, Legge KL, Turchin V, Zaghouani H, Kuchroo VK. High frequency of autoreactive myelin proteolipid protein–specific T cells in the periphery of naive mice: Mechanisms of selection of the self-reactive repertoire. J Exp Med 2000;191:761–770.
29. Klein L, Klugmann M, Nave K-A, Tuohy VK, Kyewski BA. Shaping of the autoreactive T cell repertoire by a splice variant of self protein expressed in thymic epithelial cells. Nature Med 2000;6(1):56–61.
30. Targoni OS, Lehmann PV. Endogenous myelin basic protein inactivates the high avidity T cell repertoire. J Exp Med 1998;187(12):2055–2063.
31. Wekerle H. T-cell autoimmunity in the central nervous system. Intervirol 1993;35:95–100.
32. Pribyl TM, Campagnoni CW, Kampf K, Kashima T, Handley VW, McMahon J et al. Expression of the myelin proteolipid protein gene in human fetal thymus. J Neuroimmunol 1996;67:125–130.
33. Adelmann M, Wood J, Benzel I, Fiori P, Lassmann H, Matthieu J-M et al. The N-terminal domain of the myelin oligodoendrocyte glycoprotein (MOG) induces acute demyelinating experimental autoimmune encephalomyelitis in the Lewis rat. J Neuroimmunol 1995;63:17–27.
34. Kerlero de Rosbo N, Ben-Nun A. T cell responses to myelin antigens in multiple sclerosis: Relevance of the predominant autoimmune reactivity to myelin oligodendrocyte glycoprotein. J Autoimmun 1998;11:287–289.
35. Wallström E, Khademi M, Andersson M, Weissert R, Linington C, Olsson T. Increased reactivity to myelin oligodendrocyte glycoprotein peptides and epitope mapping in HLA DR2 (15)[+] positive multiple sclerosis. Eur J Immunol 1998;28(10):3329–3335.
36. Genain CP, Cannella B, Hauser SL, Raine CS. Identification of autoantibodies associated with myelin damage in multiple sclerosis. Nature Med 1999;5(2):170–175.
37. Linington C, Webb M, Woodhams PL. A novel myelin-associated glycoprotein defined by a mouse monoclonal antibody. J Neuroimmunol 1984;6:387–396.
38. Litzenburger T, Fässler R, Bauer J, Lassmann H, Linington C, Wekerle H et al. B lymphocytes producing demyelinating autoantibodies: Development and function in gene-targeted transgenic mice. J Exp Med 1998;188(1):169–180.
39. Litzenburger T, Blüthmann H, Morales P, Pham-Dinh D, Dautigny A, Wekerle H et al. Development of MOG autoreactive transgenic B lymphocytes: Receptor editing in vivo following encounter of a self-antigen distinct from MOG. J Immunol 2000;165:5360–5366.
40. Chan OTM, Madaio MP, Shlomchik MJ. The central and multiple roles of B cells in lupus pathogenesis. Immunol Rev 1999;169:107–121.
41. Korganow A-S, Ji H, Mangialaio S, Duchatelle V, Pelanda R, Martin T et al. From systemic T cell self reactivity to organ specific autoimmune disease via immunoglobulins. I 1999;10:451–461.
42. Eugster H-P, Frei K, Kopf M, Lassmann H, Fontana A. IL-6 deficient mice resist myelin oligodendrocyte glycoprotein induced autoimmune encephalomyelitis. Eur J Immunol 1998;28(7):2178–2187.
43. Wolf SD, Dittel BN, Hardardottir F, Janeway CA. Experimental autoimmune encephalomyelitis induction in genetically B cell-deficient mice. J Exp Med 1996;184:2271–2278.
44. Bourquin C, Iglesias A, Berger T, Wekerle H, Linington C. Myelin oligodendrocyte glycoprotein-DNA vaccination induces autoantibody mediated autoaggression in experimental autoimmune encephalitis (EAE). Eur J Immunol 2000;30(12):3663–3671.
45. Lassmann H, Brunner C, Bradl M, Linington C. Experimental allergic encephalomyelitis: The balance between encephalitogenic T lymphocytes and demyelinating antibodies determines size and structure of demyelinated lesions. Acta Neuropathol 1988;75:566–576.
46. Linington C, Bradl M, Lassmann H, Brunner C, Vass K. Augmentation of demyelination in rat acute allergic encephalomyelitis by circulating mouse monoclonal antibodies directed against a

myelin/oligodendrocyte glycoprotein. Am J Pathol 1988;130:443–454.

Genetic susceptibility to multiple sclerosis

Professor Alastair Compston*
University of Cambridge Neurology unit, Addenbrooke's Hospital, Hills Road, Cambridge CB2 2QQ, UK

Introduction

The analysis of recurrence risk within families of probands has defined the contribution made by genetic factors to the aetiology of multiple sclerosis. Thereafter, slow progress has been made in identifying and locating the genes responsible for increased susceptibility using either the candidate approach or systematic screening of the entire genome with polymorphic microsatellite markers. Even if the status of genomic regions provisionally identified as encoding susceptibility genes is confirmed, these could only account for a proportion of heritability implicated by the classical genetic studies. But continued effort aimed at solving the genetics of multiple sclerosis is worthwhile because this knowledge will eventually inform our understanding of the pathogenesis and perhaps generate novel treatments.

Recurrence within families

Family studies illustrate the relative contribution of genetic factors and extrinsic triggers in determining susceptibility to multiple sclerosis. Debate on which is the more important is some extent contrived since modern concepts on the nature of complex traits necessarily involve an interplay between genes and the environment. Fifteen percent of probands with multiple sclerosis have an affected relative. Age adjusted recurrence risk is highest for siblings (3%), parents (2%) and children (2%) but lower in second and third degreee relatives [1–3]. Concordance in identical twins is at least 35% [4–6] and the age-adjusted risk may be nearly as high for children of conjugal pairs with multiple sclerosis [7]. Conversely, half siblings show approximately half the rate observed in full siblings [8]; and the frequency of multiple sclerosis in non-biological parents, siblings and children of adoptees matches the lifetime risk for unrelated individuals [9]. Population based studies show increased recurrence risk for a variety of autoimmune diseases in the relatives of patients with multiple sclerosis – implicating in particular autoimmune thyroid disease [10].

Correspondence address: University of Cambridge, Addenbrooke's Hospital, Neurology unit, Hills Road, Cambridge CB2 2QQ, UK. E-mail: alastair.compston@medschl.cam.ac.uk

Susceptibility and clinical course

Multiplex families can be used either as a resource for molecular analyses or to assess features of the disease. Genetic factors may affect susceptibility and/or variations in the clinical phenotype and course. Comparison of affected sibling pairs provides evidence for concordance of eventual disability and handicap but not initial presentation – suggesting that genetic factors not only determine susceptibility to multiple sclerosis but also the course of the disease. Comparisons between concordant parent-child pairs show no distortion in the random distribution of sex pairings and neither sex nor line of inheritance influence disability, age at onset or the clinical course; disability is highest in the male offspring of affected fathers who more commonly manifest the primary progressive phenotype [11].

Strategies for the identification of susceptibility genes

The analysis of complex traits

The concept of a complex trait assumes poly- or oligogenic models in preference to trivial contributions from an infinite number of genes in determining disease susceptibility. Unlike monogenic disorders, the genes responsible for complex disorders are not mutations which code for aberrant gene products but normal polymorphisms. They are several, acting independently or through epistasis, and each may exert a very small contributory effect on some as yet undefined function. Collectively their attributable risk determines a relative risk of around 20. When eventually in place, the application of this knowledge for improved understanding of the pathogenesis, in counselling and in designing novel treatments is potentially considerable.

Two methods underpin the analysis of complex traits. Linkage analysis aims to show co-segregation between disease and the passage of markers for loci which encode susceptibility genes. Even though the marker allele present in each family may differ, they are nevertheless linked to the same genetic locus between families. Association describes the different situation in which a specific allele, which may also be a marker not the susceptibility gene itself, is over-represented in cases compared to controls because this specific allele is in linkage disequilibrium with the disease susceptibility gene in apparently unrelated individuals within the at-risk population. Linkage and association therefore provide different information. Each has its limitations and advantages. Assuming a reasonable sample size, any marker which is well distant from the true susceptibility locus will be linked but not associated. A marker which is close to the susceptibility locus will not be linked in sibling pairs if the sample size is inadequate (as are most currently available sibling pair studies) but it may nevertheless be associated. Only the combination of a close marker tested in an adequate sample will demonstrate both linkage and association.

Population based case-control studies

The major limiting factor for population based case-control studies is confounding through ethnic mis-matching. This is most problematic in large outbred populations (north America) and reduced in genetic isolates (such as Finland and Sardinia). In fact, the problem is probably more theoretical than real and main reason for poor performance of case-control studies is lack of stringent statistical limits. Originally, population based association studies comparing the frequency of polymorphisms at a given genetic locus in groups of unrelated individuals with and without multiple sclerosis were limited by the number of polymorphic systems which could be explored, the paucity of the hypotheses which directed the search to a particular region of interest, and the confounding effects of inappropriate controls. However, an important principle to emerge was that (to a varying extent) alleles encoded at neighbouring loci tend to co-segregate. It seems that many adjacent markers are not yet randomly distributed by genetic recombinations either because the number of generations over which mitoses have occurred is still too few to establish genetic equilibrium, or because there is evolutionary pressure to maintain these linkage disequilibria.

Family-based linkage studies

The advent of laboratory methods for distinguishing short sections of DNA, accurate mapping of the many available microsatellites and single nucleotide polymorphisms (SNPs) across the genome, availability of extended pedigrees in which affected status could reliably be determined, and deployment of appropriate statistical methods for assessing genome wide significance have collectively made it possible to tackle the genetics of complex traits in more detail [12,13]. Linkage analysis in affected family members either selects candidates based on *a priori* considerations of disease pathogenesis or performs systematic screens of the whole genome. Where genome screens fail to provide evidence for linkage, rare genes with large effects probably do not exist for that disease. But since the power of a linkage genome screen is critically dependent on the frequency of susceptibility alleles in the population studied, and is thus expected to vary between studies, linkage analysis may still be worth pursuing in places where the frequency of susceptibility alleles is more favourable for historical reasons. When a small family population expands numerically in relative isolation, genetic drift can influence the frequency of mutant and polymorphic alleles. As a consequence, some genetically determined diseases may occur at higher than expected frequency. Isolated populations such as Finns, Sardinians, Icelanders and Tasmanians could in theory provide special opportunities for resolving the issue of genetic susceptibility to complex traits such as multiple sclerosis.

Meta-analysis has been explored in the expectation that this will reduce the evidence for false positive peaks and strengthen the candidature of those which are genuine [14]. However, this assumes that genes conferring susceptibility to multiple sclerosis are shared between populations especially if data-sets are combined from ethnically diverse groups. In reality, it is probable that ubiquitous genes do exist which increase susceptibility to core aspects of the pathological process in multiple sclerosis whereas others are restricted to particular populations. These considerations are clearly important when looking at meta-analyses but the exercise has proved useful in selecting regions for the further study of positional candidates and establishing the probability that, in addition to the major histocompatibility complex, certain regions may encode genes which confer susceptibility across the range of autoimmune diseases [15,16].

Family-based association studies

An alternative approach has been the use of family-based association methods in which the test is to show that a particular allele of a candidate gene, or a polymorphic marker in linkage disequilibrium with that gene, is transmitted to affected individuals more often than expected by chance. Transmission disequilibrium testing (TDT) uses trios consisting of single affecteds with both parents who are usually (but need not be) unaffected. Each parental allele has a 1:2 chance of being transmitted. The test provides evidence for an allelic association by demonstrating excess transmission and identifies that allele which is responsible for disequilibrium in the sample. Although not every family will be informative for this marker, it follows that those which are in linkage disequilibrium with susceptibility genes for multiple sclerosis will be adequately represented. One further factor limiting the extent to which the sample of family trios may not prove fully informative for a particular polymorphic locus is the presence of homozygous parents. These are rejected in the analysis because the transmitted allele cannot be identified. Segregation distortion arises when a particular allele confers a survival advantage and therefore appears to be disproportionately transmitted. This allele will be preferentially transmitted to all surviving offspring whether or not they have the disease trait in question. The issue can be resolved by studying affected and unaffected children in family quartets. But a new problem then arises which is offset by the affected family member approach and that is difficulty in assigning unaffected status to a young adult in a disease which may not manifest clinically until late in life – if at all, given the prevalence of pathologically verified but clinically silent disease in autopsy series. Because it is very unlikely that an investigator will chose to test for transmission disequilibrium at exactly the locus which encodes a disease susceptibility gene contributing to a complex trait, the dividend from family based association methods depends crucially on the existence of linkage disequilibrium. This issue is not resolved [17,18] but it is probable that ancient populations will show less, and younger populations more, linkage disequilibrium. As a result, complex traits which are more common in European than African populations (such as multiple sclerosis) should offer better prospects for detection of disease susceptibility genes using (case-control or family-based) association studies.

If the extent of linkage disequilibrium is very low or highly variable across the genome, systematic genome screening will be less rewarding than the candidate gene approach; conversely, if it is rather more extensive, an opportunity will exist for linkage disequilibrium dependent (case-control or family-based) association studies. Selecting probands for transmission disequilibrium testing from amongst families in which there are multiple cases increases the power and reduces the scale of the study especially in the context of rare alleles [13]. Associations which are reproducible between populations carry conviction with respect to their causal relevance but population differences do not exclude causality. Where there is reasonable but variable linkage disequilibrium and several markers are found to be associated in separate populations, cross cultural studies (exploiting different degrees of linkage disequilibrium) may narrow the region of interest and select the closest marker; for example, Caucasians and Asians show association between narcolepsy and DR-DQ alleles but only the DQ association is found in Africans [19].

Combined linkage and association studies

One strategy is first to screen the genome with linkage to establish regions of interest and then to refine the map with transmission disequilibrium testing. SNP mapping, at a density of one per 10,000–30,000 base pairs, is a proven method for narrowing a linked region of interest and more efficient than existing candidate gene selection although the opportunity for matching knowledge of the disease pathogenesis to the location of defined genes has altered radically with completion of the human genome project. However, in situations where regions of interest cannot confidently be identified through linkage analysis, whole genome linkage disequilibrium screening is an option. Power calculations indicate that the resource needed reliably to demonstrate transmission disequilibrium is much more attainable than that required for linkage. This opens the debate on preferred choice of genetic marker – SNPs *versus* microsatellites – and the required density of these markers. SNPs occur every 100 base pairs and it is estimated that 800,000 will be mapped by 2001. At worst, it may be necessary to use 500,000 SNPs and still accept the possibility of false negatives; at best, a linkage disequilibrium screen may yield results with 6000 microsatellites. The density of markers must be sufficient to fill the intervals between regions which are not in linkage disequilibrium (probably extending to no more than 0.25cM) whereas linkage provides much greater topographical information for a given genotyping effort. Selecting those SNPs which are within coding or promotor regions especially of functional and candidate genes and picking the 60,000–100,000 type 1 (leading to non-conservative amino acid substitutions and less heterozygosity for reasons of evolutionary pressure) may reduce the genotyping effort and increase the dividend of such screens. Thus, a trade-off exists between the ease of collecting a definitive clinical resource and the laboratory work needed to screen the genome using family linkage and association methods.

Genotyping effort can be reduced during the screening phase through the use of samples in which equal amounts of DNA from a large number of cases, controls or parents are pooled into a single sample and screened against a panel of microsatellites [20]. Pooling can detect allelic frequency differences of c5%.

Candidate genes in multiple sclerosis

Much effort has gone into the assessment of candidate susceptibility genes selected on the basis of prevailing ideas concerning the pathogenesis of multiple sclerosis and which therefore provide a logical interpretation of genetic susceptibility. In the past, these studies were necessarily limited by the availability of polymorphic markers and constrained by poorly developed concepts of disease aetiology and mechanisms. Population studies comparing unrelated cases and controls show an association between the class II MHC alleles DR15 and DQ6 and their underlying genotypes (DRB1*1501, DRB5*0101 and DQA1*0102, DQB2*0602 [21]. In Sardinians there is a specifically different association with DR4 (DRB1*0405-DQA1*0301-DQB1*0302) [22]. In the Canaries, the primary association is with DR15 and DQ6 but a secondary association exists with DR4 [23]. In Turkey, there is an allelic association with both DR2 (DBR1*1501, DQA1*0102, DQB1*0602) and DR4 (DRB1*04, DQA1*03, DQB1*0302) [24].

Outside the major histocompatibility complex, association and linkage studies identify the VH2-5 immunoglobulin heavy chain and the T cell receptor ß chain variable regions as possible candidates [25–27] but neither has been repeatedly confirmed [28,29]. The list of immunologically related candidates now screened includes adhesion molecules, immune receptors or accessory molecules, cytokines and their receptors or antagonists, and chemokines [30–36]. In the isolated population of Finns, association and linkage of susceptibility to multiple sclerosis exists for the myelin basic protein gene [37,38]. Studies of structural genes of myelin have otherwise been been uninformative [39–42] as have efforts have been directed at implicating growth factors which determine oligodendrocyte development and their receptors in susceptibility [43].

Linkage genome screens

Disappointingly, the situation is not materially advanced by six whole genome linkage analyses carried out in the United States [44], Canada [45], United Kingdom [46,47], Finland [48], Sardinia [49] and Italy [50]. These linkage analyses reveal several new genomic regions which may encode genes conferring susceptibility to multiple sclerosis. Some will be true and others false positives. Genotyping was completed on cohorts each of between 21–225 families, together involving in excess of 1000 individuals, for each of between 257–443 microsatellite markers chosen to have an average spacing of around 10 centiMorgans giving enough power to identify regions encoding a major susceptibility gene and with sufficient polymorphism to make a high proportion of the available families fully informative. Although the role of the major histocompatibility complex is confirmed (providing a positive control for the screens), in common with most other complex traits no major susceptibility gene has been identified and, superficially, the results show a disappointing lack of overlap. The regions of interest emerging from the United Kingdom genome screen are 1cen, 5 cen, 6p, 7p, 14q, 17q, 19q, Xp. They are 2p, 3p, 5p, 11q and Xp in the Canadian series. The United States/French consortium identifies 6p, 7q, 11p, 12q and 19q. There are no statistically significant regions of interest in the relatively small Finnish screen although positive lod scores were obtained for 6p21 (MHC) and 5p14-p12; increasing the density of markers raised the lod scores in several other regions (4cen, 11tel and 17q) whereas others (2q32 and 10q21) were unchanged; when all 21 families were typed across the regions of interest, the highest lod score was at 17q22-q24 – as in the previously reported United Kingdom screen. The recent genome-wide linkage screen in the largest available sample of Sardinian families with multiple sclerosis, whilst identifying three regions of interest (1q31, 10q23 and 11p15) nevertheless continues to show only modest evidence for linkage and fails to support the hypothesis that susceptibility genes might more easily be identified in a small but genetically informative isolated population [51].

The likely reason for failure of linkage analysis to solve the nature of genetic susceptibility to multiple sclerosis is that susceptibility is conferred by many genes each of small effect which are therefore not detectable by under-powered linkage studies. A major gene could yet exist which has been missed by all five studies. Alternatively, genetic complexity (ie heterogeneity) has obscured the picture. This failure of linkage genome screens reliably to identify regions of interest conferring disease susceptibility both in mixed and isolated European populations, each having a high prevalence of multiple sclerosis, has obvious implications for future genetic analyses of complex traits. Linkage data should be added from new populations with the aim of confirming and excluding regions of interest and continued assessment of positional candidates may shorten the search for genes that determine susceptibility and influence the course of multiple sclerosis but other approaches may yet be needed (see below).

Positional candidate screens

Overlap between two or more screens allows certain candidate regions to be identified with reasonable confidence and, from this selection, follows the opportunity for studying new positional candidates based on mechanistic hypotheses and genetic mapping – an exercise which has gathered momentum with the availability of a first provisional map of the human genome. In a study from Scandinavia, 15 markers from regions of interest identified in one or more full genome screens were assessed using affected relative pairs (from 46 multiplex families, 28 trios, and 190 cases and 148 unrelated controls): depending on the method of analysis positive results were obtained for 5p15, 6p21, 7ptr15 and 12q23; the latter two also showed transmission distortion in families and an association in the case-control series. The remaining markers (2p23, 5p, 5q, 6q22-25, 7q, 11q21-23, 12q24-qtr,

13q33-34, 16p12, 18p11 and Xp21) did not identify regions of interest in the Scandinavian series [52]. An Italian group used 67 markers covering regions of interst from the three full genome screens and candidates (HLA-DRB1, CTLA-4, IL-9, CSF-1R, APOE, Bcl-2 and TNF-R2) in 69 multiplex families (28 from Sardinia and 41 from mainland Italy [53]. Some positive scores were obtained, especially for the region at 5q14 in the Sardinian subset, and three regions (2p11.2, 7p15.2 and 17q12) also showed positive TDT scores. This study has since been extended to a full genome screen ([50]: see above).

Heterogeneity in multiple sclerosis

These genetic analyses are predicated on the assumption that multiple sclerosis is one disease. A major part of future studies in the genetics of multiple sclerosis will be to resolve the question of disease heterogeneity.

Clinical heterogeneity

One good example does exist in which a specific clinical phenotype almost certainly represents the expression of a genuinely different pathological process associated, in turn, with a specifically different genetic background. A minority of patients who meet the clinical criteria for definite multiple sclerosis and in whom there are associated magnetic resonance imaging abnormalities and cerebrospinal fluid oligoclonal bands have an illness in which there is disproportionate involvement of the anterior visual pathway. These are commonly women with male relatives already known to be affected by Leber's hereditary optic neuropathy and they have pathological mutations of mitochondrial DNA. This disorder (Harding's disease) therefore represents true heterogeneity in that a specific genotype determines a characteristic phenotype [54,55]. What remains unresolved is whether the mutation of mitochondrial DNA focuses the non-specific process of brain inflammation onto a particular site – constituting selective tissue vulnerability – or (as some have suggested) this merely represents the chance occurrence of relatively mild multiple sclerosis and Leber's hereditary optic neuropathy. The latter interpretation seems highly improbable given the number of cases of Harding's disease already described and the rarity of isolated Leber's hereditary optic neuropathy in women.

Another probable example of true heterogeneity is the distinct phenotype of demyelinating disease seen in Orientals and Africans. In Japan, multiple sclerosis shows either a so-called Western phenotype in which a number of sites are involved, or an optico-spinal pattern in which the clinical picture is dominated by involvement of visual and spinal cord pathways [56]. In its extreme form this mimics Devic's disease – a disorder which also occurs in Europeans, albeit rarely. Demyelinating disease is considered to be extremely rare in Africans but a number of cases are described and clinical experience suggests that the phenotype is typically a severe illness dominated by one or more episodes usually affecting the anterior visual pathway and spinal cord – again combining the anatomical features of Devic's disease with the clinical course of moderately severe relapsing remitting multiple sclerosis [57]. Unlike Harding's disease, the evidence that this represents disease heterogeneity arising from specific genetic or environmental modification of a core pathological process remains circumstantial.

Heterogeneity in the pathogenesis

Variations in imaging protocol can be used to delineate some but not all features of the pathological process leading to demyelination and astrocytosis. There is preliminary evidence for differences in the regional distribution of lesions defined by T_2-weighted imaging (which essentially depicts water accumulation in association with astrocytosis) and T_1-weighted gadolinium enhancement which images the inflammatory component [58]. The fact that these two principal types of lesion show significant differences in their anatomical clustering is at least consistent with the hypothesis that inflammation, demyelination and astrocytosis do not invariably follow in an orderly sequence but may arise independently as the expression of specifically different sequences of events. The same interpretation arises from histological comparison of the distribution of inflammation and axon injury [59].

The concept of pathogenic heterogeneity is further developed in the recent pathological studies using biopsy and autopsy material in which four distinct types of disease process are proposed [60]. Type 1 constitutes peri-venous inflammation with a sharp definition to the edge of the lesion and significant remyelination; this occurs in patients with clinically definite multiple sclerosis usually in the relapsing phase and in only a small number of progressive cases. Type 2 consists of perivenous demyelination with local deposition of immunoglobulin and terminal complement components within sharply defined lesions also having remyelination; these patients have clinically definite multiple sclerosis with an equal distribution of relapsing and progressive cases. Type 3 lacks peri-venous inflammation and the lesions are ill-defined with evidence for oligodendrocyte apoptosis; these features are also seen in areas of tissue ischaemia and the clinical course is usually an acute monophasic illness with fewer cases of relapsing remitting definite multiple sclerosis. Type 4 consists of perivenous inflammation with sharply defined lesions but oligodendrocyte loss in the normal appearing white matter; these cases typically have primary progressive involvement of the cerebrum, cerebellum or brainstem. As with variations in the clinical phenotype, these histopathological patterns *per se* are better described as complexity rather than heterogeneity unless and until specifically different aetiological factors can be linked to each type. In these preliminary assessments, types 3 and 4 are considered to represent a primary viral or ischaemic pathogenesis whereas the hypothesis is that types 1 and 2 depend on specifically different combinations of genetic predisposing factors. This is of course entirely speculative but indicates the need to link aetiology to pathogenesis and phenotype before patterns of true heterogeneity can reliably be defined. More compelling is the evidence that these histopathological lesions cluster within patients. Each lesion from a given case offers the same histological features arguing against the interpretation that these types represent phases in the temporal evolution of tissue injury and not aetiologically determined heterogeneity.

Genetic [locus and allelic] heterogeneity

Allelic heterogeneity does not complicate linkage analysis and it actually provides evidence for the functional relevance of that locus since the probability of several different mutations occurring at the same (irrelevant) locus in unrelated individuals is very low. The probability of locus heterogeneity is suggested by detailed analysis of the most secure finding with respect to genetic susceptibility in multiple sclerosis – association with alleles of the major histocompatibility complex. As discussed above, in addition to the primary association with the DR15/DQ6 phenotype in Northern Europeans, there is a subsidiary association with DR3 (DR17)-DQ2 (and its associated DRB1*0301-DRB5*0101-DQA1*0501-DQB1*0201 genotype) and specifically different associations in some Mediterranean populations which cannot be explained by site specific similarities in sequence of the crucial elements binding peptide. The implication must be that either these alleles are in linkage disequilibrium with another (shared) susceptibility gene encoded within the major histocompatibility complex (which may or may not have a primary immunological function); or that the environmental trigger which initiates the disease process in multiple sclerosis varies and so selects specifically different at-risk populations.

This evidence for genetic heterogeneity is strengthened by the probability that the primary progressive form of multiple sclerosis seen in Northern Europeans is also DR4 associated. The evidence is not entirely secure and the original finding from Scandinavia [61] has not consistently been reproduced. Although there is support from a more recent study of primary progressive multiple sclerosis in Spain [31], definitive studies are needed. The best present position would be that this distinct phenotype not only has an individual pathological profile (less inflammation and more focal axon degeneration) but may also occur on a specifically different genetic background. Furthermore, the optico-spinal but not the Western type of multiple sclerosis is associated with DPB1*0501 in northern and southern Japan whereas the western type is DRB1*1501 associated [56,62].

Conditioning the United Kingdom genome screen DR15 (or an extended DR15 linked haplotype also encoding alleles of TNF and the DQ locus) shows that the regions of interest on 1p, 17p, 17q and X cluster in families which are identical by state for DR15 whereas the non-sharing group is associated with 1cen, 3p, 5cen, 7p, 14q and 22q; in addition new regions of interest are found at 5q and 13p (DR15 sharing) and 16p and 20p (DR15 non-sharing) [47,63]. As knowledge accumulates, conditioned analyses may routinely be needed in order to suggest or exclude new regions of interest or positional candidate susceptibility genes.

Phenocopies of multiple sclerosis

The presence of phenocopies is a major concern in the analysis of complex traits where diagnosis depends on pattern recognition of symptoms, signs and laboratory investigations occurring in the absence of a test for the disease. Reassuringly, the cohort of cases included in the United Kingdom linkage genome screen was not contaminated by cases of CADASIL, spinocerebellar degenerations, or adrenoleukodystrophy in male-male pairs. However, this search did identify a potentially important aspect of genetic heterogeneity which may have implications for selective tissue vulnerability. Although there were no individuals having an excess of triplet repeats for SCA2, the 22kb allele occurred at a higher frequency in cases than controls reported in the literature. This result prompted an assessment of transmission disequilibrium in family trios which supported an association between multiple sclerosis and the 22kb allele [64]. A second data set showed similar findings although the evidence for association weakened (Stephen Sawcer: unpublished observations). If confirmed, one interpretation of this finding would be that, in individuals who have a tendency for auto-immunity as a result of the interplay between genetic susceptibility and environmental factors, the inflammatory process may be focused onto a particular system or pathway within the brain and spinal cord in those who have genetic polymorphisms exposing that pathway for tissue injury. In the case of SCA2, individuals with the 22kb allele polymorphism may have disproportionate inflammatory demyelination of the spino-cerebellar pathways – similar by analogy to involvement of the anterior visual pathway in Harding's disease [54,55].

Genetic Analysis of Multiple Sclerosis in the European Society (GAMES)

The principle of GAMES

A European origin for genetic susceptibility in multiple sclerosis underlines the logic of studying susceptibility amongst the present populations of Europe and neighbouring regions. A collaborative study assumes that some or all of the susceptibility genes for multiple sclerosis are concentrated in Europeans and were distributed by their migrations. More specifically, the varying frequency of multiple sclerosis within Europe reflects the dilution, but uneven diffusion, of factors originating in the north. The power of genetic analysis is influenced by the regional frequency of each susceptibility allele. The probability of detection is reduced where susceptibility factors are over represented in the at-risk population or (conversely) very infrequent since, although these fluctuations determine disease frequency, they also act to minimise the difference between cases and controls. GAMES assumes that several genes contribute to susceptibility in multiple sclerosis. Since it is not possible to anticipate in advance which regions will offer the greatest power to identify each susceptibility factor, a pan-European collaboration is necessary and can be expected to be highly informative.

Based on existing results, linkage disequilibrium mapping will be needed in order to advance knowledge on genetic susceptibility in multiple sclerosis. GAMES is predicated on the hypothesis that some at least of the susceptibility factors responsible for increasing susceptibility to multiple sclerosis in different parts of Europe will be identical by descent through linkage disequilibrium. The experiment assumes that linkage disequilibrium in the outbred European population averages around 0.25cM making it necessary to screen the genome with 6000 microsatellites – approximately 20 times more markers than has been used collectively in linkage genome screens. Within regions of interest, map resolution will be significantly improved by comparing different regions. Identification of the same susceptibility haplotypes in different populations shown to be identical by descent will provide tighter resolution for the location of each susceptibility locus than can be achieved by studying these populations in isolation. Using microsatellites, GAMES is planned to take forward the existing evidence from linkage to full linkage disequilibrium genome screens in the interval before new technologies for high resolution mapping of susceptibility genes using SNPs become generally available.

The GAMES protocol

GAMES will: use 6000 microsatellite markers to screen the genome; compare cases and controls (n≥200 of each) using pooled DNA to identify regions of interest; confirm their status in pooled DNA from >200 additional cases and their parents; type these (or ideally a third set of 200) families individually across all emerging regions of interest; preserve investigator independence but commit to a meta-analysis, posted at *http://www.mrc-bsu.cam.ac.uk/MSgenetics*; create separate pools for primary progressive and relapsing-secondary progressive forms of multiple sclerosis; use strict diagnostic criteria and follow-up of individual cases in order later to stratify the analysis for genotypic and phenotypic sub-groups reclassified using revised criteria [65]; where more than 100 affected sibling pairs are available, a conventional linkage genome screen will be carried out in order to identify provisional regions of interest since cumulative experience of linkage analysis in Europe should eventually cross the threshold for reliably identifying regions of interest; but the number of families required for linkage analysis in genetic isolates will be reduced to 40.

GAMES predicts that some regions of interest will disappear – the *false positives*; others will be shared, to a varying extent, between players – the *ubiquitous* genes; some regions of interest will be confined to selected participants – the *domestic* genes – providing evidence both for the effects of single quantitative traits and true heterogeneity which may be expressed as distinct clinical or laboratory phenotypes.

Conclusions

Resolving the issues of complexity and heterogeneity in multiple sclerosis, and other complex traits, has practical dividends. Without knowledge linking aetiology to pathogenesis and phenotype, putative new treatments – selected on the basis of an imperfect understanding of the pathogenesis – will continue to be screened in cohorts who may or may not have an appropriate pathological substrate for that particular intervention.

References

1. Sadovnick AD, Baird PA, Ward RH. Multiple sclerosis; updated risks for relatives. American Journal of Medical Genetics 1988;29:533–541.
2. Robertson NP, Fraser M, Deans J, Clayton D, Compston DAS. Age adjusted recurrence risks for relatives of patients with multiple sclerosis. Brain 1996;119:449–455.
3. Carton H, Vlietinck R, Debruyne J et al. Recurrence risks of multiple sclerosis in relatives of patients in Flanders, Belgium. Journal of Neurology Neurosurgery and Psychiatry 1996;62:329–333.
4. French Research Group on Multiple Sclerosis. Multiple Sclerosis in 54 twinships: concordance rate is independent of zygosity. Annals of Neurology 1992;32:724–727.
5. Sadovnick AD, Armstrong H, Rice GPA et al. A population-based study of multiple sclerosis in twins: update. Annals of Neurology 1993;33:281–285.
6. Mumford, CJ, NW Wood, HF Kellar-Wood et al. The British Isles survey of multiple sclerosis in twins. Neurology 1994;44:11–15.
7. Robertson NP, O'Riordan JI, Chataway J et al. Clinical characteristics and offspring recurrence rates of conjugal multiple sclerosis. Lancet 1997;349:1587–1590.
8. Sadovnick AD, Ebers GC, Dyment DA, Risch N, the Canadian Collaborative Study Group. Evidence for genetic basis of multiple sclerosis. Lancet 1996;347:1728–1730.
9. Ebers GC, Sadovnick AD, Risch NJ. A genetic basis for familial aggregation in multiple sclerosis. Nature 1995;377:150–151.
10. Broadley S, Deans J, Sawcer SJ, Clayton D, Compston DAS. Autoimmune disease in first degree relatives of patients with multiple sclerosis in the United Kingdom. Brain 2000;123:1102–1111.
11. Hupperts R, Broadley S, Mander A et al. Patterns of disease in concordant parent-child pairs with multiple sclerosis (in press).
12. Risch N. Linkage strategies for genetically complex traits. American Journal of Human Genetics 1990;46:222–253.
13. Risch NJ. Searching for genetic determinants in the new millennium. Nature 2000;405:847–856.
14. Transatlantic Multiple Sclerosis Genetics Cooperative. A meta-analysis of genome screens in multiple sclerosis. J of Multiple Sclerosis 2001;7:3–11.
15. Becker KG, Simon RM, Bailey-Wilson JE. Clustering of non-major histocompatibility complex susceptibility candidate loci in human autoimune disease. Proceedings of the National Academy of Sciences of the USA 1998;95:9979–9984.
16. Merriman A, Cordell HJ, Eaves IA et al. Suggestive evidence for association of human chromosome 18q12-q21 and its orthologue on rat and mouse chromosome 18 with several autoimmune diseases. Diabetes 2001;50:184–194.
17. Kruglyak L. Prospects for whole-genome linkage disequilibrium mapping of common disease genes. Nature Genetics 1999;22:139–144.
18. Taillon-Miller P, Bauer-Sardina I, Saccone NL et al. Juxtaposed regions of exctensive and mini-

mal linkage disequilibrium in human Xq25 and Xq28.Nature Genetics 2000;25:324–328.
19. Rogers AE, Meehan J, Guilleminault C et al. HLA DR15 (DR2) and DQB1*0602 typing studies in 188 narcoleptic patients with cataplexy. Neurology 1997;48:1550–1556.
20. Barcellos LF, Klitz W, Field LL et al. Association mapping of disease loci using a pooled DNA genomic screen. American Journal of Human Genetics 1997;61:734–747.
21. Olerup O, Hillert J. HLA class II-associated genetic susceptibility in multiple sclerosis: a critical evaluation. Tissue Antigens 1991;38:1–15.
22. Marrosu MG, Muntoni F, Murru MR et al. HLA-DQB1 genotype in Sardinian multiple sclerosis: evidence for a key role of DQB1. 0201 and DQB1.0302 alleles. Neurology 1992;42:883–886.
23. Coraddu F, Reyes-Yanez MP, Aladro Y et al. HLA associations with multiple sclerosis in the Canary Islands. Journal of Neuroimmunology 1998;87:130–135.
24. Saruhan-Direskeneli G, Esin S, Baykan-Kurt B et al. HLA-DR and -DQ associations with multiple sclerosis in Turkey. Human Immunology 1997;55:59–65.
25. Wood N, Sawcer SJ, Kellar-Wood H, Holmans P, Clayton D, Robertson N, Compston DAS. Susceptibility to multiple sclerosis and the immunoglobulin heavy chain variable region. Journal of Neurology 1995;242:677–682.
26. Wood N, Sawcer SJ, Kellar-Wood H, Holmans P, Clayton D, Deans J. Fraser M, Compston DAS. The T-cell receptor beta locus and susceptibility to multiple sclerosis. Neurology 1995;45:1859–1863.
27. Seboun E, Robinson MA, Doolittle TH et al. A susceptibility locus for multiple sclerosis is linked to the T cell receptor beta chain complex. Cell 1989;57:1095–1100.
28. Ligers A, He B, Fogdell-Hahn A, Olerup O, Hillert J. No linkage or association of a VNTR marker in the junction region of the immunoglobulin heavy chain genes in multiple sclerosis. European Journal of Immunogenetics 1997;24:259–264.
29. Droogan AG, Kirk CW, Hawkins SA et al. T cell receptor alpha, beta, gamma and delta chain microsatellites show no association with multiple sclerosis. Neurology 1996;47:1049–1053.
30. Mycko MP. Kwinkowoski M, Tronczynska E et al. Multiple sclerosis: increased frequency of the ICAM-1 exon 6 gene point mutation genetic type K469. Annals of Neurology 1998;44:70–75.
31. De La Concha EG, Arroyo R, Crusius JB et al. Combined effect of HLA-DRB1*1501 and interleukin-1 receptor antagonist gene allele 2 in susceptibility to relapsing/remitting multiple sclerosis. Journal of Neuroimmunology 1997;80:172–178.
32. Wansen K, Pastinen T, Kuokkanen S, Wokstrom J et al. Immune system genes in multiple sclerosis: genetic association and linkage analyses on TCR-ß, IgH, IFN-gamma and IL-12ra/IL-1ß loci. Journal of Neurology 1997;79:29–36.
33. Vanderbroeck K, Martino G, Marrosu M et al. Occurrence and clinical relevance of an interleukin-4 gene polymorphism in patients with multiple sclerosis. Journal of Neuroimmunology 1997;76:189–192.
34. He B, Xu C, Yang B, Landtblom A-M, Fredrikson S, Hillert J. Linkage and association analysis of genes encoding cytokines and myelin proteins in multiple sclerosis. Journal of Neuroimmunology 1998;86:13–19.
35. Bennets BH, Teutsch SM, Buhler MM et al. The CCR5 deletion mutation fails to protect against multiple sclerosis. Human Immunology 1997;58:52–59.
36. Reboul J, Mertens C, Levillayer F et al. Cytokines in genetic susceptibility to multiple sclerosis: a candidate gene approach. Journal of Neuroimmunology 2000;102:107–112.
37. Tienari P, Wikstrom J, Sajantila A, Palo J, Peltonen L. Genetic susceptibility to multiple sclerosis linked to myelin basic protein gene. Lancet 1992;340:987–991.
38. Tienari P, Kuokkanen S, Pastinen T et al. Golli-MBP gene in multiple sclerosis. Journal of Neuroimmunology 1998;81:158–167.
39. Nellemann LJ, Frederiksen J, Morling N. PCR typing of two short tandem repeat (STR) structures upstream of the human myelin basic protein (MBP) gene; genetic susceptibility in multiple sclerosis and monosymptomatic idiopathic optic neuritis in Danes. Multiple Sclerosis 1995;1:186–189.

40. Price SE, Sharpe G, Boots A et al. Role of myelin basic protein and proteolipid protein genes in multiple sclerosis: single strand conformation polymorphism analysis of the human sequences. Neuropathology and Applied Neurobiology 1997;23:457–467.
41. Thompson RJ, Mason CR, Douglas AJ et al. Analysis of polymorphisms of the 2'3'-cyclic nucleotide-3'-phosphodiesterase gene in patients with multiple sclerosis. Multiple Sclerosis 1996;2:215–221.
42. Rodriguez D, Della Gaspera B, Zalc B et al. Identification of a Val 145 Ile substitution in the human myelin oligodendrocyte glycoprotein: lack of association with multiple sclerosis. Multiple Sclerosis 1997;3:377–382.
43. Mertens C, Brassat D, Reboul J et al. A systematic study of oligodendrocyte growth factors as candidates for genetic susceptibility to multiple sclerosis. Neurology 1998;51:748–753.
44. Haines JL, Ter-Minassian M, Bazyk A et al A complete genomic screen for multiple sclerosis underscores a role for the major histocompatibility complex. Nature Genetics 1996;13:469–471.
45. Ebers GC, Kukay K, Bulman D et al. A full genome search in multiple sclerosis. Nature Genetics 1996;13:472–476.
46. Sawcer S, Jones HB, Feakes R et al. A genome screen in multiple sclerosis reveals susceptibility loci on chromosome 6p21 and 17q22. Nature Genetics 1996;13:464–468.
47. Chataway J, Feakes R, Coraddu F et al. The genetics of multiple sclerosis: principles, background and updated results of the United Kingdom systematic genome screen. Brain 1998;121:1869–1887.
48. Kuokkanen S, Gschwend M, Rioux JD et al. Genomewide scan of multiple sclerosis in Finnish multiplex families. American Journal of Human Genetics 1997;61:1379–1387.
49. Coraddu F, Sawcer S, D'Alfonso S et al. A genome screen for multiple sclerosis in Sardinian multiplex families (submitted).
50. Broadley S, Sawcer S, D'Alfonso S et al. A genome screen for multiple saclerosis in Italian families (submitted).
51. Eaves IA, Merriman TR, Barber RA et al. The genetically isolated populations of Finland and Sardinia may not be a panacea for linkage disequilibrium mapping of common disease genes. Nature Genetics 2000;25:320–323.
52. Xu C, Dai Y, Fredrikson, Hillert J. Association and linkage analysis of candidate chromosomal regions in multiple sclerosis: indication of disease genes in 12q23. European Journal of Human Genetics 1999;7:110–116.
53. D'Alfonso S, Nistico L, Zavattari P et al. A linkage analysis of multiple sclerosis with candidate region markers in Sardinian and continental Italian families. European Journal of Human Genetics 1999;7:377–385.
54. Harding AE, Sweeney MG, Brockington M et al. Occurrence of a multiple sclerosis-like illness in women who have a Leber's hereditary optic neuropathy mitochondrial DNA mutation. Brain 1992;115:989–989.
55. Kellar Wood H, Robertson N, Govan GG et al. Leber's hereditary optic neuropathy mitochondrial DNA mutations in multiple sclerosis. Annals of Neurology 1994;36:109–112.
56. Yamasaki K, Horiuchi I, Minohara M et al. HLA-DPB1*0501-associated opticospinal multiple sclerosis: clinical, neuroimaging and immunogenetic studies. Brain 1999;122:1689–1696.
57. Dean G, Bhighee AIG, Bill PLA et al. Multiple sclerosis is black South Africans and Zimbabweans. Journal of Neurology Neurosurgery and Psychiatry 1994;57:1064–1069.
58. Lee MA, Smith S, Palace J et al. Spatial mapping of T_2 and gadolinium-enhancing T_1 lesion volumes in multiple sclerosis: evidence for distinct mechanisms of lesion genesis. Brain 1999;122:1261–1270.
59. Bitsch A, Schuchardt S, Bunkowski S, Kuhlmann T, Bruck W. Acute axonal injury in multiple sclerosis: correlation with demyelination and inflammation. Brain 2000;123:1174–1183.
60. Lucchinetti C, Bruck W, Parisi J et al. Heterogeneity for multiple sclerosis lesions: implications for the pathogenesis of demyelination. Annals of Neurology. 2000;47:707–717.

61. Hillert J, Gronning M, Hyland H, Link H, Olerup O. Immunogenetic heterogeneity in multiple sclerosis. Journal of Neurology Neurosurgery and Psychiatry 1992;55:887–890.
62. Ma, JJ, Nishimura M, Mine H et al. HLA-DRB1 and tumor necrosis factor gene polymorphisms in Japanese patients with multiple sclerosis. Journal of Neuroimmunology 1998;92:109–112.
63. Coraddu F, Sawcer S, Feakes R et al. HLA typing in the United Kingdom multiple sclerosis genome screen. Neurogenetics 1999;2:24–33.
64. Chataway J, Sawcer S, Coraddu F et al. Allelic variants of the spinocerebellar ataxia genes contribute to multiple sclerosis susceptibility. Neurogenetics 1999;2:91–96.
65. McDonald WI, Compston DAS, Edan G et al. International panel on the diagnosis of multiple sclerosis: new diagnostic criteria for multiple sclerosis. Annals of Neurology (in press).

HLA controlled susceptibility to multiple sclerosis – a survey

Arne Svejgaard*

Department of Clinical Immunology, University Hospital of Copenhagen (Rigshospitalet), Denmark

Introduction

One of the most important developments in understanding the etiology and pathogenesis of multiple sclerosis (MS) was the discovery in 1972 that this disease preferentially occurs in individuals who have inherited certain factors belonging to the HLA (Human Leukocyte Antigen) system. This discovery had two important implications. First, it definitely proved that the predisposition to MS is inherited. Second, it provided a new tool to study how the inheritance of the susceptibility is inherited. Third, because the significance of the HLA system in the immune response was becoming apparent at the same time, it provided evidence for the involvement of immune abnormalities in MS, and because most autoimmune diseases have subsequently been found to be HLA associated, the mere existence of an HLA association indicates the involvement of autoimmunity.

In this survey, we first summarize some basic information on the HLA system and its biological importance, review the HLA associations in MS including a few historical notes, and finally discuss some of the possible mechanisms behind the associations with special reference to very recent observations throwing new light over these mechanisms. It is impossible to cite all relevant publications within the space available and accordingly, references are mainly made to the most recent publications and to surveys.

The HLA complex

The HLA system is the Major Histocompatibility Complex (MHC) of man and contains three classes of molecules: HLA Class I (e.g. the HLA-A, B, and C cell surface molecules), Class II (e.g. HLA-DR, DQ, and DP cell surface molecules), and the HLA Class III molecules which comprise a variety of factors described below. The HLA genes are located on the short arm of chromosome 6 where they occupy about 3.6 megabases at position 6p21. This entire region has recently been cloned [1] and shown to contain 128 expressed genes, about half of which have immune function. A detailed description of the system is given in [2] and the nomenclature is summarized in [3]. The relative positions of some selected genes

---

Correspondence address: Department of Clinical Immunology, Section 7631, Rigshospitalet, Blegdamsvej 9, DK-2100 Copenhagen O, Denmark. Tel: +(45) 35 45 76 30. Fax: +(45) 35 39 87 66. E-mail: arnesvej@post4.tele.dk

HLA Complex

Fig. 1. The HLA Complex on Chromosome 6. The location of some critical loci and regions are shown relative to each other. The distances between the loci/regions are not to scale.

in the HLA complex are shown in Fig. 1. The Class II genes are located in the centromeric end of the complex, while the Class I genes are in the telomeric end, and most of the Class III genes lie in the middle of the complex Classes.

Both Class I and Class II molecules are anchored to the cell surface, and they have related biological function because they both bind antigenic peptides in their so-called antigen-binding cleft located in the part of the molecule pointing away from the cell membranes. These antigenic peptides are presented to the T-cell receptor (TCR) on T lymphocytes. If the TCR is sufficiently specific for the HLA/peptide complex on the antigen-presenting cell, the T cell is induced to respond. Class I molecules present peptides to CD8-positive cytotoxic T-cells, while Class II molecules present their peptides to CD4-positive T-helper lymphocytes. The LMP (large multifunctional protease) and TAP (Transporter of Antigenic Peptides) molecules are involved in the loading of HLA Class I molecules with antigenic peptides, but the genes are located in the Class II region.

HLA Class I and Class II molecules are thus deeply involved in the thymus-dependent, T-cell mediated immune response including the development of tolerance to self antigens, and this is probably the explanation for the associations between various HLA factors and most autoimmune disease [4]: the tolerance development have failed in some way.

The term HLA Class III factors is generally used for all other molecules (apart form Class I and II) controlled by genes in the HLA complex. Some of these molecules have immune functions, e.g. the complement factors C2 and C4 and properdin factor Bf, the cytokines TNFα and β, and heat shock protein 70 (hsp70) and some are without immunological function, e.g. 21-hydroxylase.

The HLA system is characterized by pronounced linkage disequilibrium, i.e. alleles at neighboring loci tend to occur more or less frequently in the same haplotype (i.e. the HLA genes inherited from one parent) than should be expected from the frequency of the individual alleles. This phenomenon is on the one hand an advantage because it helps identifying the mere existence of polymorphic disease susceptibility genes within or near the HLA complex, while it on the other complicates the precise identification of the genes actually responsible for the various disease associations. Typically, one or a few alleles at a locus is in strong positive

Table 1. Some Alleles in the HLA region involved in the MS association. The involvement of markers in brackets is probably due to linkage disequilibrium.

Locus	DQB1	DQA1	DRB1	DRB5	TNFa	HLA-B
Allele	0602	(0102)	1501	0101	(121-bp)	(B7)

linkage disequilibrium with one or a few alleles at a nearby locus. The phenomenon can be exemplified by the some MS associated HLA genes in Caucasoids (Table 1). For example, the *HLA-B7,DR2* haplotype has a frequency of about 8.2% in Denmark against an expected frequency of only 0.22% if there were linkage equilibrium.

HLA and MS – the first studies

The first studies of HLA in MS were reported in 1972 by Naito et al. [5], Jersild et al. [6], and Bertrams et al. [7] who found positive associations between MS and HLA-A3 and HLA-B7. At that time, only HLA Class I antigens were known, but in 1973, the Class II determinant's were discovered by means of the so-called MLC (Mixed Leukocyte Culture) test and cellular typing by means of MLC homozygous typing cells was developed. It soon became clear that there is strong linkage disequilibrium between HLA-B7 and a specific MLC determinant, which is now called HLA-D2. Accordingly, Jersild et al. [8] investigated a series of MS patients and controls both for HLA-B7 and for HLA-D2 and found that the association between MS and HLA-D2 is stronger than the association with HLA-B7. When applying the adequate statistical test [9] as illustrated in Table 2, it appears that this conclusion is actually correct.

The detailed HLA associations in MS

Several years later when serological typing for HLA Class II determinants became possible, it was not surprisingly found that the serologic equivalent of HLA-D2, i.e. the HLA-DR2 antigen, is also strongly associated with MS. Subsequently, the DR2 determinant was split in two serological subtypes, HLA-DR15 and 16, and when the corresponding genes were finally cloned and sequenced, it became clear that each of these subtypes can again be split in two: HLA-DR1*1501 and 1502 for DR15 and DRB1*1601 and 1602 for DR16. It is now clear that it is the DRB1*1501 determinant which is associated with MS (cf. [10]).

However, the situation becomes further complicated because there are a number of additional HLA-DRB loci, and one of these is particularly interesting in terms of MS: the HLA-DRB5 locus where the allele DRB5*0101 is in virtually absolute linkage disequilibrium with HLA-DRB1*1501, i.e. these two genes are always present together in the same haplotype.

The final complication is due to the fact that in the DQ series of HLA Class II factors, the DQB1*0602 alleles is also in almost absolute linkage disequilib-

Table 2. Testing the HLA-B7 vs. the HLA-D2 associations in MS.

Comparison	Odds Ratio (OR)	Deviation of OR from Unity: Fisher's exact p-value
HLA-B7 association in MS	1.4	0.28
HLA-D2 associations in MS	5.6	0.0014*
B7 association in D2-positives	0.5	0.28
B7 association in D2-negatives	0.2	0.15
D2 association in B7-positives	20	0.034*
D2 association in B7-negatives	7.3	0.052*

*Calculation from [9] on data from Jersild et al. [8]. OR-values above or below unity indicate positive or negative associations, respectively. P-values marked with asterisks have been corrected for the number of comparisons. It appears that the D2 marker is positively associated with MS both in B7-positive patients as compared with B7-positive controls and in B7-negative patients compared with corresponding controls. Conversely, the weak B7 association remains negative when stratified for the presence and absence of D2.

rium with DRB1*1501 and DRB5*0101. The strong linkage disequilibria between these Class II alleles makes it extremely difficult to decide which one of them shows the strongest association with MS.

One clue came from a study by molecular typing of large numbers of Norwegian DR2 positive MS patients and controls [11]. It was noted that two of 181 patients carried DQB1*0602 in the absence of DRB1*1501, while this haplotype was not observed in 294 controls. Although these figures are too small to allow meaningful significance testing, this observation indicates that DQB1*06202 may by itself predispose to MS. However, another interesting observation appears from this report because it was noted that seven of the patients but only one of the controls carried DRB1*1501 in the absence of DQB1*0602 indicating that DRB1*1501 apparently can contribute to the disease without DQB1*0602 being involved. Most recently, a very important study [12] on MS in African Brazilians showed that DQB1*0602 was present in 45% of 44 MS patients, but in only 17% of 88 ethnically matched controls (Fisher's p = 0.001). Importantly, DRB1*1501 was only present in three patients and one control (S. Leon personal communication), i.e. not significantly increased. Accordingly, in this special population, HLA-DQB1*0602 is definitely more strongly associated with MS than is DRB1*1501.

The HLA association for MS in the special population of Sardinia [13] is extraordinary because it mainly involves HLA-DR4 and 3 in the following haplotypes: *DRB1*0405,DQA1*0501,DQB1*0301* and *DRB1*0301,DQA1*0501,DQB1*0201*. Recently, however, it has become clear that HLA-DR2 is actually also associated with MS in this population [14]. HLA-DR2 is also associated with MS in both Ashkenazi and non-Ashkenazi Israeli Jews [15], in MS patients from Spain [16] and the Canary Islands [17].

The situation in Japanese patients was unclear until recently when distinction between two clinical forms was made: the so-called Western form characterized by disseminated central nervous system involvement appears to have a DR2 association [18,19], while this is not the case for the so-called optico-spinal or Asian form, which has no clear HLA association.

Thus, the DRB1*1501 and/or DQB1*0602 associations are almost universal with Sardinians as notable exceptions having additional, stronger associations with DR4 and DR3. In general, it seems quite plausible that both the DRB1*1501 and DQB1*0602 molecules (and perhaps the DRB5*0101 molecule) can confer susceptibility to MS, and it is even possible that they may act in concert.

Impact of HLA on the genetic susceptibility to MS

The HLA association in Northern Europe may be exemplified by the situation in Denmark, where DRB1*1501 is present in 65% of MS patients versus 28% of controls giving a relative risk of 3.8, i.e. DRB1*1501-positives have about 3.8 times higher risk of developing MS compared with the risk in DRB1*1501-negatives. From these data, the so-called etiologic fraction can be estimated to be about 48% for DRB1*1501 in MS indicating that this genetic marker is involved in the pathogenesis of MS in about half of the cases.

This figure is in striking contrast to the low estimates of HLA involvement obtained from linkage studies. For example, in our studies [20], the λs for HLA was 1.6, which is only about 10% of the total λs of about 15 for full sibs. Indeed, two of the three complete genome screens in affected sib pairs did not detect linkage to the HLA region (cf. [21] for references), but these studies did not include HLA Class genes in the typing. An obvious reason for the discrepancy between results obtained from association and linkage studies is the fact that linkage studies include all sib pairs, also those without DR2, and if DR2 is responsible for the association, non-DR2 MS cases may not show linkage to MS, i.e. these cases will dilute the influence of HLA. This actually seems to be the case: when HLA Class II genes are included in the sib pair studies, linkage can de demonstrated [20,21], and it also appears that the strongest evidence for linkage comes from the HLA-DR genes as compared with micro satellite markers located very closely to the DR locus. Moreover, when including information on DR2, Haines et al. [21] observed that linkage were limited to families segregating for DR2, and they estimated that the HLA system explains between 17 and 62% of the genetic etiology of MS, which include the above estimate of 48% for the etiologic fraction derived solely from association data in unrelated individuals.

Frequently, both dominant and recessive models for MS are used in linkage studies including HLA. However, it is worth noting that the distribution between DR2 homo- and heterozygosity in unrelated MS patients is not compatible with a dominant model, but does not reject a recessive model (Svejgaard et al. unpublished). Thus, the HLA controlled susceptibility to MS seems to be recessive. This may be explained by a gene dose effect of, for example, the HLA-DR2 molecule in antigen presentation, but theoretically, it is also compatible with the involvement of another as yet unknown gene in linkage disequilibrium with DR2 because *DR2/2* homozygotes would be more likely to carry such genes than would *DR2/X* heterozygotes.

Influence of other (non Class II) HLA genes on the susceptibility to MS

As described above, the HLA association in MS was originally detected as a Class I association with HLA-B7 and A3 in particular, but these were secondary and due to linkage disequilibrium as shown in Table 2 for HLA-B7. Some investigators report other Class I associations which apparently cannot be explained by linkage disequilibrium (Hillert, personal communication). However, such associations need very careful matching of controls in terms of the DR2 association. A number of other genetic markers controlled by genes in the HLA complex have been studied in MS. Some show association, but whenever it has been investigated, these associations have been shown to be due to linkage disequilibrium with the above Class II alleles. Thus, allelic variants at the following loci do not appear to separately influence the susceptibility to MS: Complement variants, LMP [22], TAP [22–24], HLA-DM [25], TNF ([19,20] and referenced in [26]), or MOG [27]. The assumption that the genetic predisposition to MS (and narcolepsy as described below) is due to another as yet unknown gene in strong linkage disequilibrium with DQB1*0602 and DRB1*1501 has been weakened considerably after the complete sequencing of the HLA complex [1] which gave little evidence of new, expressed genes close to these Class II genes.

Other HLA-DR2 associated disorders

When considering the HLA associations of MS, the situation in another neurological disorder comes to mind: narcolepsy is associated with precisely the same HLA Class II alleles as MS, and again, the association seems to be stronger with DQB1*0602 than with DRB1*1501 [28]. Although narcolepsy, contrary to MS, is often not considered an autoimmune disease, this possibility cannot be excluded. For example, it may be speculated that hypocretin (=orexin) or its receptor recently identified in dog and mouse models of narcolepsy (reviewed in [29]) may be the targets of autoantibodies, which would make narcolepsy a homologue of myasthenia gravis.

It should also be brought in mind, that DRB1*1501 and/or DQB1*0602 (or associated genes) is strongly negatively associated with insulin-dependent diabetes and thus have a strong dominant, protective effect against this disease, but again the mechanisms is unknown and possibly unrelated to that conferring susceptibility to MS.

Clinical heterogeneity and HLA

Many attempts have been made to clarify whether different disease courses (e.g. relapsing remitting vs. primary progressive) have different HLA associations. Several claims have been made, but none have been confirmed in subsequent studies (see e.g. [30]). Already in 1973, we found evidence [8] that D2-postive patients have a more rapidly progressing MS compared with other patients, but this has been difficult to confirm. If DR2 were associated with rapid progression, it might be expected that DR2-positive patients with isolated optic neuritis would have a higher risk of developing MS subsequently and this may actually be the case [30–32], although the situation may be complicated [33]. An early suggestion [34] that HLA-DR3 may actually protect against rapid progression of MS still needs confirmation but gained recently some support from animal studies [35].

Animal models of MS

Experimental Allergic Encephalitis (EAE) has for many years been used as an animal model of MS in mice and rats, in whom the disease is induced by immunization of susceptible strains with nervous tissue together with Freund's adjuvant. However, EAE may in several respects differ from MS, and it would be surprising if the pathogenesis of EAE completely mirrored that of MS.

A significant step towards understanding the pathogenesis of MS was made recently by Fugger and coworkers [36] who described a humanized murine model of MS using HLA-DR2 and a human T-cell receptor. Three different human molecules possibly involved in the pathogenesis of MS were expressed in transgenic mice: (i) the DR2 molecule controlled by the *DRA*0101/DRB1*1501* genes; (ii) a TCR from an MS derived T-cell clone specific for the immunodominant myelin basic protein (MBP) peptide (84-102) bound to HLA-DR2; (iii) the human CD4 molecule, which reacts better with the human Class II molecule than does the murine counterpart. Lymphocytes from these animals (but not from control mice) responded to the MBP peptide presented by DR2-positive cells in vitro by proliferation and IL-2 and IFNγ production. When immunized with the MBP 84-102 peptide together with adjuvant and pertussis toxin, the mice developed CNS inflammation and demyelinization leading to a disease resembling MS. Actually, four per cent of non-immunized animals developed spontaneous disease, and this incidence increased dramatically when the mice were backcrossed twice to Rag2-deficient mice, which cannot recombine their TCR or immunoglobulin genes. Taken together, these experiments provide strong evidence that the HLA-DR2 molecule can mediate an MS-like disease by presenting an MBP self-peptide to T cells. Thus, it seems very likely that the HLA-DR2 molecule can confer susceptibility to MS. However, this does not exclude the possibility that some of the other MS associated HLA Class II molecules (e.g. DQB1*0602 and/or DRB1*0101) may also be involved in autoimmune responses leading to MS. The strongly increased incidence in Rag-2-deficient mice indicates that suppressor cells partially protect non-Rag-2-deficient mice from the development of myelitis as previously observed in a pure murine model of EAE [37]. The humanized murine model is now being used to test various immune interventions aimed at diminishing the demyelinization involved in MS.

Final remarks on the mechanism(s) behind the HLA association

If the association between MS and HLA is indeed due to HLA-DR2 (or one of the other Class II molecules listed in Table 2), a critical question concerns the mechanism behind the occurrence of MS in DR2-negative individuals. Several, not mutually exclusive mechanisms may be at work. First, if DR2 is merely a marker in linkage disequilibrium with a gene truly responsible for the association, this gene may also be present in some DR2-negative haplotypes, and it may have different linkage disequilibria in different populations, which could explain the different Class II associations, seen. Second, there may be different triggering events and/or agents (e.g. virus) in DR2-positive versus DR2-negative individuals, i.e. etiologic and/or pathogenic heterogeneity. Third, if antigen presentation is always involved in the pathogenesis (which may actually not be the case), it is quite possible that other Class II molecules (e.g. DR3 and/or DR4 in Sardinians) can bind and present other peptides from MBP or other proteins in the nervous system, e.g. from MOG, which could induce the disease process. Conceivably, if the triggering event is sufficiently forceful, any Class II molecule may bind and present at least one of the peptides from one of proteins present in the nervous system. However, the plausible key role of the DR2 molecule was strengthened by the recent demonstration of myelin basic protein (MBP) T-cell epitopes in human multiple sclerosis lesions by means of a monoclonal antibody specific for the HLA-DR2-MBP 85-99 complex [38]. Although HLA Class I associations in MS independent of linkage disequilibrium with Class II alleles remain to be proven, their absence does not exclude that Class I molecules may be involved in the pathogenesis, e.g. as targets in T-cell cytotoxicity. It is quite possible that the combined repertoire of peptides which can bind to Class I molecules may be so diverse that all individuals have at least one Class I molecule which can bind and present one of the peptides.

A very recent review on the HLA System [39] in the prestigious New England Journal of Medicine illustrates how even distinguished scientists can make serious errors. First, a subsequent correction [40] was required to inform the readers that critical data given in the review on HLA and disease associations, but not referenced, were actually taken from reference 4 in the present survey.

Second, the authors postulate that the very strong association between HLA-DQB1*0602 and narcolepsy is due to the location of the *HCRTR2* gene coding for the hypocretin (=orexin) 2 receptor in the vicinity of the HLA-DQB1*0602 allele. This assumption is based solely on references to observations in dogs [41] and mice [42]. Apparently, the authors imply that the HLA association of narcolepsy is due to linkage disequilibrium between an *HCTTR2* allele and HLA-DQB1*0602. However, the human *HCRTR2* gene was already in 1998 shown by Sakurai et al. [43] to be located between 6p11 and 6q11, *i.e.* close to the centromere of chromosome 6, which also contains the HLA complex. However, the HLA complex is located at 6p21.33–6p21.31, *i.e.* quite a long way from the centromere. Indeed the distance between *HCRTR2* and HLA is so large that it definitely excludes linkage disequilibrium between an *HCRTR2* allele and the HLA-DQB1*0602 allele as an explanation for the very strong association between narcolepsy and HLA-DQB1*0602. In addition, Peyron et al. [44] more recently found only one mutation in the *HCRTR2* receptor gene among 74 patients with narcolepsy. As mentioned above, these authors are probably right when assuming that the absence of hypocretin in the cerebrospinal fluid of almost all narcoleptics is probably due to autoimmunity, which probably explains the HLA Class II association also in this disorder.

Acknowledgments

This report was aided by grants from the Danish Medical Research Council, the Danish Multiple Sclerosis Society, and the EU (contract No. DG12 SSMJ).

References

1. The MHC sequencing consortium. Complete sequence and gene map of a human histocompatibility complex. Nature 1999;401:921–923.
2. Parham P (Ed). Genomic organization of the MHC: structure origin and function. Immunological Review 1999;vol. 167.
3. Bodmer JG, Marsh SG, Albert ED, Bodmer WF, et al. Nomenclature for factors of the HLA system, 1998. Tissue Antigens 1999; 53:407–446.
4. Svejgaard A. MHC and Disease Associations, Chapter 37, pp.37.1–37.13. In:Weir's Handbook of Experimental Immunology in Four Volumes, 5th Edition (Herzenberg LA, Herzenberg LA, eds), New York:Blackwell Scientific Publications, 1996; 37.1–37.13.
5. Naito S, Namerow N, Mickey MR, Terasaki PI. Multiple sclerosis: association with HL-A3. Tissue Antigens 1972; 2:1–4.
6. Jersild C, Svejgaard A, Fog T. HL-A antigens and multiple sclerosis. Lancet 1972;1:1240–1241.
7. Bertrams J, Kuwert E, Liedtke U. HL-A antigens and multiple sclerosis. Tissue Antigens 1972;2:405–408.
8. Jersild C, Fog T, Hansen GS, Thomsen M, et al. Histocompatibility determinants in multiple sclerosis, with special reference to clinical course. Lancet 1973; 2:1221–1225.
9. Svejgaard A, Ryder LP. HLA and disease associations: detecting the strongest association. Tissue Antigens 1994; 43:18–27.
10. Olerup O, Hillert J. HLA class II-associated genetic susceptibility in multiple sclerosis: a critical evaluation. Tissue Antigens 1991;38:1–15.
11. Spurkland A, Celius EG, Knutsen I, Beiske A. The HLA-DQ(alpha 1*0102, beta 1*0602) heterodimer may confer susceptibility to multiple sclerosis in the absence of the HLA-DR(alpha 1*01, beta 1*1501) heterodimer. Tissue Antigens 1997;50:15–22.
12. Caballero A, Alves-Leon S, Papais-Alvarenga R, Fernandez O. DQB1*0602 confers genetic susceptibility to multiple sclerosis in Afro- Brazilians. Tissue Antigens 1999;54:524–526.
13. Marrosu MG, Murru MR, Costa G, Murru R, et al. DRB1-DQA1-DQB1 loci and multiple sclerosis predisposition in the Sardinian population. Hum Mol Genet 1998;7:1235–1237.
14. Marrosu MG, Murru MR, Costa G, Murru R, et al. Relative predispositional effects (RPEs) of HLA-DRB1-DQA1-DQB1 haplotype in Sardinian multiple sclerosis. ECTRIMS Abstract Book 1999 (Abstract).
15. Kwon OJ, Karni A, Israel S, Brautbar C. HLA class II susceptibility to multiple sclerosis among Ashkenazi and non-Ashkenazi Jews. Arch Neurol 1999;56:555–560.

16. Clerici N, Hernandez M, Fernandez M, Rosique J. Multiple sclerosis is associated with HLA-DR2 antigen in Spain. Tissue Antigens 1989;34:309–311.
17. Coraddu F, Reyes-Yanez MP, Parra A, Gray J. HLA associations with multiple sclerosis in the Canary Islands. J Neuroimmunol 1989;87:130–135.
18. Ono T, Zambenedetti MR, Yamasaki K, Kawano Y, et al. Molecular analysis of HLA class I (HLA-A and -B) and HLA class II (HLA- DRB1) genes in Japanese patients with multiple sclerosis (Western type and Asian type). Tissue Antigens 1998;52:539–542.
19. Ma JJ, Nishimura M, Mine H, Saji H, et al. HLA-DRB1 and tumor necrosis factor gene polymorphisms in Japanese patients with multiple sclerosis. J Neuroimmunol 1998;92:109–112.
20. Oturai A, Larsen F, Ryder LP, Madsen HO, et al. Linkage and association analysis of susceptibility regions on chromosomes 5 and 6 in 106 Scandinavian sibling pair families with multiple sclerosis [In Process Citation]. Ann Neurol 1999;46:612–616.
21. Haines JL, Terwedow HA, Burgess K, Pericak-Vance MA, et al. Linkage of the MHC to familial multiple sclerosis suggests genetic heterogeneity. The Multiple Sclerosis Genetics Group. Hum Mol Genet 1998;7:1229–1234.
22. Liblau R, van Endert PM, Sandberg-Wollheim M, Patel SD, et al. Antigen processing gene polymorphisms in HLA-DR2 multiple sclerosis. Neurology 1993;43:1192–1197.
23. Bell RB, Ramachandran S. The relationship of TAP1 and TAP2 dimorphisms to multiple sclerosis susceptibility. J Neuroimmunol 1995;59:201–204.
24. Vandevyver C, Stinissen P, Cassiman JJ, Raus J. TAP 1 and TAP 2 transporter gene polymorphisms in multiple sclerosis: no evidence for disease association with TAP. J Neuroimmunol 1994;54:35–40.
25. Ristori G, Carcassi C, Lai S, Fiori P, et al. HLA-DM polymorphisms do not associate with multiple sclerosis: an association study with analysis of myelin basic protein T cell specificity. J Neuroimmunol 1997;77:181–184.
26. Chataway J, Feakes R, Coraddu F, Gray J, et al. The genetics of multiple sclerosis: principles, background and updated results of the United Kingdom systematic genome screen. Brain 1998;121:1869–1887.
27. Roth MP, Dolbois L, Borot N, Pontarotti P, et al. Myelin oligodendrocyte glycoprotein (MOG) gene polymorphisms and multiple sclerosis: no evidence of disease association with MOG. J Neuroimmunol 1995;61:117–122.
28. Mignot E, Lin X, Arrigoni J, Macaubas C, et al. DQB1*0602 and DQA1*0102 (DQ1) are better markers than DR2 for narcolepsy in Caucasian and black Americans. Sleep 1994;17:S60–7.
29. Siegel JM. Narcolepsy: a key role for hypocretins (orexins). Cell 1999; 98:409–412.
30. McDonnell GV, Mawhinney H, Graham CA, Hawkins SA, et al. A study of the HLA-DR region in clinical subgroups of multiple sclerosis and its influence on prognosis. J Neurol Sci 1999;165:77–83.
31. Sandberg-Wollheim M, Bynke H, Cronqvist S, Holtas S. A long-term prospective study of optic neuritis: evaluation of risk factors. Ann Neurol 1990;27:386–393.
32. Soderstrom M, Ya-Ping J, Hillert J, Link H. Optic neuritis: prognosis for multiple sclerosis from MRI, CSF, and HLA findings. Neurology 1998;50:708–714.
33. Frederiksen JL, Madsen HO, Ryder LP, Larsson HB, et al. HLA typing in acute optic neuritis. Relation to multiple sclerosis and magnetic resonance imaging findings. Arch Neurol 1997;54:76–80.
34. Engell T, Raun NE, Thomsen M, Platz P. HLA and heterogeneity of multiple sclerosis. Neurology 1982;32:1043–1046.
35. Drescher KM, Nguyen LT, Taneja V, Coenen MJ, et al. Expression of the human histocompatibility leukocyte antigen DR3 transgene reduces the severity of demyelination in a murine model of multiple sclerosis. J Clin Invest 1998;101:1765–1774.
36. Madsen LS, Andersson EC, Jansson L, Krogsgaard M, et al. A humanized model for multiple sclerosis using HLA-DR2 and human T-cell receptor. Nature Genetics 1999;23:343–347.

37. Lafaille JJ, Nagashima K, Katsuki M, Tonegawa S. High incidence of spontaneous autoimmune encephalomyelitis in immunodeficient anti-myelin basic protein T cell receptor transgenic mice. Cell 1994;78:399–408.
38. Krogsgaard M, Wucherpfennig KW, Canella B, Hansen BE, Svejgaard A, Pyrdol J, Ditzel H, Raine C, Engberg J, Fugger L. Visualization of myelin basic protein (MBP) T cell epitopes in multiple sclerosis lesions using a monoclonal antibody specific for the human histocompatibility leukocyte antigen (HLA) DR2-MBP 85-99 complex. J Exp Med 2000;191:1395–1412.
39. Klein J, Sato A. The HLA system. Second of two parts. N Engl J Med 2000; 343:782–786.
40. Correction. The HLA system. Second of two parts. N Engl J Med 2000;343:1504.
41. Lin L, Faraco J, Li R, Kadotani H, Rogers W, Lin X, Qiu X, de Jong PJ, Nishino S, Mignot E. The sleep disorder canine narcolepsy is caused by a mutation in the hypocretin (orexin) receptor 2 gene. Cell 1999;98:365–376.
42. Chemelli RM, Willie JT, Sinton CM, Elmquist JK, Scammell T, Lee C, Richardson JA, Williams SC, Xiong Y, Kisanuki Y, Fitch TE, Nakazato M, Hammer RE, Saper CB, Yanagisawa M. Narcolepsy in orexin knockout mice: molecular genetics of sleep regulation . Cell 1999;98:437–451.
43. Sakurai T, Amemiya A, Ishii M, Matsuzaki I, Chemelli RM, Tanaka H, Williams SC, Richardson JA, Kozlowski GP, Wilson S, Arch JR, Buckingham RE, Haynes AC, Carr SA, Annan RS, McNulty DE, Liu WS, Terrett JA, Elshourbagy NA, Bergsma DJ, Yanagisawa M. Orexins and orexin receptors: a family of hypothalamic neuropeptides and G protein-coupled receptors that regulate feeding behavior. Cell 1998;92:573–585.
44. Peyron C, Faraco J, Rogers W, Ripley B, Overeem S, Charnay Y, Nevsimalova S, Aldrich M, Reynolds D, Albin R, Li R, Hungs M, Pedrazzoli M, Padigaru M, Kucherlapati M, Fan J, Maki R, Lammers GJ, Bouras C, Kucherlapati R, Nishino S, Mignot E. A mutation in a case of early onset narcolepsy and a generalized absence of hypocretin peptides in human narcoleptic brains. Nat Med 2000;6:991–997.

Nordic contributions in multiple sclerosis genetics

Jan Hillert*

Division of Neurology, Karolinska Institutet at Huddinge University Hospital, Huddinge, Sweden

Introduction

According to a colourful latter day Viking "saga", the world distribution of multiple sclerosis (MS) may be explained by the spread of "Scandinavian genes" [1]. Even though this remains a vaguely supported hypothesis, still in need of support from epidemiology and molecular genetics, it may serve to increase the interest in MS genetics from an anthropological point of view. At the same time, the widely used argument for the search for MS genes, an increased understanding of the underlying pathogenic mechanisms, is, however also somewhat shaky due to lack of a good track record for this approach in complex, i.e. polygenetic and multifactorial disorders.

Compared with practically all comparable diseases, the case of genes in MS, as indicated by genetic epidemiology, is extremely well documented. In addition to a number of twin studies we also have access to well-performed studies of adoptees, half-siblings and conjugal families. In addition, after a phase of some confusion following the publication of the first generation of genomic screens [2,3,4] we are now beginning to see a picture of increasing clarity and coherence. In fact, we may already have roughly localised a number of the most importance genes to specific chromosomal regions.

Still, it is important to remember that even though we have known of one "MS-gene", or at least specific chromosomal region, namely the HLA DR-DQ subregion, for more than two decades, we are still far from having a good understanding of its role in MS.

HLA: class II and beyond

There is a long history of MS genetic studies from the Nordic countries. Indeed, two of the original reports of association with HLA-A, -B and -D were indeed reported from Denmark [5,6]. Ever since, a large number of HLA-related studies have been performed by various groups in both Denmark, Norway, Finland and Sweden reviewed in [7].

It is well established that the HLA class II genes are of importance for the susceptibility to MS. In the 80:ies and 90:ies, the association with serological specifi-

---

Correspondence address: Department of Neurology, Karolinska Institutet at Huddinge Hosptial, R54, SE-141 86 Huddinge, Sweden. Tel: +46-8-58582056. Fax: +46-8-58587080. E-mail: jan.hillert@neurotec.ki.se

cities was translated into genomic terminology and defined in detail by genomic typing. The MS-associated HLA-DR-DQ haplotype is now established to be HLA-DRB1*1501,DRB5*0101,DQA1*0102,DQB1*0602 [8]. In addition, both Swedish and Finnish studies have demonstrated significant linkage for HLA-DR and -DQ genes [9,10].

Recently, we have identified additional associations in MS with the HLA class I alleles HLA-A2 (negative association) and HLA-A3 (positive) in both Swedish and Norwegian MS patients [11]. These associations were clearly independent of the HLA-DR15,DQ6 haplotype. An increased frequency of HLA-B7, on the other hand, was found to be completely secondary to the DR15,DQ6 association. This observation may help to explain why, in other populations, it has been so relatively difficult to find evidence for linkage in spite of a well-established and often strong association. These findings also raise interesting questions regarding the mechanisms behind HLA molecules in MS in particular and in autoimmune diseases in general.

Recently, we have re-addressed the importance of HLA-DRB alleles in a large number of sporadic Swedish MS patients [12]. A database of 948 MS patients typed for HLA-DRB revealed the lack of importance of any allele on severity as well as on clinical course. The most striking effect was a significantly earlier onset of disease (approximately 2.5 years). In addition to an association with DR15, a relative predispositional effect analysis revealed an independent significant association with HLA-DR3(17).

Candidate gene analysis

Efforts to identify susceptibility genes have generally been frustrating since most positive reports have been followed by lack of confirmation in subsequent studies. Maybe the most promising recent candidate gene in autoimmune disorders in general is the CTLA-4 gene. In MS, both a Norwegian and a Swedish study have found case-control associations [13,14] with the G allele of a biallelic polymorphism at position 49 of the first exon, supported further by a significant transmission distortion of this allele in a family study [14].

The gene coding for myelin basic protein (MBP) has been much in focus after the initial reports of first an association [15] and subsequently both association and linkage in a Finnish study [16]. However, most later studies have been clearly negative including a Swedish study employing a technique allowing definite allele identification [17]. In addition, the MBP locus failed to reveal evidence for linkage in the three genome-wide screens [2-4].

Candidate locus analysis

In 1996, three genomic screens in MS failed to identify single genetics factors of great importance in MS [2,3,4]. On the other hand, a number of loci of possible importance were found. At first sight, the evidence for each of these loci was regarded as too weak to allow great hope for confirmation. The exception from this was a segment on the short arm and centromerically on chromosome 5, which gave positive indications in all three studies, and in a Finnish study published in parallel [18]. Recently, in a joint Nordic study of 117 affected sibling pairs, we have gained further support for this region [19]. In 1997, an independent Finnish genomic screen [20] added support to a locus on 17q first indicated to be of importance in the genomic screen by Sawcer [4]. Recently, this locus has also shown slight evidence for linkage in the Nordic affected sib-pair families [21].

In 48 Swedish multiplex families and large numbers of sporadic MS patients and controls, we have investigated these loci for presence of linkage and/or association. In fact, we have observed suggestive linkage and/or significant association for markers in the following loci: 3p12-13, 5p, 7pter-15, 7q35, 12p12-13, 12q23 and 17q25 [22,23]. At least three of these loci are syntenic to loci of importance for experimental autoimmune encephalomyelitis as well as for experimental arthritis which increases the likelihood that these loci harbour genes with importance for autoimmunity. This is also well in line with the observation of clustering of genes for autoimmune disorders as pointed out recently by Becker et al. [24]. In summary, it seems likely that these loci may indeed in the end reveal genetic factors of importance not only for MS but for autoimmune disorders in general.

Isolated population analysis

We are currently investigating 30 individuals from a small Swedish community of which 25 belong to an extended pedigree originating from a common ancestral couple in the eighteenth century. So far, we have found that most of patients share an 8 cM haplotype on 17p identical by descent. This illustrates a way to identify chromosomal regions of relevance in sedentary populations of a type still widely available in the Nordic countries and suggests that a focus on such populations should be performed before an increased mobility of populations occur.

References

1. Poser CM. Viking voyages: the origin of multiple sclerosis? An essay in medical history. Acta Neurol Sc and Suppl 1995;161:11–22.
2. Ebers GC, Kukay K, Bulman DE et al. A full genome search in multiple sclerosis. Nature Genetics 1996;13: 472–476.
3. Haines JL, Ter-Minassian M, Bazyk A, et al. A complete genomic screen for multiple sclerosis underscores a role for the major histocompatability complex. The Multiple Sclerosis Genetics Group. Nature Genetics 1996;13:469–471.
4. Sawcer S, Jones HB, Feakes R, et al. A genome screen in multiple sclerosis reveals susceptibility loci on chromosome 6p21 and 17q22. Nature Genetics 1996;13:464–468.
5. Jersild C, Svejgaard A. and Fog T. HL-A antigens and multiple sclerosis. Lancet 1972;1:1240–1241.
6. Jersild C, Fog T, Hansen GS et al. Histocompatibility determinants in multiple sclerosis, with special reference to clinical course. Lancet 1973;2:1221–1225.
7. Hillert J. Human leukocyte antigen studies in multiple sclerosis. Ann Neurol 1994;

36(suppl):15–17.
8. Fogdell A, Hillert J, Sachs C and Olerup O. The multiple sclerosis- and narcolepsy-associated HLA class II haplotype includes the DRB5*0101 allele. Tissue Antigens 1995;46:333–336.
9. Tienari PJ, Wikstrom J, Koskimies S, et al. Reappraisal of HLA in multiple sclerosis: close linkage in multiplex families. European Journal of Human Genetics 1993;1:257–268.
10. Fogdell A, Olerup O, Fredrikson S et al. Linkage analysis of HLA class II genes in Swedish multiplex families with multiple sclerosis. Neurology 1997;48:758–762.
11. Fogdell-Hahn A, Ligers A, Gronning M, Hillert J, and Olerup O. Multiple sclerosis. A modifying influence of HLA class I genes in an HLA class II associated autoimmune disease. Tissue antigens 2000;55:140–8.
12. Masterman T, Ligers A, Olerup O, Hillert J. HLA-DRB1 alleles do not influence course or outcome in multiple sclerosis, but DR15 is associated with lower age of onset. Annals of Neurology 2000;48:211–219.
13. Flinstad Harbo H, Celius EG, Vartdal F, Spurkland A. CTLA-4 promoter and exon 1 dimorphisms in multiple sclerosis. Tissue Antigens 1999;53:106–110.
14. Ligers A, Xu C, Saarinen S, Hillert J and Olerup O. The CTLA-4 gene is associated with multiple sclerosis. J Neuroimmunol 1999;97:182–190.
15. Boylan KB, Takahashi N, Paty DW et al. DNA length polymorphism 5′ to the myelin basic protein gene is associated with multiple sclerosis. Ann Neurol 1990:27:291–297.
16. Tienari PJ, Wikström J, Sajantila A, Palo J, Peltonen L. Genetic susceptibility to multiple sclerosis linked to myelin basic protein gene. Lancet 1992:340:987–991.
17. He B, Yang B, Lundahl J, Fredrikson S, and Hillert J. The myelin basic protein gene in multiple sclerosis: Identification of discrete alleles of a 1.3 kb tetranucleotide repeat sequence. Acta Neurologica Scandinavica 1998;97:46–51.
18. Kuokkanen S, Sundvall M, Terwilliger JD, et al. A putative vulnerability locus to multiple sclerosis maps to 5p14-p12 in a region syntenic to the murine locus Eae2 [see comments]. Nature Genetics 1996;13:477–480.
19. Oturai A, Larsen F, Ryder L, Madsen H, Hillert J, Fredrikson S, Sandberg-Wollheim M, Laaksonen M, Koch-Henriksen N, Sawcer S, Fugger L, Sørensen PS, Svejgaard A. Annals of Neurology 1999;46:612–616.
20. Kuokkanen S, Gschwend M, Rioux JD, Daly MJ, Terwilliger JD, Tienari PJ, Wikstrom J, Palo J, Stein LD, Hudson TJ, Lander ES, Peltonen L. Genomewide scan of multiple sclerosis in Finnish multiplex families. Am J Hum Genet 1997 Dec;61(6):1379–87 1997;61:1379–1387.
21. Larsen F, Oturai A, Ryder L, Madsen H, Hillert J, Fredrikson S, Sandberg-Wollheim M., Laaksonen M, Harboe H, Sawcer S, Fugger L, Sorensen PS, Svejgaard A. Linkage analysis of a candidate region in Scandinavian sib pairs with multiple sclerosis reveals linkage to chromosome 17q. Genes and Immunity 2000;1:456–9.
22. Xu C, Dai Y, Fredrikson S, and Hillert J. Association and linkage analysis of candidate chromosomal regions in multiple sclerosis: indication of disease genes in 12q23. Eur J Human Genet 1999:7:110–116.
23. Xu C. Candidate genes and chromosomal loci in multiple sclerosis (Academic thesis). Karolinska Institutet 1999.
24. Becker KB, Simon RM, Bailey-Wilson JE, Freidlin B, Biddison WE, McFarland HF, Trent JM. Clustering of non-major histocompatibility complex susceptibility candidate loci in human autoimmune diseases. Proc Natl Acad Sci USA 1998;95(17):9979–9984.

Genomic screening in multiple sclerosis

Françoise Clerget-Darpoux on behalf of the French Multiple Sclerosis Genetic Group*

INSERM U535, Bâtiment Gregory Pincus, Le Kremlin Bicêtre, France

Introduction

Although the risk to develop Multiple Sclerosis (MS) is low for first degree relatives of an affected (only 2 to 4% for full sibs), this risk is 20 to 60 times greater than for an individual of the general population [1]. Familial correlation may result from both genetic and familial environmental factors but further evidence for a genetic contribution is provided by the difference between the concordance rates for MS, 24% versus 3%, in monozygotic and dizygotic twins respectively [2].

MS is a multifactorial disease and the number of susceptibility genes is unknown. The involvement of HLA factor(s) in the susceptibility is no more disputable but it does not explain the whole familial concentration. Search for other genetic factors have been undertaken by different teams who screened the genome by linkage tests.

Linkage tests

Linkage between a genetic marker and a disease means that the disease and marker transmission in families is not independent. Linkage with a marker implies the presence of a disease susceptibility gene in the marker region. The principle of a systematic screening of the genome by linkage test is to search among markers regularly spaced on the genome those which are linked with the disease. Different statistics may be used for linkage tests.

For a long time, the most widely applied method was the lod score method, proposed by Morton [3]. Computation of Morton's lod scores requires the specification of an underlying disease model which assigns the probabilities of genotypes from the observed phenotypes. If the specification is incorrect, the recombination fraction which is the key parameter will not be correctly estimated [4]. A wrong specification may even lead to an exclusion of the true location of a risk factor [5],

---

*The French Multiple Sclerosis Genetic Group coordinated by Michel Clanet (Toulouse): Cécile Azaïs-Vuillemin (Toulouse), Marie-Claude Babron (Paris), Jean-François Bureau (Paris), Michel Brahic (Paris), David Brassat (Paris), Michel Clanet (Toulouse), Françoise Clerget-Darpoux (Paris), Hélène Copin (Toulouse), Isabelle Cournu (Paris), André Dautigny (Paris), Gilles Edan (Rennes), Sophie Eichenbaum-Voline (Paris), Bertrand Fontaine (Paris), Emmanuelle Génin (Paris), Roland Liblau (Paris), Olivier Lyon-Caen (Paris), Caroline Mertens (Paris), Jean Pelletier (Marseille), Danièle Phan Dinh (Paris), Erwan Quelvennec (Rennes), Jocelyne Reboul (Paris), Marie-Paule Roth (Toulouse), Gilbert Semana (Rennes), Jacqueline Yaouanq (Rennes).

Table 1.

Recombination fraction θ between HLA and MS disease susceptibility locus	Recessive	Dominant
	q = 0.03 f=0.43	q = 0.18 f=0.02
0.00	−31.34	3.11
0.02	−14.08	3.00
0.04	−7.59	2.90
0.06	−3.87	2.72
0.08	−1.49	2.56

as illustrated by the following example. HLA typing was available on 58 nuclear families of Multiple Sclerosis (MS) patients. Lod Score analyses between HLA and a putative MS susceptibility locus were performed under two different models at the disease locus. As shown in Table 1, when assuming a rare recessive allele (q=0.03) with a penetrance of 0.43 [6], the lod scores are extremely negative for small recombination fraction. Thus, under this model, the presence of a risk factor in the HLA region is strongly excluded. In contrast, evidence for a risk factor within the HLA region was obtained (lod score greater than 3 for θ=0) when a frequent dominant allele (q=0.18) with low penetrance (f=0.02) was assumed [7].

In multifactorial diseases, the underlying model is unknown. Morton's lod score statistics should not be applied and alternative statistics, which are model free, should be preferred. The principle of these methods, stated in 1935 [8], is to test if relatives with the same status for the disease are more similar for the marker than expected under independent segregation of the marker and the disease. A very popular strategy, developed for the study of HLA associated diseases [9–11], is to consider a variable IBD measuring the number of marker alleles shared Identical By Descent (2, 1 or 0) by relatives and to compare the observed distribution to what is expected under independent segregation of the marker and the disease. Different statistics are based on the IBD variable. This is the case of the Maximum Lod Score (MLS) proposed by Risch [12] which applies to Affected Sib Pairs (ASP). In the initial method, linkage was tested using a single marker information (software : ASPEX). The method was extended [13,14] for using the simultaneous information given by a set of linked markers (software : MAPMAKER/SIBS). The Non Parametric Linkage (NPL) statistics may be applied to more extended familial structures than ASP and is implemented in the software GENEHUNTER [15]. Other methods proposed to regress the disease trait (which may be a quantitative trait) on the marker genotype [16] (software : SIBPAL) or the WPC method [17].

Genome searches on multiple sclerosis

For Multiple Sclerosis, three full genome searches using information on ASPs have been performed and published in a same issue of Nature Genetics ([18–20]: (G1, G2, G3)). A fourth genome search has been performed not on ASP but on

Table 2.

Reference	Population	# families	# ASP	# independent ASP	# markers
G1	American	52	81	68	443
G2	British	129	143	136	311
G3	Canadian	61	100	76	257
G4	Finnish	16	–	–	328

extended Finnish pedigrees ([21]: (G4)). Description of these four studies is given Table 2.

Although the real statistical significance of results is difficult to assess, it seems that linkage conclusion cannot be drawn at the 5% level in any of the four studies. Indeed, interpretation and comparison come up against several difficulties. Different statistics and different marker information have been used among studies. Some studies used information on each marker separately (single point analyses), others on all markers simultaneously (multipoint analyses). Most reported results have been obtained on non-independent affected sib pairs. It is important to note that using non independent ASPs artificially inflates linkage statistics and thus the false positive rate. The increase may be particularly high when the ASP parents are untyped [22]. In addition, when parents are untyped, the single-point analyses are very sensitive to the specification of marker allele frequencies [23]. This problem is well illustrated by the Canadian study (G3) with a marker of chromosome 5 (D5S406) for which the MLS value is 4.24 in single point analysis on the 100 non independent ASP. The MLS drops to 3.1 in multipoint analysis or to 2.72 when the analysis is performed only on the 76 independent sib pairs. The result of multipoint analysis on independent sib-pairs is not provided.

In a genome screen, faced to multiple testing (many linkage tests), a high threshold for linkage conclusion is required to limit the rate of false positives to 5%. Consequently the power to detect the true positive in a single study is low.

In a first step, it may be of interest to compare the regions suggested at a less stringent significance level than 5%. Table 3 shows the region suggested in the different studies. In bold are the results expected to occur once at random in a genome scan, the other results indicate smaller hits.

The more consistent results seem to be on chromosome 6p and correspond to the HLA region, although the individual results are not highly significant. Three studies also obtained a small hit in the 19q region.

A more formal meta-analysis of the four genome screens was recently performed [24]. Its principle was :
1) to split the chromosomes into bins of equal genetic length;
2) to rank, for each scan, the results according to their significance;
3) to sum, for each bin, the ranks across scans;
4) and lastly to compute the probability of the summed ranks under the null hypothesis (no susceptibility locus).

Five regions of interest can be retained corresponding to a probability lower than 5%:

Table 3.

US	UK	Canada	Finland
	1p	1p	
2p	**2p**	2p	
	3p	3p	
	3q	3q	
4q		4q	
		5p	
5q	**5q**		
6p	**6p**	6p	6p
6q	**6q**		
7q		7q	
	17q		17q
18p		18p	
19q	**19q**	19q	

the regions 6p (p= 0.0004) ; 19q (0.002) ; 5p (0.025) ; 17q (0.029) ; 2p (0.033).

Although such an approach is clearly more promising than the interpretation of separate studies, it also has inherent weaknesses. An evident one is the quality of the classification in step 2 when, as emphasized above, the significance of the individual results are disputable. The second one is in the choice of the bin lengths and cutpoints. Indeed, a susceptibility locus should contribute to the summed rank through the same bin accross studies. Lastly, such an approach cannot use the results of studies which are not full scans such as partial replication studies.

Indeed, additional linkage studies were performed focusing only on regions already suggested. In particular, linkage was re-tested in 4 regions 2p, 5q, 17q, 19q in new samples of 82 American ASPs [18], of 108 British ASPs [19] and of 94 French independent ASPs sampled by the French Multiple Sclerosis group [25]. The results are all negative, except a small hit on 19q for US data, on 17q for UK data and on 5q for French data. The Canadian group re-tested the 5p region on two additional samples of 44 and 78 ASP and obtained a small hit in only one of the two samples [20].

Discussion

Overall, genome screens of MS provide poor evidence of linkage. In addition, large variations of results from one sample to another are observed. This is what is expected in presence of risk factors with moderate effect [26]. The involvement of genetic risk factors with an important effect in MS is unlikely but the existence of factors with moderate effect cannot be excluded in particular in the regions revealed by meta-analysis.

The HLA region is a good illustration of the difficulty to detect a risk factor with moderate effect by a genome scan. Linkage tests did not reach significance in the HLA region in any of the four genome scans although the presence of a risk factor in this region is clearly evidenced by a candidate gene approach [27].

A candidate gene strategy focuses on specific genes called «candidate genes», chosen either for their functionality (i.e. HLA genes, oligodendrocytes, cytokines can be considered as candidates for MS) or for their position in a region showing linkage or in a region homologous to another species (murine EAE loci). A candidate gene strategy has several tangible benefits over systematic linkage tests on the whole genome. First, focusing only on a few genes limits the number of tests. Second, information may be gained through gametic disequilibrium between the marker alleles and those of the functional factor, the ideal situation being complete gametic disequilibrium, i.e. working on the functional factor itself. Lastly, one gets rid of the uncertainty of the susceptibility gene location. However, the choice of good candidate genes as well as markers showing gametic disequilibrium is not that simple and a tight collaboration between physiopathologists and geneticists is a prerequisite for success.

Faced to the difficulty of detecting the genetic factors involved in a multifactorial disease, the geneticist aims at designing the most efficient strategies [28]. In that respect, meta-analysis approaches are very much promising and large collaborations are set up with the intent to homogenize the clinical data sampling, the marker typing and the strategy of analysis. It is also important to take into account already known co-variables and risk factor(s) (genetic and/or environmental). This may be fundamental in case of interaction. For example, conditioning on HLA may greatly increase the power to detect other non-HLA risk factors in an autoimmune disease. When studying the susceptibility to an infectious disease, the degree of exposure to the infectious agent is clearly necessary. For MS, the environmental agent(s) are unfortunately unknown. Genetic and environmental heterogeneity increases even more the complexity. A possible way to minimise heterogeneity is to study the disease model in animal species. Such an approach may provide candidate genes or regions which can be further studied for humans.

Acknowledgments

We would like to thank "Association pour la Recherche sur la Sclérose en Plaques" (ARSEP) for their active support.

References

1. Sadovnick AD, Ebers GC. Epidemiology of multiple sclerosis: a critical overview. Can J Neurol Sci 1993;20:17–29.
2. Fontaine B, Clerget-Darpoux F, le réseau INSERM de Recherche Clinique sur la susceptibilité à la SEP. La prédisposition génétique à la Sclérose en plaques. Pathologie 2000;48:87–92.
3. Morton NE. Sequential tests for the detection of linkage. Am J Hum Genet 1955;7:277–318.
4. Clerget-Darpoux F, Bonaïti-Pellié C, Hochez J. Effects of misspecifying genetic parameters in lod score analysis. Biometrics 1986;42:393–399.
5. Clerget-Darpoux F, Bonaïti-Pellié C. An exclusion map covering the whole genome : a new challenge for genetic epidemiologists ? Am J Hum Genet 1993;52:442–443.
6. Tiwari JL, Hodge SE, Terasaki PI, Spence MA. HLA and the inheritance of multiple sclerosis : linkage of 72 pedigrees. Am J Hum Genet 1980;31:103–111.

7. Clerget-Darpoux F, Govaerts A, Feingold N. HLA and susceptibility to Multiple Sclerosis. Tissue Antigens 1984;24:160–169.
8. Penrose LS. The detection of autosomal linkage in data which consist of pairs of brothers and sisters of unspecified parentage. Ann Eugenics1935;6:133–138.
9. Day NE, Simons MJ. Disease susceptibility genes-their identification by multiple case family studies. Tissue Antigens 1976;8:109–119.
10. Thomson G, Bodmer W. The genetics of HLA and disease associations. In: Dausset J, Svejgaard A, (eds) HLA and Disease. Copenhagen: Munksgaard, 1977.
11. Suarez BK. The affected sib pair IBD distribution for HLA-linked disease susceptibility genes. Tissue Antigens 1978;12:87–93.
12. Risch N. Linkage strategies for genetically complex traits. III. The effect of marker polymorphism on analysis of affected relative pairs. Am J Hum Genet 1990;46:242–253.
13. Kruglyak L, Lander ES. High resolution genetic mapping of complex traits. Am J Hum Genet 1995;56:1212–1223.
14. Kruglyak L, Daly MJ, Reeve-Daly MP, Lander ES. Parametric and non-parametric linkage analysis: a unified multipoint approach. Am J Hum Genet 1996;58:1347–1363.
15. Kruglyak L, Lander ES. Complete multipoint sib pair analysis of qualitative and quantitative traits. Am J Hum Genet 1995;57:439–454.
16. Haseman J, Elston RC. The investigation of linkage between a quantitative trait and a marker locus. Behav Genet 1972; 2:3–19.
17. Commenges D. Robust genetic linkage analysis based on a score test of homogeneity : the weighted pairwise correlation statistic. Genet Epidemiol 1994;11:189–200.
18. The Multiple Sclerosis Genetics Group. A complete genome screen for multiple sclerosis underscores a role for the major histocompatibility complex. Nature Genet 1996;13:469–471.
19. Sawcer S, Jones HB, Feakes R, Gray J, Smaldon N, Chataway J, Roberston N, Clayton D, Goodfellow PN, Compston A. A genome screen in multiple sclerosis reveals susceptibility loci on chromosome 6p21 and 17q22. Nature Genet 1996;13:466–468.
20. Ebers GC, Kukay K, Bulman DE, Sadovnick AD, Rice G, Anderson C, Armstrong H, Cousin K, Bell R B, Hader W, Paty DW, Hashimoto S, Oger J, Duquette P, Warren S, Gray T, O'Connor P, Nath A, Auty A, Metz L, Francis G, Paulseth JE, Murray TJ, Pryse-Phillips W, Nelson R, Freedman M, Brunet D, Bouchard JP, Hinds D, Risch N. A full genome search in multiple sclerosis. Nature Genet 1996;13:472–476.
21. Kuokkanen S, Gschwend M, Rioux JD, Daly MJ, Terwilliger,JD, Tienari PJ, Wikström J, Palo J, Stein LD, Hudson TJ, Lander ES, Peltonen L. Genomewide scan of multiple sclerosis in Finnish multiplex families. Am J Hum Genet 1977;61:1379–1387.
22. Babron MC, Truy F, Eichenbaum-Voline S, Genin E, Clerget-Darpoux F. Behaviour of the maximum likelihood score when applied to dependent affected sib-pairs. Ann Hum Genet 1997;61:532.
23. Eichenbaum-Voline S, Génin E, Babron MC, Margaritte-Jeannin P, Prum B, Clerget-Darpoux F. Caution in the interpretation of MLS. Genet Epidemiol. 1997;14:1079–1084.
24. Wise LH, Lanchbury JS, Lewis CM. Meta-analysis of genome searches. Ann Hum Genet 1999; 63:263–272.
25. Fontaine B, Clanet M, Babron MC, Rimmler JB, Woodruff LK, Lincoln R, Garcia ME, Roth MP, Coppin H, Cournu I, Lyon-Caen O, Edan G, Hauser SL, Oksenberg JR, Pericak-Vance MA, Haines JL. Analysis of six previously identified regions of linkage in multiple sclerosis (MS). Am J Hum Genet 1998;63:1670.
26. Eichenbaum-Voline S, Babron MC, Prum B, Clerget-Darpoux F. Designing a linkage replication study in affected sib pairs. Genet Epidemiol 1998;15:524.
27. Yaouanq J, Semana G, Eichenbaum S, Quelvennec E, Roth MP, Clanet M, Edan G, Clerget-Darpoux F. Evidence for linkage disequilibrium between HLA-DRB1 gene and multiple sclerosis. Science 1997;276:664–665.
28. Clerget-Darpoux F. Overview of strategies for complex genetic diseases. Kidney Int

1998;63:1441–1445.

… 2001 Elsevier Science B.V. All rights reserved.
Genes and Viruses in Multiple Sclerosis.
O.R. Hommes, H. Wekerle, M. Clanet, editors.

Cytokine gene polymorphisms in multiple sclerosis

Bernard M.J. Uitdehaag*

Department of Neurology, Academic Hospital 'Vrije Universiteit', P.O. Box 7057, 1007 MB Amsterdam, The Netherlands

Introduction

Multiple sclerosis (MS) is a chronic inflammatory demyelinating disorder of the central nervous system which has a substantial genetic component. In recent years a large number of reports has been published concerning the role of cytokine gene polymorphisms in MS.

Research on cytokine gene polymorphisms in MS is based on several assumptions. First of all the assumption that the immune system plays an important role in MS. Although the exact pathogenesis of the disease is not yet cleared, there is little doubt that the immune system is significantly involved in the disease process in MS.

A second assumption involves the role of cytokines in the pathogenesis of MS. Cytokines are molecules that are predominantly secreted by immunocompetent cells. They take part in the regulation of the activity of the immune system. On the basis of their action they can be divided into pro-inflammatory and anti-inflammatory cytokines. The balance between these two kinds of cytokines is believed to be of importance in MS.

A third assumption is that cytokine gene polymorphisms are in one way or another related to the function of the cytokine. This may be the result of a change in biological activity or a difference in production [1].

The most often studied parameter with respect to cytokine gene polymorphisms is disease susceptibility. However, also differences in disease course and progression may be linked to genetic factors. Finally, response to treatment, especially when a therapy is believed to change a specific balance in the immune system, may depend on the genetic make up of immunological factors.

'Intelligent choices'

In contrast to a complete genomic screen, investigation of a specific gene polymorphism starts with the choice of that particular gene. Such a choice is usually based on the hypothesis that the product of that gene (e.g. a cytokine or its antagonist) is relevant in the disease process. Since in practice only a limited number of genes can be studied this way it is even more important to carefully consider your

---

Correspondence address: Department of Neurology, Academic Hospital 'Vrije Universiteit', P.O. Box 7057, 1007 MB Amsterdam, The Netherlands. E-mail: bmj.uitdehaag@azvu.nl

choice.

Due to a variety of reasons tumor necrosis factor-α (TNF-α) is one of the most frequently investigated cytokines in MS. TNF-α expression is associated both with the activity of relapsing remitting MS and with the development of chronic progressive MS. The TNF-α gene is located on the short arm of chromosome 6 between class I and II regions of the HLA complex. Based on the results of the full genome search this is an area that is likely to be relevant for MS [2–4]. Moreover an association between HLA and MS has been shown before. It can not be excluded that these associations are due to linkage disequilibrium between the HLA genes and closely located genes like the TNF-α gene.

The gene encoding for lymphotoxin (TNF-β) is located within the same region and polymorphisms of this gene have also often been investigated in MS.

Genes for interleukin-1 (IL-1) and its natural occurring antagonist IL-1 receptor antagonist (IL-1RA) are located on the short arm of chromosome 2. This region has not been identified as being important in the full genomic search. However IL-1 and IL-1RA are potentially important in the disease process. Therefore it seems to be relevant to look into more detail at the genes encoding for these cytokines.

Interleukin-4 and -10 (IL-4 and IL-10) and transforming growth factor-β (TGF-β) are all important anti-inflammatory cytokines. Their down regulatory effect in the immune process is believed to counterbalance the stimulatory effects of pro-inflammatory cytokines like TNF-α and IL-1.

Interferon-γ (IFN-γ) is one of the cytokines that has been associated with disease activity in MS. The gene is located on chromosome 12q, a region that showed a slightly positive lod score in the genomic screen study. Genes for the type I interferons (IFN-α and IFN-β) are located on chromosome 9. IFN-β is known for its beneficial effect in MS, whereas IFN-γ caused a clear deterioration. These intriguing observations in combination with many unresolved issues in interferon treatment make these cytokines a challenging topic for MS research.

Effect on disease susceptibility

One of the most frequently studied aspects of cytokine gene polymorphisms is their impact on disease susceptibility. Approaching the problem of genetically determined disease susceptibility by choosing a limited number of cytokine gene polymorphisms comes close to looking for a needle in a haystack. Especially in a disease that is probably determined by a combination of several genes such an approach seems likely to fail. However the majority of studies focusses on susceptibility, perhaps because this seems to be the most appealing aspect of genetic research. For the purpose of detecting disease susceptibility genes a full genomic screen may proof te be a more valuable technique.

TNF-α and TNF-β

Because of their presumed importance as indicated above many studies have evaluated polymorphisms of the genes encoding for TNF-α and TNF-β [5–14]. Many reports dealing with a variety of different mutations have been published so far. In none of them data were presented that would advocate a significant role of these polymorphisms in disease susceptibility.

IL-1 and IL-1RA

Allele 2 of the IL-1RA gene is associated with several autoimmune diseases. Also in MS a role of this gene in disease susceptibility was shown in some studies [15,16]. An increased frequency of carriage of allele 2 was found in a Dutch population (odds ratio 2.2) [15]. Subsequently this was not confirmed in other populations including a larger Dutch patient sample studied by the same group [11,17–20]. In contrast the A1/A1 genotype was more frequent in a group of Italian MS patients compared to healthy controls (odds ratio 1.83) [21]. However in the majority of studies no significant increase or decrease in frequency of a certain IL-1 and/or IL-1RA gene polymorphism was found [22].

IL-4, IL-10 and TGF-

A limited number of studies have been published concerning the possible association between MS and the genes encoding for the most familiar anti-inflammatory cytokines IL-4, IL-10 and TGF-β [23–26]. Therefore it is hard to draw definite conclusions, however so far no indications of a relation with disease susceptibility have been found for the known mutations of these genes.

IFN

Despite the promising theoretical background with respect to genetic research of IFN in MS, the results published to date concerning IFN-γ and IFN-β are rather disappointing [19,23,27–29]. However just recently functionally relevant polymorphisms in the region encoding for IFN-α were shown to be associated with the risk to develop MS [28].

Effect on disease type

Careful clinical observations have led to the distinction of several clinical phenotypes of the disease. Most patients initially experience a relapsing remitting (RR) course followed by a second phase with gradual deterioration (secondary progressive; SP). A minority of the patients shows a gradual deterioration from disease onset (primary progressive; PP). Magnetic resonance imaging (MRI) studies revealed differences of imaging characteristics between these clinical phenotypes.

Also data from pathological studies argue in favour of a pathogenetic heterogeneity of the disease. The background of these differences is unknown but may be related to immunogenetic variability. To detect these factors one must either divide the patients into subgroups or select patients to be included in the study based on specific disease characteristics.

Several investigators have approached the problem by dividing the group of MS patients in accordance to their disease phenotype [11,13,16,20,30]. A number of different ways of dividing the group have been chosen. RRMS has been compared to SPMS and PPMS. However since the disease course of most of the RR patients will eventually become SP these patients are often combined and described as 'bout onset' in contrast to PPMS. Also the comparison of 'Asian type' as opposed to 'European type' have been made [12]. So far the different studies did not result in the identification of a certain polymorphism related to a specific clinical phenotype of MS. However it must be stressed that up to now the proven value of MRI in the identification of subgroups have not been fully appreciated with respect to immunogenetic studies.

The age of onset can also be considered as a clinical features of MS. One could argue that in some patients this is an expression of disease activity whereby a late onset is considered to be representative for a long subclinical phase, however this is certainly not always the case. In one study IL-4 B1 allele was found to be associated with late onset of MS but not with disease progression [31].

Effect on disease progression

A change in the delicate balance within the immune system may affect the activity of the system and thereby change clinical disease activity in immune mediated diseases. One can imagine that this may be the case in cytokine regulation or production. Therefore investigating the relations between cytokine gene polymorphisms and disease activity in MS is potentially meaningful.

There are many ways disease activity in MS is defined and there are shortcomings in any of them. From a clinical point of view several definitions have been used in immunogenetic research. Some have used the Expanded Disability Status Scale (EDSS) score as such and only checked for significant differences in disease duration. Others have calculated a progression index by dividing the EDSS score by disease duration. Also the time to reach a certain EDSS score can be used as a measure of disease activity.

The value of MRI in the determination of disease activity is well known. Compared to clinical parameters progression of lesion load over time or brain atrophy are more objective measures that may turn out to be more closely related to the pathogenesis of the disease. Despite this, so far there are very few immunogenetic reports that deal with these parameters.

There are a limited but increasing number of studies who address the issue of disease progression in relation to cytokine gene polymorphisms [13,16,20–22,25,26,30]. For TNF-α and TNF-β so far no clear association with disease progression could be made for the chosen parameters. In one of the studies it was suggested that the -238 G to A transition in the TNF-α promoter may be protective in MS since it was found to be less common in patients with severe neurological damage [10].

We found a relation between the time to reach EDSS 6 and a specific combination of genes encoding for IL-1β and IL-1RA. The presence of allele 2 of the IL-1RA gene in the absence of allele 2 of the IL-1β gene was associated with a higher rate of progression [20]. In an Italian group of patients the presence of allele 1 of the IL-1RA gene was more often found in patients with an EDSS score above 3 at several cutoff times [21]. In the latter study the IL-1 genotype was not taken into account.

In two studies IL-10 promoter polymorphism was not found to be associated with disease progression [25,26].

Effect on treatment result

Apart from influencing disease susceptibility and clinical course and progression of the disease the immunogenetic make-up may have an impact on the effect of a given treatment. This can especially be true in case therapy is aimed at influencing the immune system as a whole and more specifically the cytokine network. In the latter situation it is imaginable that the presence or absence of certain cytokine gene polymorphisms affect the result of the treatment. The impact may not be restricted to the overall positive or negative effect but also the magnitude of the effect. In addition to that, production of neutralising antibodies in response to cytokine treatment, an important and not yet solved issue in MS treatment, may be related to the genetic background.

These considerations are not hypothetical or purely theoretical. Already in other immune mediated diseases a relation between cytokine gene polymorphism and treatment effect has been reported [32,33]. In MS so far no such study has been published.

Conclusion

Research on cytokine gene polymorphisms in MS is an rapidly expanding field. Many new mutations have been discovered and described in recent years. One might expect that this development will continue in the coming years. This implicates a enormous task in the evaluation of the importance of these polymorphisms in MS, especially when also certain combinations must be taken into account.

Whether the susceptibility of the disease is associated with these polymorphisms remains to be seen. It is possible that the approach chosen by most investigators may not reveal a clear answer to that question in the near future. However this approach may be very suitable for detecting the influence of cytokine gene polymorphisms on disease phenotype and disease progression. For that purpose integration of clinical and laboratory data with data obtained by MRI is potentially of great importance.

Another issue that should be addressed in the near future concerns the role of cytokine gene polymorphisms in the effects of (immune) therapy.

References

1. Wilson AG, Symons JA, McDowell TL, McDevitt HO, Duff GW. Effects of a polymorphism in the human tumor necrosis factor alpha promoter on transcriptional activation. Proc Natl Acad Sci USA 1997;94:3195–3199.
2. Sawcer S, Jones HB, Feakes R, et al. A genome screen in multiple sclerosis reveals susceptibility loci on chromosome 6p21 and 17q22. Nature Genet 1996;13:464–468.
3. Haines JL, Ter-Minassian M, Bazyk A, et al. A complete genomic screen for multiple sclerosis underscores a role for the major histocompatability complex. Nature Genet 1996;13:469–471.
4. Ebers GC, Kukay K, Bulman DE, et al. A full genome search in multiple sclerosis. Nature Genet. 1996;13:472–476.
5. Fugger L, Morling N, Sandberg-Wollheim M, Ryder LP, Svejgaard A. Tumor necrosis factor alpha gene polymorphism in multiple sclerosis and optic neuritis. J Neuroimmunol 1990;27:85–88.
6. Roth MP, Nogueira L, Coppin H, Clanet M, Clayton J, Cambon-Thomsen A. Tumor necrosis factor polymorphism in multiple sclerosis: no additional association independent of HLA. J Neuroimmunol 1994;51:93–99.
7. He B, Navikas V, Lundahl J, Söderström M, Hillert J. Tumor necrosis factor alpha-308 alleles in multiple sclerosis and optic neuritis. J Neuroimmunol 1995;63:143–147.
8. Garcia-Merino A, Alper CA, Usuku K, et al. Tumor necrosis factor (TNF) microsatellite haplotypes in relation to extended haplotypes, susceptibility to diseases associated with the major histocompatibility complex and TNF secretion. Hum Immunol 1996;50:11–21.
9. Wingerchuck D, Liu Q, Sobell J, Sommer S, Weinshenker BG. A population-based case-control study of the tumor necrosis factor alpha-308 polymorphism in multiple sclerosis. Neurology 1997;49:626–628.
10. Huizinga TW, Westendorp GRJ, Bollen ELEM, et al. TNF-alpha promoter polymorphisms, production and susceptibility to multiple sclerosis in different groups of patients. J Neuroimmunol 1997;72:149–153.
11. Epplen C, Jäckel S, Santos EJM, et al. Genetic predisposition to multiple sclerosis as revealed by immunoprinting. Ann Neurol 1997;41:341–352.
12. Ma JJ, Nishimura M, Mine H, et al. HLA-DRB1 and tumor necrosis factor gene polymorphisms in Japanese patients with multiple sclerosis. J Neuroimmunol 1998;92:109–112.
13. Mycko M, Kowalski W, Kwinkowski M, et al. Multiple sclerosis: the frequency of allelic forms of tumor necrosis factor and lymphotoxin-alpha. J Neuroimmunol 1998;84:198–206.
14. Trojano M, Liguori M, De Robertis F, et al. Comparison of clinical and demographic features between affected pairs of Italian multiple sclerosis multiplex families; relation to tumour necrosis factor genomic polymorphisms. J Neurol Sci 1999;162:194–200.
15. Crusius JBA, Peña AS, Van Oosten BW, et al. Interleukin-1 receptor antagonist gene polymorphism and multiple sclerosis. Lancet 1995;346:979–980.
16. De la Concha EG, Arroyo R, Crusius JBA, et al. Combined effect of HLA-DRB1*1501 and interleukin-1 receptor antagonist gene allele 2 in susceptibility to relapsing/remitting multiple sclerosis. J Neuroimmunol 1997;80:172–178.
17. Huang WX, He B, Hillert J. An interleukin 1-receptor-antagonist gene polymorphism is not associated with multiple sclerosis. J Neuroimmunol 1996;67:143–144.
18. Semana G, Yaouanq J, Alizadeh M, Clanet M, Edan G. Interleukin-1 receptor antagonist gene in multiple sclerosis. Lancet 1997;349:476.

19. Wansen K, Pastinen T, Kuokkanen S, et al. Immune system genes in multiple sclerosis: genetic association and linkage analyses on TCRβ, IGH, IFN-γ and IL-1Ra/IL-1β loci. J Neuroimmunol 1997;79:29–36.
20. Schrijver HM, Crusius JBA, Uitdehaag BMJ, et al. Association of interleukin-1beta and interleukin-1 receptor antagonist genes with disease severity in MS. Neurology 1999;52:595–599.
21. Sciacca FL, Ferri C, Vandenbroeck K, et al. Relevance of interleukin 1 receptor antagonist intron 2 polymorphism in Italian MS patients. Neurology 1999;52:1896–1898.
22. Feakes R, Sawcer S, Broadley S, et al. Interleukin 1 receptor antagonist (IL-1ra) in multiple sclerosis. J. Neuroimmunol 2000;105:96–101.
23. He B, Xu C, Yang B, Landtblom AM, Fredrikson, Hillert J. Linkage and association analysis of genes encoding cytokines and myelin proteins in multiple sclerosis. J Neuroimmunol 1998;86:13–19.
24. McDonnell GV, Kirk CW, Hawkins SA, Graham CA. Lack of association of transforming growth factor (TGF)-β1 and β2 gene polymorphisms with multiple sclerosis (MS) in Northern Ireland. Multiple Sclerosis 1999;5:105–109.
25. Pickard C, Mann C, Sinnot P, et al. Interleukin-10 (IL10) promoter polymorphisms and multiple sclerosis. J Neuroimmunol 1999;101:207–210.
26. Mäurer M, Kruse N, Giess R, Toyka KV, Rieckmann P. Genetic variation at position -1082 of the interleukin 10 (IL10) promoter and the outcome of multiple sclerosis. J Neuroimmunol 2000;104:98–100.
27. Giedraitis V, He B, Hillert J. Mutation screening of the interferon-gamma gene as a candidate gene for multiple sclerosis. Eur J Immunogenet 1999;26:257–259.
28. Miterski B, Jaeckel S, Epplen JT, et al. The interferon gene cluster: a candidate region for MS predisposition? Genes Immunity 1999;1:37–44.
29. Goris A, Epplen C, Fiten P, et al. Analysis of an IFN-γ gene (IFNG) polymorphism in multiple sclerosis in Europe: effect of population structure on association with disease. J Interferon Cytokine Res 1999;19:1037–1046.
30. Weinshenker BG, Wingerchuk DM, Liu Q, Bissonet AS, Schaid DJ, Sommer SS. Genetic variation in the tumor necrosis factor alpha gene and the outcome of multiple sclerosis. Neurology 1997;49:378–385.
31. Vandenbroeck K, Martino G, Marrosu MG, et al. Occurrence and clinical relevance of an interleukin-4 gene polymorphism in patients with multiple sclerosis. J Neuroimmunol 1997;76:189–192.
32. Czaja AJ, Cookson S, Constantini PK, Clare M, Underhill JA, Donaldson PT. Cytokine polymorphisms associated with clinical features and treatment outcome in type 1 autoimmune hepatitis. Gastroenterology 1999;117:645–652.
33. Facklis K, Plevy SE, Vasiliauskas EA, Kam L, et al. Crohn's disease-associated genetic marker is seen in medically unresponsive ulcerative colitis patients and may be associated with pouch-specific complications. Dis Colon Rectum 1999;42:601–605.

Genetic control of Theiler's virus infection

S. Aubagnac, L. Behrens, F. Bihl[1], M. Brahic*, J.-F. Bureau, S. Vigneau

Unité des Virus Lents, CNRS URA 1930, Institut Pasteur, 28, rue du Dr. Roux, 75724 Paris Cedex 15, France

Theiler's virus, a murine picornavirus, causes a persistent infection of the central nervous system accompanied by focal inflammation and demyelination. The virus and the persistent infection of the CNS were first described by Max Theiler in 1934 [1]. Chronic demyelination in the white matter of spinal cord was reported later by Daniels et al. [2]. In nature, the virus is transmitted by the fecal-oral route and reaches the CNS only occasionally. In the laboratory, CNS infection is obtained reproducibly after intracranial inoculation. In the mid 1970s, Howard Lipton showed that the disease that follows intracranial inoculation is biphasic, with an initial polio encephalomyelitis with destruction of neurons followed by a persistent infection of spinal cord with chronic inflammation and demyelination [3]. This disease is a now a classical model for multiple sclerosis [4,5].

Although all inbred mouse strains are susceptible to the early polio encephalomyelitis, they vary greatly in their susceptibility to persistent infection and to demyelination. The control of viral persistence is multigenic, with a strong effect of the MHC. The H-2^b haplotype confers dominant resistance. Within the MHC, the H-$2D$ gene plays a major role in resistance, as shown with congenic mouse strains [6,7]. This observation was confirmed using susceptible FVB mice transgenic for the H-D^b gene and resistant C57BL/6 mice in which the D^b or the K^b gene had been inactivated [8,9]. H-$2D^b$ confers resistance through a Class I restricted CTL response mediated by an immunodominant epitope [10–12].

Non H-2 genes controlling susceptibility have been mapped by scanning the genome of an (SJL/J x B10.S) x B10.S backcross. Susceptibility loci have been found close to the *Ifng* locus on chromosome 10 and close to the *Mbp* locus on chromosome 18 [13]. A series of H-2^s mice congenic for the chromosome 10 region has been constructed and characterized. The analysis showed that *Ifng* which codes for interferon gamma, is not the gene responsible for the susceptibility of the SJL/J strain. Instead, two susceptibility loci, *Tmevp2* and *Tmevp3*, have been located in the region [14]. *Tmevp3* does not affect viral persistence by modulating cytokine expression in the CNS, in particular the Th1/Th2 balance [15], or the specific immune responses in general, as suggested by experiments in which

---

[1]*Present address:* Center for the Study of Host Resistance, Montreal General Hospital, 1650 Cedar Ave., Room L11-144, Montreal, Quebec H3G 1A4, Canada.
*Correspondence address: Unité des Virus Lents, CNRS URA 1930, Institut Pasteur, 28, rue du Dr Roux, 75724 Paris Cedex 15, France, Phone: 33(0)1 45 68 87 70, Fax: 33(0)1 40 61 31 67. E-mail: mbrahic@pasteur.fr

H-2^S parental or congenic animals were lethally irradiated and grafted with bone marrow from parental mice with a different phenotype (Aubagnac et al. in preparation).

The existence of a susceptibility locus on chromosome 18, close to the gene which codes for the myelin basic protein (*Mbp*), led to the study of susceptible C3H mice with mutations in the *Mbp* or the *Plp* gene (the latter codes for another major component of myelin, PLP). Surprisingly, mutations in either gene made the C3H strain totally resistant to persistent infection [16]. This resistance is not immune mediated and may indicate that myelin, as a structure, is necessary for persistence of the infection.

The interferon gamma pathway plays a central role in resistance to persistence of the infection and to demyelination, as shown by the susceptible phenotype of 129/Sv mice whose *Ifngr* gene has been inactivated. These mice present with extensive demyelinating lesions and severe clinical disease [17]. Surprisingly, although they are persistently infected, C57BL/6 mice whose *Ifng* gene has been inactivated do not present with clinical symptoms. By comparing F2 crosses between these mice and their parental strains the difference of phenotype was linked to the difference of genetic background. Screening the entire genome of these crosses led to the description of *Tmevd5*, a locus on chromosome 11 which affects the severity of clinical disease for mice with a high viral load [18].

Several conclusions can be drawn from these genetic studies. First, genetic analysis is a powerful tool to uncover important steps in the pathogenesis of complex traits such as susceptibility to a viral infection and to a demyelinating disease. The approach has the important feature that it does not require a priori hypotheses. As a result, the findings may be totally unexpected and all the more interesting. Second, the genetic control of an infectious, demyelinating disease can be extremely complex. Sorting out the multiple genes involved, and their respective effects, is a serious enterprise for a model disease for which inbred strains are available. With a disease occurring in an outbred population, such as humans, it becomes a formidable task. Lastly, if the usual viro-epidemiological techniques used in medicine, such a serology or nucleic acid detection by PCR, were the only tools used to look for the cause of the demyelinating disease described in this review, it would be almost impossible to incriminate Theiler's virus. Looking for an infectious cause in a multifactorial disease of man can be a major challenge.

References

1. Theiler M. Spontaneous encephalomyelitis of mice: a new virus disease. Science 1934;80;122–123.
2. Daniels JB, Pappenheimer AM, Richardson S. Observations on encephalomyelitis of mice (DA strain). J Exp Med 1952;96;22–24.
3. Lipton HL. Theiler's virus infection in mice: an unusual biphasic disease process leading to demyelination. Infect Immun 1975;11;(5);1147–1155.
4. Brahic M, Bureau J-F. Genetics of susceptibility to Theiler's virus infection. Bioessays 1998;20;(8);627–633.

5. Monteyne P, Bureau J-F, Brahic M. The infection of mouse by Theiler's virus: from genetics to immunology. Immunol Rev 1997;159;163–176.
6. Rodriguez M, Leibowitz JL, David CS. Susceptibility to Theiler's virus-induced demyelination. Mapping of the gene within the H-2D region. J Exp Med 1986;163;620–631.
7. Bureau J-F, Montagutelli X, Lefebvre S, Guénet J-L, Pla M, Brahic M. The interaction of two groups of murine genes determines the persistence of Theiler's virus in the central nervous system. J Virol 1992;66;4698–4704.
8. Azoulay A, Brahic M, Bureau J-F. FVB mice transgenic for the H-$2D^b$ gene become resistant to persistent infection by Theiler's virus. J Virol 1994;68;4049–4052.
9. Azoulay-Cayla A, Dethlefs S, Pérarnau B, et al. H-$2D^{b-/-}$ mice are susceptible to persistent infection by Theiler's virus. J Virol 2000;74;(12);5470–5476.
10. Dethlefs S, Brahic M, Larsson-Sciard EL. An early, abundant cytotoxic T-lymphocyte response against Theiler's virus is critical for preventing viral persistence. J Virol 1997;71;8875–8878.
11. Borson ND, Paul C, Lin X, et al. Brain-infiltrating cytolytic T lymphocytes specific for Theiler's virus recognize H2D^b molecules complexed with a viral VP2 peptide lacking a consensus anchor residue. J Virol 1997;71;5244–5250.
12. Dethlefs S, Escriou N, Brahic M, van der Werf S, Larsson-Sciard E-L. Theiler's virus and Mengo virus induce cross-reactive cytotoxic T lymphocytes restricted to the same immunodominant VP2 epitope in C57BL/6 mice. J Virol 1997;71;5361–5365.
13. Bureau J-F, Montagutelli X, Bihl F, Lefebvre S, Guénet J-L, Brahic M. Mapping loci influencing the persistence of Theiler's virus in the murine central nervous system. Nat Genet 1993;5;87–91.
14. Bihl F, Brahic M, Bureau J-F. Two loci, *Tmevp2* and *Tmevp3*, located on the telomeric region of Chromosome 10, control the persistence of Theiler's virus in the central nervous system. Genetics 1999;152;385–392.
15. Monteyne P, Bihl F, Levillayer F, Brahic M, Bureau J-F. The Th1/Th2 balance does not account for the difference of susceptibility of mouse strains to Theiler's virus persistent infection. J Immunol 1999;162;7330–7334.
16. Bihl F, Pena-Rossi C, Guénet J-L, Brahic M, Bureau J-F. The shiverer mutation affects the persistence of Theiler's virus in the central nervous system. J Virol 1997;71;5025–5030
17. Fiette L, Aubert C, Müller U, et al. Theiler's virus infection of 129Sv mice that lack the interferon a/b or interferon g receptors. J Exp Med 1995;181;2069–2076.
18. Aubagnac S, Brahic M, Bureau J-F. Viral load and a locus on chromosome 11 affect the late clinical disease caused by Theiler's virus. J Virol 1999;73;(10);7965–7971.

© 2001 Published by Elsevier Science B.V.
Genes and Viruses in Multiple Sclerosis.
O.R. Hommes, H. Wekerle, M. Clanet, editors.

MHC proteins and human diseases: a tale of recognition in two immune systems*

Dr. Jack L. Strominger**
Higgins Professor of Biochemistry, Harvard University, USA

The discovery of transplantation antigens

Transplantation antigens were named by Peter Gorer in the 1930's during studies of organ graft and tumor transplant acceptance or rejection in mice to describe the putative chemical substances that were recognized as foreign by the graft recipient. The recipient, however, must also have these substances since its tissues were in turn rejected by other mice. Many different variants of transplantation antigens differing in small but important ways occur in each species, i.e., they are polymorphic. These studies were the impetus for some of the early mouse breeding experiments which led to the definition of a single genetic region in mice that encoded the polymorphic chemical substance(s) responsible for graft acceptance or rejection. This region was named the major histocompatibility complex (MHC) because differences in it led to very rapid graft rejection in contrast to differences in many other regions which led to delayed graft rejection (minor histocompatibility loci).

The discovery of antibodies raised in humans after whole blood transfusion (against white blood cells which gave rise to their original name, HLA, Human Leucocyte Antigens) or during pregnancy (against paternal substances present in the fetus), and of genetic factors in mice which controlled the immune response, was followed by a period of intensive immunogenetic study and of molecular studies, which together defined two principal classes of proteins (now called class I and class II MHC proteins). The differences in these proteins control not only graft acceptance or rejection, but also the ability of the host to respond immunologically to a very large number of foreign proteins derived from viruses, bacteria and other environmental agents.

The isolation and structures of human histocompatibility proteins

The task which was undertaken in my laboratory, beginning in the early 1970's, was to isolate and characterize these proteins, to understand how they are involved in immune responsiveness and most importantly, to understand the nature of the biological force(s) which led to the evolution of their polymorphism. It could not

---
*This material was reprinted by the courtesy of The Science and Technology Foundation of Japan, from the booklet "Japan Prize 1999". Copyright: The Science and Technology Foundation of Japan.
**Correspondence address: Harvard University, Department of Molecular and Cellular Biology, 7 Divinity Avenue, Cambridge, MA 02138, USA. E-mail: jlstrom@fao.harvard.edu

have been the exchange of surgical grafts which led to their discovery.

At the time we began our studies, several laboratories had attempted to purify these substances. The task of purifying them was, however, complicated by several obstacles.

1. Most importantly, a quantitative assay for the activity of this material that could be used in following its purification was not available. The development of a quantitative assay was of key importance. Previous studies had shown that some materials containing histocompatibility antigens would inhibit the lysis of human lymphocytes by pregnancy alloantisera in the presence of complement as observed by trypan blue exclusion, a procedure that had been developed for tissue typing. We were able to develop this simple procedure into a quantitative method and, as the result, an assay for following purification was available. The purification was carried out much like purification of an enzyme, which I had learned carefully during the period I had spent at Washington University (St. Louis) and at the National Institutes of Health in the 1950's and early 1960's.

2. Histocompatibility proteins were present in very small amounts on the membranes of cells and, moreover, a source of human material for large scale purification was not immediately obvious. Just at this time the human B lymphotropic virus, Epstein-Barr virus, the causative agent of an important tumor in African children and of infectious mononucleosis in Western civilizations, was discovered. This virus was also able to transform human B lymphocytes *in vitro* which then multiplied rapidly and could be grown in the laboratory. These transformed cells were found to express large amounts of histocompatibility proteins and, although it represented a logistic problem, they could be grown in tissue culture in large amounts.

3. Few membrane proteins had been purified because of their insolubility in aqueous solvents. The discovery that human histocompatibility antigens, like their mouse homologues, could be solubilized by treatment with the proteolytic enzyme, papain, and retained their biological activity was a key to subsequent purification of a water-soluble form. In addition, however, a scheme was also developed for the purification of these substances in detergent solution which permitted the whole structure to be worked out. We clarified the question that had arisen when the proteins initially purified by papain solubilization were found to have two polypeptide chains. The detergent-soluble material also had two chains and thus their two chain structure was not an artifact of proteolysis during papain solubilization. The smaller chain was a protein that had been isolated from the urine of patients with nephrotic kidney disease and had been named β_2-microglobulin.

By this time immunogenetic studies had begun to reveal the complexity of histocompatibility proteins. Two genetic loci had been identified that encoded these proteins, HLA-A and HLA-B, and soon still a third locus, HLA-C, was uncovered. Purification schemes had to take account of the presence of multiple protein products, particularly since the two alleles at each locus were expressed co-dominantly, i.e., most cells in an individual express six histocompatibility proteins. The problem was somewhat simplified by the use of homozygous cell lines derived by transformation with Epstein-Barr virus, mainly from the offspring of consanguineous first cousin marriages that express only three different histocompatibility proteins. Another surprise was the discovery that a major impurity was separated at a late purification stage, which had similar properties to the histocompatibility proteins. It proved to be a second type of histocompatibility protein. The two types are now called class I MHC proteins and class II MHC proteins. The latter were also soon found to be the products of three distinct genetic loci (HLA-DR, HLA-DQ, and HLA-DP) and these too were ultimately separated to provide pure materials. By chance they turned out to be the human B lymphocyte proteins recognized by another series of human pregnancy alloantisera that were discovered at the same time.

Another important step was the cloning of DNA encoding the histocompatibility proteins and indeed the cloning of a major portion of the human major histocompatibility complex which is about four megabases in size. For a time these clones represented the largest available material from the human genome. Many previously unknown genes were identified and most of them have some function in the immune system. The availability of the genes encoding the histocompatibility proteins and their sequences contributed greatly to the elucidation of the primary structures of these proteins and later to the development of expression systems for production of the proteins in larger amounts and in a facile manner. The first amino acid sequences derived both from protein sequencing and from DNA sequencing illuminated many important features: the occurrence of immunoglobulin-like domains in the structure (the first members, in addition to immunoglobulins, of what is now called the immunoglobulin superfamily of proteins), the domain structure of the heavy chain of class I and of both chains of class II MHC proteins, the location of disulfide bonds, the positions in the sequences in which polymorphisms occurred, and finally the similarity in structure between the class I and class II MHC proteins despite the fact that the former had a heavy chain with three domains and a light chain with one domain, while in the latter the heavy and light chains each had two domains.

All of this work led up to the collaboration with my colleague Don Wiley, which resulted in the crystallization and finally elucidation of the three-dimensional structures of both class I and class II MHC proteins. An exceptionally important feature of these structures, which will be outlined in his lecture, is the occurrence of bound peptides within a groove in each of these proteins. In the case of class I proteins both ends of the peptide, usually nine amino acids long, are embedded within a closed groove in the structure, while in the case of class II proteins the peptides are 13-14 amino acids long but can extend further beyond the groove which is open at both ends. We had learned that the function of human histocompatibility antigens is to present peptides derived from foreign agents to the immune system to initiate an immune response and that the polymorphism which occurs almost entirely in the groove where the peptides are bound had evolved to ensure that peptides derived from a large variety of foreign organisms and even from mutants of these organisms would not escape recognition. Infectious agents must have been the driving force for the evolution of the polymporphism. To some extent the polymorphism protects individuals from catastrophic variations in infectious agents since each individual presents six polymorphic variants of each class of proteins that can bind and present different peptides, but to a larger extent, the extreme population polymorphism also protects the population as a whole from catastrophes that might result from variations in infectious agents, for example, viruses. Graft rejection, which led to the initial discovery of histocompatibility proteins, is a by-product of the fundamental role that they play in initiating an immune response.

Rather than provide further details of this past work, I would like to describe some ways that in the spirit of the Japan Prize, the knowledge we have gained in molecular recognition in the immune system might be used to improve human health. The past is a prologue to the future.

MHC proteins, T cells and autoimmune diseases

Autoimmune diseases result from the attack of the immune system on normal human tissues. The most frequent of these diseases are rheumatoid arthritis (affecting 1% of the Western Caucasian population), multiple sclerosis (0.1%), and type I diabetes mellitus (also about 0.1%). These three diseases result from attack by the immune system on joint synovium, the myelin sheath of neurons in the central nervous system, and pancreatic beta cells that produce insulin respectively. In addition, many other autoimmune diseases affect the population and include myasthenia gravis (neuromuscular junction), pemphigus vulgaris (skin), biliary cirrhosis (liver), celiac disease (intestinal mucosa), lupus erythematosus (systemic) and Graves' disease (thyroid), for example. In most cases the nature of the inciting agent is unknown but two examples, glomerulonephritis and rheumatic fever, were initiated by infection with streptococci and now with the advent of antibiotics have virtually been eliminated. The prevalent view is that many of these diseases are initiated by virus or bacterial infections and that the disease results from the cross-reactivity of an immune effector T cell involved in an immune response to these infectious agents with a normal host tissue, i.e., molecular mimicry, or, as it was possible to define further based on our studies of the binding of peptides to MHC proteins, degenerate molecular mimicry.

Autoimmune diseases are frequently linked to alleles of MHC genes. Linkage means that the frequency of that allele in the disease population is much greater than the frequency in the normal population. Ankylosing spondylitis was the first example of an "autoimmune" disease linked to an MHC allele, HLA-B27. The linkage is very tight: at least 95% of patients with ankylosing spondylitis carry the HLA-B27 allele. Despite this strong linkage, this disease is curious in at least two ways: it has never been proven to be autoimmune and it is the only "autoimmune" disease that is linked to a class I MHC protein. The other diseases are linked to particular alleles of class II MHC genes, e.g., rheumatoid arthritis to HLA-DR4 (DRB1*0401), multiple sclerosis to HLA-DR2 (DRB1*1501), and type I diabetes to HLA-DQ8 (DQB1*0302) and to other alleles, at least in Western populations.

The immune system has exquisite mechanisms for distinguishing self peptides bound to MHC proteins from foreign peptides similarly bound. These mechanisms need to be exquisite because the vast majority of peptides found in isolated MHC proteins are self peptides. All the proteins of the cell are under continuous degradation and resynthesis and peptides derived from them are "surveyed" by the effector T cells of the immune system through the intermediacy of their presentation at the surface of cells by MHC proteins. Both central and peripheral mechanisms exist for deleting or inactivating (anergizing) self-reactive T cells generated stochastically by the immune system. Autoimmunity results from failure of the tolerance mechanisms, which presumably results, for example, from an excess stimulation by a foreign peptide that leads to a cross-reactive T cell.

Besides improving our knowledge of how MHC proteins may be involved in autoimmune diseases, what has our present knowledge contributed to efforts to ameliorate these diseases? After all, science has several goals. One is the pursuit of knowledge for its own sake. The revelation of the beautiful way in which nature has evolved the immune system to recognize and eliminate foreign pathogens was an immense reward for having participated in this work. Another equally important goal of scientific research is to improve the lot of mankind.

Many laboratories are presently investigating the origins, mechanisms and therapy of autoimmune diseases using the information which has been obtained in the past three decades on the genetics of these diseases and in the molecular studies of the class II MHC proteins to which these diseases are linked. I will briefly describe four approaches that have been used in my laboratory in an attempt to understand and ameliorate several autoimmune diseases.

The use of structural information to predict disease-related epitopes and also to examine the possibility that peptide epitopes derived from viruses and bacteria might provide the stimulus to break self tolerance and initiate these diseases

The structure of MHC proteins and particularly the nature of "pockets" in the groove in which peptides are bound make it possible to predict which peptides derived from putative disease-related proteins could be the actual disease epitopes, and furthermore to define relatively precisely the nature of the epitope. In the case of multiple sclerosis, for example, a disease-related epitope derived from myelin basic protein was precisely defined as amino acid residues 85-99. In the case of pemphigus vulgaris (a blistering skin disease) linked to a subtype of HLA-DR4 (DRB1*0402), seven peptides derived from the disease-related skin adhesion protein desmoglein-3 were predicted to be possible disease epitopes and one of these was found to stimulate T cells from all four of the first small group of pemphigus patients studied. We could also use our knowledge of the structural information, in particular our knowledge of which of the MBP 85-99 residues were involved in binding in the pockets of the HLA-DR2 protein and which were important in contacting the T cell receptor, coupled with the knowledge that the residues that bound in the pockets could be "degenerate" in the sense that they could be substituted by other residues of the same amino acid class, to show that mimicry could occur without a perfect match of amino acid sequence between the disease epitope and, for example, a virus epitope. Thus the possibility that degenerate molecular mimicry may actually apply to the initiation of at least some autoimmune diseases was greatly enhanced.

The use of oligomerized disease-related peptide epitopes to inactivate immunopathogenic T cells and thereby to ameliorate disease

The clustering of class II MHC/peptide complexes and of T cell receptors as an important component of an effective immune response has become a focus of interest in several laboratories. In our laboratory peptide oligomers were constructed, that is, peptide epitopes linked by spacers of an equivalent size as a means of forcing clustering or oligomerization. The immune response to an oligomer containing sixteen such epitopes was enhanced by a thousand-fold. Moreover, at higher doses, the immune response was suppressed. This phenomenon has been known as high zone tolerance. In the case of the oligomers it was accompanied by T cell receptor down-regulation and T cell death. When this technique was applied in experimental models of autoimmune disease, experimental autoimmune encephalomyelitis and experimental autoimmune neuritis, induction of disease could be prevented. These experiments illustrate an approach to the therapy of autoimmune diseases that could be of great importance.

A study of the mehanism by which Copaxone (Copolymer 1), a random polypeptide comprised of four amino acids (tyrosine, Y; glutamic acid, E; alanine, A; and lysine, K), a drug newly introduced for the treatment of multiple sclerosis, exerts its beneficial effect

Copaxone was introduced into therapy because it reduced significantly the frequency of relapses in patients with multiple sclerosis. However, this drug is a random mixture of polypeptides. What is (are) the effective component(s)? Taking advantage of the technology which was developed, 80% of the polypeptides in this mixture were shown to bind to HLA-DR2 proteins within the groove, but they bind in a manner so that there are marked preferences for which residues bind at P minus 2 (E), P minus 1 (K), and P1 (A or Y). These findings have suggested ways in which the composition of the copolymer might be improved to increase its efficacy in the treatment of multiple sclerosis and have also suggested the design of a copolymer that could be effective in other autoimmune diseases such as rheumatoid arthritis.

The discovery of an abnormality in regulatory T cells in patients with type I diabetes mellitus

Monozygotic twins are often discordant for development of autoimmune diseases, i.e., in different studies and with different diseases only 20 to 60% of the twin of a diseased proband will also have disease. In our studies of monozygotic diabetic twins and one triplet set discordant for the development of type I diabetes, the diabetic twins/triplets had a greatly reduced frequency of a regulatory T cell subset know as Vα24JαQ T cells whose ligand is CD1d, a class I protein encoded *outside* of the MHC. Those cells which remained were defective in the production of IL-4, a cytokine believed to be protective for the development of autoimmune diseases. The nature and origin of the defect remain to be defined but the subject of the types and roles of regulatory T cells in immune function and dysfunction is receiving increasing attention.

Natural killer cells, MHC proteins and disease

During these three decades, while attention was focused on MHC/peptide complexes and on their receptors on T cells, another class of circulating lymphocyte, the Natural Killer cells, was discovered in Sweden and in the U.S. Until recently, however, they received relatively scant attention. Natural Killer cells are the reciprocal of T cells. Peripheral T cells are normally inactive and are activated (for proliferation, cytokine release and cytolytic activity) by the recognition of specific MHC/peptide complexes. By contrast, Natural Killer cells are normally active and are inhibited (inactivated) by the recognition of class I MHC proteins, particularly HLA-C and HLA-E. One role of Natural Killer cells is to eliminate cells that have lost expression of class I MHC proteins. Class I MHC proteins are absent from cells in at least three circumstances:

The fetal placenta

The extravillous cytotrophoblast of the fetal placenta forms the fetal-maternal interface. This cell layer of the placenta does not express the normal class I MHC proteins, a physiological regulation which must have evolved to prevent recognition by maternal effector T cells of paternal MHC proteins expressed on nearly all other fetal tissues. A novel class I MHC protein (HLA-G) is expressed only on the fetal extravillous cytotrophoblast. HLA-G was found to be a potent inhibitor of Natural Killer cells. Alternatively, however, it could possibly function as a signaling molecule involved in the development of the placenta or the maintenance of pregnancy. Futhermore, its role, and that of a special type of Natural Killer cell found in the uterine decidua, in abnormalities of pregnancy such as recurrent spontaneous abortion and pre-eclampsia is an important area for future study.

Tumor cells

Some tumor cells, for example, some colon carcinoma and melanoma cells, have also lost expression of class I MHC proteins, and should, therefore, be targets for lysis by Natural Killer cells. However, means of avoidance of this recognition event have also evolved. One possibility to explain their resistance is the loss of expression of an accessory molecule whose interaction would also be required for the function of Natural Killer cells, as some accessory molecules are for the function of T cells. Still another possibility would be the failure to express the ligands for the lysis receptors on Natural Killer cells. Natural Killer cells express two types of receptors: inhibitory receptors for class I MHC proteins which prevent the lysis of cells that express these proteins normally (for which HLA-C and HLA-E are the dominant ligands) and lysis receptors on the Natural Killer cells that recognize still unidentified ligands on target cells. At least three lysis receptors have been identified: CD16 in our laboratory and NKp44 and NKp46 in another laboratory. The ligands for these lysis receptors are differentially expressed on potential target cells. Thus, loss of their expression would lead to tumor cell resistance to Natural Killer cells.

Virus-infected cells

Viruses have evolved a variety of mechanisms to down-regulate expression of class I MHC proteins as a means of escape of infected cells from immunosurveillance. A complex interplay between the virus and the host immune defenses involving both T cells and Natural Killer cells has resulted. One example is provided by human immunodeficiency virus (HIV).

The Nef protein encoded by HIV down-regulates HLA-A and HLA-B proteins, presumably as a means of diminishing the cytotoxic T cell response to infection. However, Nef does not down-regulate HLA-C or HLA-E, the dominant ligands for protection of target cells from NK cells. The system has been shown to work during infection of cells by HIV as a means of protection of the infected cells that have been down-regulated HLA-A and –B from lysis by NK cells.

These findings provide an explanation for the fact that Nef-deleted HIV replicates normally in tissue culture. However, Nef-deleted Simian immunodeficiency virus (SIV) replicates very poorly in SIV-infected monkeys. Coincidentally, an HIV infected individual in Australia who did not develop Acquired Immune Deficiency Syndrome was found to have a Nef-deleted strain of HIV. Moreover, individuals who received transfusions of blood donated by this individual also became infected but their disease did not progress. Presumably, in the absence of down-regulation of HLA-A and HLA-B proteins, the human immune system is far better able to control infection with HIV. These findings suggest a possible therapeutic approach to the control of HIV infection.

Thus, at the end of three decades an enormous amount has been learned about the molecular basis of recognition of foreign proteins derived from viruses, bacteria and other agents by the immune system. We have also learned that the effector lymphocytes of the immune system have diverged into two groups, T cells and Natural Killer cells, each of which has a different role in protection and each of which recognizes class I and/or class II MHC proteins, although in different ways. The description of MHC molecules and their complexes with self and foreign peptides and their interaction with these two types of effector cells has revealed many aspects of their normal and abnormal functioning. The molecular knowledge gained may in the future permit many advances in understanding aberrant conditions that result from abnormalities in this exquisitely tuned system, as well as therapeutic approaches to treat these conditions.

But the problems before us now and in the future are not so much scientific or medical as they are social and economic. In order to reap the benefits of these advances we must learn to live together and to appreciate and treasure the differences among us and to also treasure and preserve the resources of this beautiful planet for ourselves in the present as well as for future generations. Perhaps we have something to learn from the Japanese people who now live in relative peace and prosperity on these crowded islands, so that in the spirit of the Prize established by the Science and Technology Foundation of Japan with the endorsement of the Japanese government, we may "further world peace and prosperity and thereby make a vital contribution to the positive development of mankind."

Acknowledgement

Finally, I acknowledge with much pleasure and appreciation the collaboration with my colleague, Don Wiley, and the members of our two laboratories who worked so intelligently to make these accomplishments possible. The work in my laboratory that was cited for the Japan Prize proceeded over a period of nearly thirty years. The isolation of class I MHC proteins and the description of its heterodimeric structure was primarily the work of Mervyn Turner and Peter Cresswell and Tim Springer, and correspondingly, of Class II of Bob Humphreys and Jim Kaufmann. Harry Orr and José Lopez de Castro did all of the amino acid sequencing and analysis of polymorphic residues in class I MHC proteins with great skill at a time when methods were very primitive. Class I and Class II MHC gene cloning was carried out by Hidde Ploegh, Alan Korman and Charles Auffray. Finally, Pamela Bjorkman (who started her Ph.D. thesis in my lab and finished in Don Wiley's lab) crystallized class I (HLA-A2). Joan Gorga solubilized and crystallized a second isotype of class I (HLA-B27). She then also separated the three isotypes of class II and finally solubilized and crystallized HLA-DR1 and related allotypes. Many other young scientists contributed in large and in small ways to the development of this work. I am deeply grateful to each one of them.

Central nervous system delivery of therapeutic genes using viral vectors as an alternative strategy in autoimmune demyelinating diseases

R. Furlan[1], P.L. Poliani[1], P.C. Marconi[2], E. Brambilla[1], F. Ruffini[1], A. Bergami[1], G. Comi[1] and G. Martino[1],*

[1]Dept. of Neuroscience, San Raffaele Scientific Institute, Milan, Italy; [2]Telethon Institute for Gene Therapy (TIGET), San Raffaele Scientific Institute, Milan, Italy

Therapeutic targets in MS

The pathological hallmark of multiple sclerosis (MS) is the presence within the central nervous system (CNS) of inflammatory infiltrates containing few autoreactive T cells and a multitude of pathogenetic nonspecific lymphocytes [1,2] determining the typical patchy CNS demyelination, ranging from demyelination with preservation of oligodendrocytes to complete oligodendrocyte and axonal loss and severe glial scarring [3]. It is currently believed that CNS antigen-specific T cells provide the organ specificity of the pathogenic process and regulate the recirculation within the CNS of non antigen-specific lymphomononuclear cells that in turn act as effector cells by directly destroying oligodendrocytes and/or by releasing myelinotoxic substances [4]. In most instances, however, oligodendrocytes or their precursors are morphologically preserved in demyelinating plaques, and remain capable of differentiating and remyelinating [3]. A successful therapeutic approach of MS should therefore be aimed to inhibit the activation of antigen- and non-antigen specific immune cells and/or to rescue the surviving oligodendrocytes within demyelinating plaques. CNS gene delivery using non-replicative viral vectors able to infect post-mitotic cells such as those resident in the CNS is becoming a useful therapeutic approach to deliver antinflammatory cytokine genes able to inhibit the activation of antigen- and non-antigen specific immune cells and/or growth factor genes able to rescue the surviving oligodendrocytes within demyelinating plaques. However, the short lasting transgene expression due to viral vector in the CNS along with some residual immunogenicity and/or toxicity of these vectors are the limiting factors of this new therapeutic approach.

Immunomodulatory cytokine-based therapies: evidence in humans

Several in vitro and in vivo evidence suggest that cytokines with a proinflammatory profile (i.e. Th1 cytokines) are crucial in the development of MS due to their

---

*Correspondence address: Dr. Gianvito Martino, Neuroimmunology Unit, San Raffaele Scientific Institute - DIBIT, Via Olgettina 58, 20132 Milano, Italy. Tel: +39-02-26434853. Fax +39-02-26434855. E-mail: martino.gianvito@hsr.it

role in activating not only myelin antigen specific T cells (regulatory cells) but also non specific effector cells such as macrophages and B cells [1,4]. Considering the counteracting activity of pro- (Th1) and anti-inflammatory (Th2) cytokines [5,6], therapies aimed at interfering with the pathogenic process leading to autoimmune demyelination should encompass administration of Th2 cytokines or cytokines able to down regulate activity of Th1 cytokines. This general view is supported by the well established clinical efficacy of systemic IFNβ treatment in patients with relapsing remitting MS [7] which has been attributed to its anti-inflammatory effects [8]. However, although IFNβ ameliorates disease course in MS, its clinical efficacy is only partial. The route of IFNβ administration can in part explain the incomplete efficacy of this cytokine. IFNβ, in fact, does not efficiently cross the blood-brain barrier and therefore when systemically administered does not accumulate within the CNS [9]. Moreover, systemic administration of IFNβ causes undesirable side effects such as the formation of anti-cytokine antibodies in 10%–40% of MS patients after one year of treatment thus limiting the immunomodulatory activity of the drug [10]. A further limiting factors of systemic therapeutic delivery of "antinflammatory" cytokines in CNS-confined diseases, such as MS, is due to the intrinsic biological characteristics of cytokines which have a short half-life and act in an auto-/para-crine fashion. The failure of two recent trials employing systemic administration of TGFβ [11] and of a soluble chimeric form of the TNF receptor-1 [12] further stress the partial efficacy of systemically administered cytokines in MS. CNS delivery of potentially therapeutic cytokines using gene therapy approaches might overcome some of the above mentioned limitations.

Cytokine-based gene therapy approaches: experimental evidence

Gene therapy delivery of anti-inflammatory cytokines in MS patients has never been attempted so far. However, some experiments have been performed in rodents affected by experimental autoimmune encephalomyelitis (EAE), the animal model for MS. While several gene therapy devices have been used (i.e. plasmids, naked DNA, viral vectors) only two different gene delivery strategies have been experienced so far. The first one encompasses the delivery of cytokine genes directly into the peripheral circulation or into encephalitogenic T cells. The second approach encompasses the delivery of cytokine genes directly into the CNS.

Systemic delivery

IL-4 [13], IL-10 [14] or TGFβ1 [15] genes incorporated into autoreactive encephalitogenic T cells using plasmids or retroviral vectors ameliorated EAE in mice while the opposite occurred when the TNFα gene was used [16]. EAE was also ameliorated when IL-6, TNFα, IL-1β, IL-2, IL-10 but not IL-4 or IFNγ genes were intravenously (i.v.) administered to EAE mice using vaccinia virus (VV)-derived vectors [17]. The intramuscular (i.m.) injection of naked plasmid DNA expression vectors encoding either TGFβ1 or IL-4-IgG(1) chimeric protein resulted in protection from myelin basic protein (MBP)-induced EAE [18]. However, regulatory sequences used in plasmids for naked DNA vaccination can modulate cytokine production in vivo and injection of plasmid DNA can suppress EAE by itself via induction of IFNγ production [19].

CNS delivery

DNA-liposome constructs injected into the CNS parenchima was a successful strategy to ameliorate EAE when IL-4, TGFβ, IFNβ, p75TNF receptor but not IL-10 were incorporated into the construct [20]. CNS gene delivery of the costimulatory molecule CTLA4-human Ig using a non-replicative adenoviral vector was effective in ameliorating EAE course [21].

CNS gene delivery using non-replicative herpetic vectors

We developed a novel system to deliver cytokine genes into the CNS which is based on the use of Herpex Simplex Virus (HSV) type 1-derived non-replicative vectors. These vectors are considered as an alternative to classical retroviral and adenoviral vectors mainly because HSV is able to accomodate multiple foreign genes and to infect post-mitotic cells, such as neurons [22–24]. We used the HSV-1-derived vector named d120 [25–27] which is an HSV type-1 virus (Kos) strain containing a deletion in both copies of the immediate early (IE) infected cell protein (ICP)4 gene and the IL-4 gene inserted into the TK locus under the ICP4 promoter. In vitro and in vivo preliminary experiments indicated that the d120 is easily transferred within the CNS, diffuses consistently in all ventricular spaces, is able to efficiently infect the layer of ependymal cells surrounding the ventricles and, within the CNS, redirects the cell machinery of the infected CNS cells to produce discrete amount of the cytokine into the cerebrospinal fluid (CSF) of the mice until day 28 post injection [28]. We then used the IL-4 containing vector as therapeutic gene in EAE mice; IL-4 was administered i.c. in Biozzi AB/H mice immunized with myelin oligodendrocyte glycoprotein (MOG)40-55 before [29] and after [30] the appearance of EAE signs. No toxic reactions have been observed. A significant amelioration of clinical and pathological CNS features of EAE was observed with both therapeutic protocols. The protective effect of this therapy was mostly due to the ability of IL-4 to inhibit macrophage trafficking and activation in the CNS via downregulation of monocyte chemoattractant chemokines (MCP-1, RANTES) and proinflammatory cytokines (i.e. IL-1β, TNFα). Furthermore, we found that lymph node cells from IL-4 treated vs. non treated mice were able to process and proliferate in response to the encephalitogenic antigen as well as to drive an appropriate Th1 response thus indicating a lack of interference of the IL-4 gene therapy approach we used on the proper functioning of the immune system. Finally, we have recently shown that the same therapeutic strategy (i.e. delivery of the human IL-4 gene in the CNS of Rhesus monkeys affected by a fulminant and hyperacute form of EAE induced using whole myelin) applies also for bigger animals such as EAE-affected non-human primates (P.L.P., in press) [41].

The main advantages of our approach are: (a) the availability of high cytokine levels in the CNS; (b) the persistent therapeutic effect (i.e. 4 weeks) after a single vector administration, and (c) the absence of unpredictable/undesirable side effects on the peripheral immune system.

Fostering remyelination with gene therapy: future applications

For long time it was believed that repair of myelin sheaths does not occur in MS. However, more recently a detailed analysis of MS pathology provided evidence for extensive remyelination [31,32]. Remyelination is prominent during the early stage of disease evolution and apparently depends upon the availability of oligodendrocytes or their progenitor cells within the lesions [33,34]. In the late chronic stage of the disease, repair of myelin is sparse and, if present at all, restricted to a small rim at the plaques edge. It is as yet unclear what cells in the CNS accomplish remyelination. They could in part be recruited from oligodendrocytes that have survived the acute phase of demyelination [35]. However, recent experimental data suggest that mature, terminally differentiated oligodendrocytes are incapable of synthesizing new myelin [36]. In contrast, in most experimental situations, remyelinating cells are derived from the pool of undifferentiated glial precursor cells, which are present even in the adult CNS tissue and can also be found in low numbers in demyelinated MS plaques. Thus, it is suggested that the failure of myelin repair in late chronic MS lesions is due to a depletion of this progenitor cell pool, which is likely to occur in areas of repeated demyelinating episodes [31,36,37]. On the other hand, some oligodendrocyte progenitor cells can be found even in old demyelinated scars of MS patients.

All these findings suggest that myelin repair in MS lesions could be therapeutically approached either by introducing in the pathologic tissue growth factors which prevent destruction of oligodendrocytes and stimulate undifferentiated glial precursor cells to divide, differentiate and remyelinate axons. The use of growth factors in humans should, however, take into consideration that: (a) growth factors may drive terminally differentiated mature oligodendrocytes into apoptosis [38]; (b) rescue of oligodendrocytes and/or their progenitor cells in demyelinating conditions depends upon the mechanisms of myelin injury. Nevertheless, gene therapy safety study using a growth factor named ciliary neurotrophic factor (CNTF) has been already performed in humans affected by amyotrophic lateral sclerosis (ALS) a neurodegenerative disease of the motor neurons [39]. This study showed that introduction of heterologous genes coding for CNTF into the cerebrospinal fluid using encapsulated genetically modified CNTF-producing myoblasts is feasible and non toxic. Moreover, experiments aimed to deliver genes coding for other neurotropic factors have been also performed in non human primates affected by an experimental model of Parkinson's disease in which therapeutic effects have been

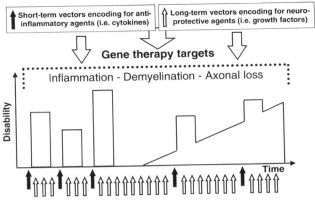

Fig. 1. Synopsis of the possible future scenario of gene therapy approaches in multiple sclerosis. Considering the heterogeneity of the disease, the future gene therapy approaches should be enough flexible to be used in the different phases of the disease as indicated. Short-term expressing vectors coding for anti-inflammatory cytokines (black arrows) might be useful in the relapsing phases of the disease in which the inflammatory component is prominent while long-term expressing vectors coding for growth factors (white arrows) fostering oligodendrocyte progenitor proliferation and differentiation might be more useful when chronically administered due to the continuos accumulation of axonal loss during the disease evolution.

observed without any side effects related to the procedure or to the vector toxicity. While no attempts have been made for transferring growth factor genes in MS patients in order to induce oligodendrocyte progenitor proliferation, few experimental evidence are available in EAE animals. It has been recently shown that autoreactive T cells isolated from SWXJ mice immunized with the PLP139-151 and transfected with an antigen-inducible transgene for platelet-derived growth factor-A (PDGF), a growth factor important in regulating the development of oligodendrocytes, upon adoptive transfer, migrated to the CNS and ameliorated ongoing EAE [40]. We also found that the introduction of HSV-1-derived vectors coding for FGF-II, another growth factor inducing differentiation of oligodendrocyte progenitors, into the intrathecal compartment of C57BL/6 mouse after the onset of MOG35-55-induced EAE is feasible and non toxic and might induce myelin restoration (F.R. manuscript submittes). A significantly higher number of oligodendroctye precursor cells (PDGFR$^+$ cells) as well as of PLP-producing oligodendrocytes have been, in fact, found in areas surrounding demyelinating plaques in FGF-II-treated EAE mice compared to control mice indicating that stimulation of differentiation of oligo precursor cells could be a useful therapeutic approach in autoimmune demyelination. In conclusion, all together these studies, although preliminary, indicate that delivery of neurotrophic factor genes into the CNS can be approached using gene therapy systems without overt undesirable toxic effects.

Conclusions

The gene therapy results obtained so far, including our own, are encouraging due to the wide range of therapeutic flexibility that biological vectors possess (Fig. 1). However, rodent immuno-gene therapy cannot be easily transferred to humans since (a) autoreactive T cells can be isolated from animal models of MS but not from patients with MS where the antigen(s) is still unknown, (b) DNA-liposomes as well as VV have limited transfection efficiency when introduced into post-mitotic cells such as those resident in the CNS, (c) VV and first-generation adenoviral vectors are still toxic and immunogenic, and, (d) HSV-1-derived vectors are short lasting. Nevertheless, the new improvements of viral vector technology should overcome the above mentioned limitations and lead soon to the development of: (a) 'inducible' vectors in which the transcription of the heterologous gene contained in the vector could be exogenously induced or inhibited, (b) chimeric vectors able to integrate into host DNA of post mitotic cells (i.e. retroviruses combined with herpesviruses), and (c) short- vs. long-term (i.e. "gutless" adenoviral vectors) expressing vectors.

In conclusion, a possible and desirable future scenario of gene therapy in immune mediated demyelination might include different "human-grade" vectors which can be used to deliver anti-inflammatory molecules as well as neuroprotective agents into the CNS (Fig. 1) in a flexibile and useful way since the new potential "therapeutic" vectors will have different life-span, tissue tropism and infectivity rate.

Acknowledgement

This work has been in part supported by Telethon (Italy), Associazione Italiana Sclerosi Multipla (AISM), and MURST.

References

1. Martino G, Hartung HP. Immunopathogenesis of multiple sclerosis: the role of T cells. Curr Opin Neurol 1999;12:309–321.
2. Steinman L. A few autoreactive cells in an autoimmune infiltrate control a vast population of nonspecific cells: a tale of smart bombs and the infantry. Proc Natl Acad. Sci USA 1996;93:2253–2256.
3. Lucchinetti CF, Brück W, Rodriguez M, Lassmann H. Distinct patterns of multiple sclerosis pathology indicates heterogeneity on pathogenesis. Brain Pathol 1996;6:269–274.
4. Kieseier BC, et al. Effector pathways in immune mediated central nervous system demyelination. Curr Opin Neurol. 1999;12:323–336.
5. Romagnani S. Human Th1 and Th2 subsets: doubt no more. Immunol Today 1991;12:256–257.
6. Abbas AK, Murphy KM, Sher A. Functional diversity of helper T lymphocytes. Nature 1996;383:787–793.
7. The IFNB Multiple Sclerosis Study Group, Interferon beta-1b is effective in relapsing-remitting multiple sclerosis. I. Clinical results of a multicenter, randomized, double-blind, placebo-con-

trolled trial. Neurology 1993;43:655–661.
8. Weinstock-Guttman B, Ransohoff RM, Kinkel RP, Rudick RA. The interferons: biological effects, mechanisms of action, and use in multiple sclerosis. Ann Neurol.1995;37:7–15.
9. Khan OA, et al. Interferon beta-1b serum levels in multiple sclerosis patients following subcutaneous administration. Neurology 1996;46:1639–1643.
10. The IFNB Multiple Sclerosis Study Group and the University of British Columbia MS/MRI Analysis Group, Neutralizing antibodies during treatment of multiple sclerosis with interferon beta-1b: experience during the first three years. Neurology 1996;47:889–894.
11. Calabresi PA, et al. Phase 1 trial of transforming growth factor beta 2 in chronic progressive MS. Neurology 1998;51:289–292.
12. The Lenercept Multiple Sclerosis Study Group and University of British Columbia MS/MRI Analysis Group. TNF neutralization in MS: results of a randomized, placebo-controlled multicenter study. Neurology 1999;53:457–465.
13. Shaw MK, et al. Local delivery of interleukin 4 by retrovirus-transduced T lymphocytes ameliorates experimental autoimmune encephalomyelitis. Local delivery of interleukin 4 by retrovirus-transduced T lymphocytes ameliorates experimental autoimmune encephalomyelitis. J Exp Med 1997;185:1711–1714.
14. Mathisen PM, et al. Treatment of experimental autoimmune encephalomyelitis with genetically modified memory T cells. J Exp Med 1997;186:159–164.
15. Chen LZ, et al. Gene therapy in allergic encephalomyelitis using myelin basic protein-specific T cells engineered to express latent transforming growth factor-beta1. Proc Natl Acad Sci USA 1998;95:12516–12521.
16. Dal Canto RA, et al. Local delivery of TNF by retrovirus-transduced T lymphocytes exacerbates experimental autoimmune encephalomyelitis. Clin Immunol 1999;90:10–14.
17. Willenborg DO, Fordham SA, Cowden WB, Ramshaw IA. Cytokines and murine autoimmune encephalomyelitis: inhibition or enhancement of disease with antibodies to select cytokines, or by delivery of exogenous cytokines using a recombinant vaccinia virus system. Scand J Immunol 1995;41:31–40.
18. Piccirillo CA. Prud'homme GJ. Prevention of experimental allergic encephalomyelitis by intramuscular gene transfer with cytokine-encoding plasmid vectors. Hum Gene Ther 1999;10:1915–1922.
19. Boccaccio GL, Mor F, Steinman L. Non-coding plasmid DNA induces IFN-gamma in vivo and suppresses autoimmune encephalomyelitis. Intern Immunol 1999;11:289–296.
20. Croxford JL, et al. Cytokine gene therapy in experimental allergic encephalomyelitis by injection of plasmid DNA-cationic liposome complex into the central nervous system. J Immunol 1997;160:5181–5187.
21. Croxford JL, et al. Local gene therapy with CTLA4-immunoglobulin fusion protein in experimental allergic encephalomyelitis. Eur J Immunol 1998;28:3904–3916.
22. Glorioso JC, et al. Gene transfer to brain using herpes simplex virus vectors. Ann Neurol 1994;35:S28–34.
23. Krisky DM, et al. Rapid method for construction of recombinant HSV-1 gene transfer vectors. Gene Ther 1997;4:1120–1125.
24. Marconi P, et al. Replication-Defective Herpes Simplex Virus Vectors For Gene Transfer In Vivo. Proc Natl Acad Sci USA 1996;93:11319–11320.
25. Krisky DM, et al. Development of Herpes simplex virus replication defective multigene vectors for combination gene therapy applications. Gene Ther 1998;5:1517–1530.
26. DeLuca NA, McCarthy AM, Schaffer PA. Isolation and characterization of deletion mutants of herpes simplex virus type 1 in the gene encoding immediate-early regulatory protein ICP4. J Virol. 1985;56:558–570.
27. Kuklin NA, et al. Modulation of mucosal and systemic immunity by enteric administration of non-replicating herpes simplex virus expressing cytokines. Virology 1998;240:245–253.
28. Martino G, et al. A gene therapy approach to treat demyelinating diseases using non-replicative

herpetic vectors engineered to produce cytokines. Mult Scler 1998;4:222–227.
29. Furlan R, et al. (1998) Central nervous system delivery of interleukin-4 by a non-replicative herpes simplex type 1 viral vector ameliorates autoimmune demyelination. Hum Gene Ther 1998;9:2605–2617.
30. Furlan R, et al. Central nervous system gene therapy with interleukin-4 inhibits progression of ongoing relapsing-remitting autoimmune encephalomyelitis in Biozzi AB/H mice. Gene Ther 2001;8:13–19.
31. Prineas JW, et al. Multiple sclerosis: remyelination of nascent lesions. Ann Neurol 1993;33:137–151.
32. Rodriguez M. Central nervous system demyelination and remyelination in multiple sclerosis and viral models of disease. J Neuroimmunol 1992;40:255–263.
33. Brück W, et al. Oligodendrocytes in the early course of multiple sclerosis. Ann Neurol 1994;35:65–73.
34. Ozawa K, et al. Patterns of oligodendroglia pathology in multiple sclerosis. Brain 1994;117:1311–1322.
35. Targett MP, et al. Failure to achieve remyelination of demyelinated rat axons following transplantation of glial cells obtained from the adult human brain. Neuropath Appl Neurobiol 1996;22:199–206.
36. Ludwin SK. Central nervous system demyelination and remyelination in the mouse: an ultrastructural study of cuprizone toxicity. Lab Invest 1978;39:597–612.
37. Linington C, Engelhardt B, Kapocs G, Lassmann H. Induction of persistently demyelinated lesions in the rat following the repeated adoptive transfer of encephalitogenic T cells and demyelinating antibody. J Neuroimmunol 1992;40:219–224.
38. Muir DA, Compston DA. Growth factor stimulation triggers apoptotic cell death in mature oligodendrocytes. J Neurosci Res 1996;44:1–11.
39. Aebischer P, et al. Intrathecal delivery of CNTF using encapsulated genetically modified xenogeneic cells in amyotrophic lateral sclerosis patients. Nature Med 1996;2:696–699.
40. Mathisen PM, et al. Th2 T cells expressing transgene PDGF-A serve as vectors for gene therapy in autoimmune demyelinating disease. J Autoimm 1999;3:31–38.
41. Poliani PL, Brok H, Furlan R, Ruffini F, Bergami A, Desina G, Marconi PC, Rovaris M, Glorioso JC, Penna G, Adorini L, Comi G, t'Hart B, Martino G. Delivery of a non-replicative herpes symplex type-1 vector engineered with the IL-4 gene to the central nervous system protects rhesus monkeys from hyperacute autoimmune encephalomyelitis. Hum Gene Ther, in press.

The contribution of axonal injury to neurologic dysfunction in theiler's virus-induced inflammatory demyelinating disease

D.R. Ure[1], D.B. McGavern[3], S. Sathornsumetee and M. Rodriguez[1,2,3]
[1]Department of Immunology, [2]Department of Neurology, and [3]Program of Molecular Neuroscience, Mayo Medical and Graduate School, Rochester, MN 55905, USA

It is well appreciated that demyelination impairs conduction [1–8], and this phenomenon contributes to clinical dysfunction in MS. Inflammatory factors such as nitric oxide can also impair conduction [9,10]. Remyelination or cessation of inflammation could therefore help to explain the periods of partial recovery seen in relapsing-remitting MS [11–14]. To explain the accumulation of permanent neurologic deficits in progressive MS, one attractive hypothesis is that the deficits result from permanent neuronal and/or axonal damage. Recent reviews have summarized findings that support this idea [15,16]. In addition to numerous accounts of axon pathology in MS tissue, additional supporting evidence comes from more recent studies using magnetic resonance spectroscopy. Notably, levels of the neuronal marker, N-acetyl aspartate, are reduced in both lesioned and normally-appearing central nervous system (CNS) tissue in MS, and the reductions have correlated with neurologic impairment [17–22]. T1-weighted magnetic resonance imaging also has been reported to be a measure of axonal pathology which correlates with disease progression [23–27].

To help understand the relationships among inflammatory demyelination, neuron damage, and neurologic dysfunction, we have studied progressive inflammatory demyelination induced by Theiler's virus infection of SJL/J mice [28,29]. Theiler's murine encephalomyelitis virus (TMEV) is a picornavirus that is naturally-occurring in mice. Experimental intracranial injection of the Daniel's strain of TMEV into SJL/J and other susceptible strains of mice results in a biphasic disease where the virus initially replicates in the brain (acute phase), resulting in a mild encephalitis that lasts for approximately 2–3 weeks. An immune response develops that is sufficient to overcome encephalitic disease and clear virus from neurons, but is incapable of completely clearing TMEV in susceptible mice. After clearance from the gray matter, TMEV establishes lifelong persistence in the brain stem and spinal cord white matter, replicating predominantly in oligodendrocytes and infiltrating macrophages / microglia. The latter is referred to as the chronic phase of disease. During the chronic phase of disease, the immune response involves infiltration of predominantly T / B lymphocytes and macrophages, and activation of resident glial cells. The cell-mediated inflammatory response is partly responsible for the

---

Correspondence address: Moses Rodriguez, Department of Immunology, Mayo Clinic, 428 Guggenheim Building, 200 1st Street SW, Rochester, MN 55905, USA.

demyelinated lesions of the spinal cord white matter, which are observed as early as 3 weeks post-infection. The lesions are concentrated in the cervical and thoracic white matter, and increase in size and number until a plateau at approximately 3–4 months post-infection [30]. Neurologic dysfunction resulting from the inflammatory, demyelinating phase of the disease (after 3 weeks post-infection) becomes apparent near the time that lesion load plateaus and then progresses toward severe spasticity, paralysis, incontinence, and a decline in sensory function. The progression of neurologic deficits beyond the time that lesion load plateaus suggests that demyelination alone may contribute but is not sufficient for development of severe deficits during the later stage of disease.

Neuronal damage resulting from inflammatory demyelination has been studied most thoroughly in SJL/J mice infected for 6–9 months (chronically infected mice). Inflammatory demyelinating lesions are well developed at this time and as mentioned have reached a plateau (Figs. 1A–D). Macrophages are the predominant inflammatory cell and are often laden with lipid vacuoles due to phagocytosis of myelin and cellular debris (Fig. 1B). Many axons are completely demyelinated or partly remyelinated (defined as having thin myelin sheaths). Morphological evidence of stressed or degenerating axons consists of dense accumulations of axoplasmic organelles and axons with dark, amorphous axoplasm (Figs. 1C, D). Consistent with axon degeneration, Bielschowski silver staining (Figs. 1E, F) and neurofilament immunocytochemistry [31] has demonstrated considerable disruption of the axon architecture within lesions. Also, the number of sodium channels is reduced, as demonstrated by immunocytochemistry and by saxitoxin binding techniques [31] (Figs. 1G, H). The reduction in sodium channels could reflect both axon loss and reduced synthesis or transport of channels to the axons.

The effects of inflammatory demyelination are evident not only within lesions but also in non-lesioned white matter. Axonal damage that results as axons traverse inflammatory lesions can be observed over long distances in the normal-appearing white matter. A similar concept was recently described in the brains of MS patients [32]. Using this concept we developed an automated method to quantify myelinated axons from transverse fields captured in the normal-appearing spinal cord white matter. We have shown using this method that the medium (4–10 μm^2) and large (>10 μm^2) axons are preferentially lost in mice infected for 195–220 days [30,33]. In the midthoracic spinal cord, a 30% reduction in medium axons and a 78% reduction in large axons was found at this time, whereas a less significant reduction was observed at earlier time points (Fig. 2C) [30]. When data from sampled fields were extrapolated to describe the entire anterior and lateral column area (which includes the lateral, anterolateral, and anterior columns) at L1/L2, total axon number was reduced by 23%, and total axon area was reduced by 37% at 270 dpi [34]. Thus, a significant degree of axon degeneration was evident in normal-appearing white matter.

Atrophy of the spinal cord is another pathologic characteristic of chronically infected mice [30,33]. Interestingly, spinal cord atrophy correlates almost perfectly with the loss of medium to large fibers (≥4 μm^2) at 192 dpi ($R = 0.94$)

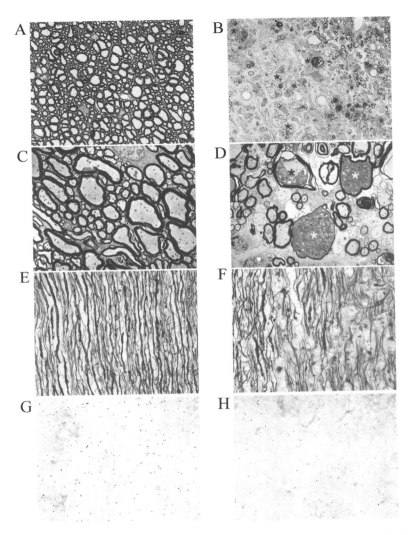

Fig. 1. Spinal cord pathology observed in susceptible SJL/J mice chronically infected with TMEV. The spinal white matter of uninfected (A,C,E,G) and chronically infected (at least 180 dpi) (B,D,F,H) SJL/J mice is illustrated above. (A,B) Demyelination plateaus by 100 dpi in TMEV infected SJL/J mice as is evident by light microscopy. The lesions are almost completely devoid of myelin when compared to normal myelin surrounding intact axons observed in uninfected white matter. Note the extensive macrophage infiltration and the lipid vacuoles within these macrophages (asterisks). (C,D) Electron micrographs of demyelinated lesions reveal evidence of axonal degeneration in chronically infected mice. Note the opaque, electron dense axoplasm of three degenerating axons (asterisks) from a demyelinating lesion when compared to the normal axons from an uninfected mouse. (E,F) Bielschowski silver staining demonstrates the significant disruption of the axonal architecture and axonal degeneration in the lesions of chronically infected mice. Note the contrast between normal and disrupted axon fibers. (G,H) Sodium channel densities detected by ^{3}H-saxitoxin are significantly reduced in lesions. The black dots represent sodium channel densities in the spinal cord white matter. Note the marked reduction in chronically infected mice as compared to uninfected mice.

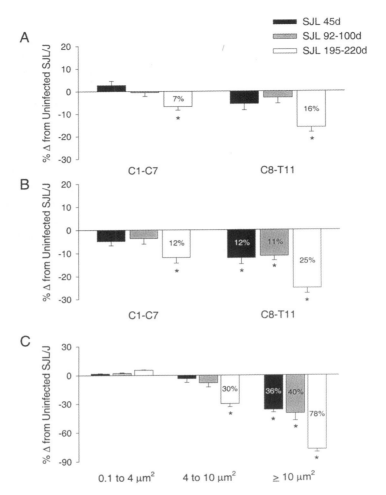

Fig. 2. Spinal cord atrophy and axonal loss in susceptible SJL/J mice following TMEV infection. (A) Total spinal cord atrophy was quantified in SJL/J mice at various time points post-infection. Data are represented as a percent change (±SEM) from a group of uninfected SJL/J mice. Total spinal cord atrophy was only observed at 195–220 dpi at the level of C1-C7 and C8-T11. (B) Quantification of lateral and anterior column area (which includes the lateral, anterolateral, and anterior columns) also revealed the most significant atrophy at 195–220 dpi. However, a lessor amount of atrophy was observed at 45 and 92–100 dpi at C8-T11. (C) Myelinated axonal area distributions were quantitatively measured from the midthoracic level of spinal cord and separated into three different size categories: small (0.1 to 4 μm^2), medium (4 to 10 μm^2), and large (greater than 10 μm^2). Data are represented as a percent change (±SEM) from a group of uninfected SJL/J mice. No reductions in small fibers were observed at any time point post-infection. The most significant reduction in axonal fibers was 195–220 dpi, the time point that also showed the most severe spinal cord atrophy. Reductions in both medium and large axon fibers were observed. Large axon fibers were also reduced at 45 and 92–100 dpi, but not to the same degree as the later time point. (In all graphs asterisks denote statistical significance by one way ANOVA when compared to uninfected SJL/J mice, and the percentages indicate the percent reduction or loss of the respective variable when compared to uninfected SJL/J mice.)

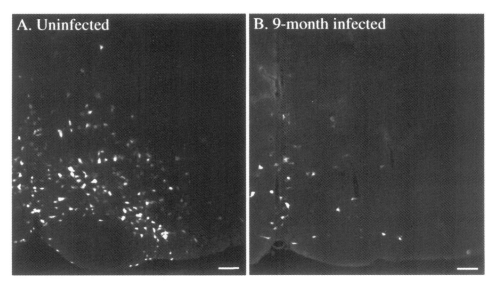

Fig. 3. Retrograde labeling of descending neurons in the brainstem. In uninfected mice and mice infected for 270 days, the retrograde tracers, Fluorogold and Fast Blue, were applied to T11/T12. After 7 days, the brains were examined for the retrograde accumulation of the fluorescent tracers in cell bodies of descending neurons. Two fields are shown from the ventral brainstem. Retrograde labeling of reticulospinal and raphespinal neurons was extensive in uninfected mice, but was greatly reduced in chronically infected mice.

(Table 1). Total spinal cord atrophy was only observed at the time point when the most severe loss of axons was observed [195–220 days post-infection (dpi)] (Fig. 2A). This total spinal cord atrophy resulted from a reduction in the anterior and lateral column area, which was found to be reduced by 12% in the cervical region (C1-C7) and 25% in the thoracic region (C8-T11) (Fig. 2B). Atrophy was also observed at earlier time points, although not to the same degree as 195–220 dpi. Total spinal cord atrophy did not result from reductions in gray matter area since this spinal cord region, which does not contain persistent virus or inflammation, was preserved at all time points post-infection. It is of note that atrophy was evident in the lateral and anterior columns despite the presence of inflammatory cells that can potentially contribute to the enlargement of the spinal cord. A combination of axon loss and demyelination are most likely responsible for the observed atrophy.

Retrograde axonal transport appears to be another neuronal function affected by chronic inflammatory demyelination. In retrograde labeling experiments using fluorescent tracers, we have found extensive reductions in the labeling of descending neurons in 270-day-infected mice, compared to uninfected mice (Figs. 3A,B). The reductions ranged from 60–93%, depending on the descending neuron population assessed. Although axon loss was partly responsible for the reduced labeling, the magnitude of the reduction in labeling suggests that retrograde organelle trafficking was also compromised [34].

Another important question is whether the destructive effects of chronic inflammatory demyelination are limited to the axon or also extend to the cell body. At least for rubrospinal and vestibulospinal neurons, the chronic axonal dysfunction and degeneration did not result in an appreciable dropout of cell bodies. Interestingly, rubrospinal cell bodies were completely preserved in number despite viral persistence and the presence of activated macrophages / microglia within the red nucleus of chronically infected mice. However, cell body pathology typical of retrograde responses to axonal transection was observed. This included cell body atrophy, nuclear indentation, and lipofuscin accumulation.

To determine how progressive inflammatory demyelination and axonal loss contribute to electrophysiologic abnormalities, motor evoked potentials (MEPs) were measured in susceptible mice at different time points post-infection [30,31]. As expected, susceptible mice showed reduced conduction velocities and amplitudes, as well as increased latencies of MEPs. However, of interest is the temporal profile for these electrophysiologic abnormalities. Increased latencies, but no reductions in MEP amplitudes were observed up through 100 dpi. This electrophysiologic pattern can best be explained by demyelination. Reductions in MEP amplitudes were only observed in mice infected for 180 days, which coincided with the most significant loss of the fastest conducting, medium and large myelinated fibers (Fig. 2C) [30].

Several methods have been used to objectively measure the progressive motor dysfunction associated with chronic disease. These methods include spontaneous activity measurements, stride length and width analyses, and rotarod assays [30,31,35]. Chronically infected mice have greatly reduced spontaneous activity in both the horizontal and vertical planes. In particular, there was nearly a complete cessation of vertical hindlimb rearing after several months of infection [31,35]. Stride length was also reduced in susceptible mice as early as 23 dpi, and further reductions were observed at later time points [35]. Interestingly, reductions in hindlimb stride width were only observed at 192 dpi, and these reductions correlated almost perfectly with axonal loss ($R = 0.97$) (Table 1). One of the most convenient and useful devices for measuring gross motor dysfunction is the rotarod, an elevated, rotating cylinder upon which the mice run. Similar to stride length studies, rotarod performance was reduced as early as 24 dpi, but worsened as the demyelinating disease progressed [35]. Since the onset of spinal cord demyelination is at approximately 21 dpi and is minimal at this time point, the mild neurologic deficits observed at this time are probably the result of the early encephalitic stage of infection, whereas the later deficits result from inflammatory demyelination and superimposed axonal injury in the spinal cord.

Correlations between measures of motor coordination / gait and various pathologic variables in chronically infected mice have brought us closer to understanding what mechanisms are primarily responsible for neurologic dysfunction [30]. Motor dysfunction is clearly related to the extent of inflammatory demyelination. Linear correlations exist between rotarod performance and both the cross-sectional areas of the largest lesions and overall lesion load in the spinal cord (Table 1). An

Table 1. Correlations between quantitative measures of neurologic function and pathology in SJL/J mice infected for 192 days.

Variable One	Variable Two	Correlation coefficient	P value
C7 combined lateral & anterior column area	frequency of medium to large fibers ($\geq 4 \cdot m^2$) measured at T6	$R = 0.94$	$P = 0.020$
C7 combined lateral & anterior column area	rotarod performance	$R = 0.92$	$P = 0.008$
percentage of spinal cord demyelination	rotarod performance	$R = -0.66$	$P = 0.110$
frequency of medium to large fibers ($\geq 4 \cdot m^2$) measured at T6	rotarod performance	$R = 0.90$	$P = 0.013$
frequency of medium to large fibers ($\geq 4 \cdot m^2$) measured at T6	change from baseline width of stance (footprint analysis)	$R = 0.97$	$P = 0.001$

Correlation coefficients and *P* values were calculated using the Pearson product moment correlation analysis. SJL/J mice (n = 6–7) were analyzed at 192 days post-infection using quantitative measures of the variables described above. This table is adapted from McGavern et al. [30].

even better predictor of motor dysfunction, however, is spinal cord atrophy. A positive correlation ($R = 0.92$) was found between rotarod performance and the cross-sectional C7 lateral and anterior column area (Table 1). Spinal cord atrophy has also correlated with disability in MS [36,37]. Dropout of medium and large axons (≥ 4 μm^2) best explains why atrophy correlates so well with function, because rotarod performance also correlated strongly with preservation of those axons ($R = 0.90$). Thus, the data suggest that preservation of the larger caliber axons is critical for maintaining normal motor function.

Informative data on the association between axon injury and neurologic dysfunction has also come from studies of β_2-microglobulin knockout mice (C57BL/6x129 background) that lack a class I MHC-induced immune response. These mice are normally resistant to TMEV infection; however, deletion of a class I MHC-induced immune response breaks resistance, resulting in viral persistence and inflammatory demyelination in the spinal cord white matter. Following TMEV infection, these mice have spinal cord demyelination similar to what is observed in susceptible SJL/J mice, which ultimately show axonal loss, electrophysiologic abnormalities, and neurologic deficits as described above. However, β_2-microglobulin knockout mice do not have electrophysiologic abnormalities or neurologic deficits at 180 dpi [31]. In addition, assessment of axons (Bielschowski and neurofilament staining) and sodium channel densities (immunocytochemistry and saxitoxin staining) in class I-deficient mice revealed "relative" axonal preservation and increased sodium channel densities when compared to susceptible SJL/J mice [31]. In addition, extensive spontaneous remyelination has previously been described in class I-deficient mice [38]. These results suggest that neurologic dysfunction can be prevented if axons are preserved, virus burden is reduced, and spontaneous remyelination occurs. This is consistent with clinical observations in MS where demyelination can be present in the absence of functional deficits [39–41].

It is clear from experimental animal models and MS that inflammatory demyelination can damage neurons and their projecting axons under certain conditions. Identifying the factors that result in axonal damage and permanent neurologic dysfunction is critical for MS patients. For example, it is not known whether axon loss results primarily from chronic demyelination or from inflammatory factors present in the milieu. It is likely that both factors contribute. Chronically demyelinated axons may become especially susceptible to damage by complement-fixing antibodies, free radicals, proteases, cytokines, or other inflammatory components. Although the CNS can compensate for a modest loss of certain neuronal populations, widespread neuronal destruction will ultimately result in permanent deficits. Therefore, aggressive measures should be taken to prevent neuronal dropout early in the disease course before clinical deficits become permanent. One therapeutic aim should focus on the maintenance of neuronal function and survival by strategies that promote remyelination or supply trophic support. Another therapeutic aim should focus on the attenuation of the pathologic mechanisms associated with inflammation. By utilizing the vast experience obtained from studying traumatic CNS injury models, many new opportunities may become available to prevent neuronal damage and progressive neurologic dysfunction in demyelinating disease.

References

1. McDonald WI, Sears TA. Effect of demyelination on conduction in the central nervous system. Nature 1969;221:182–183.
2. McDonald WI, Sears TA. The effects of experimental demyelination on conduction in the central nervous system. Brain 1970;93:583–598.
3. Rasminsky M, Sears TA. Internodal conduction in undissected demyelinated nerve fibres. J Physiol 1972;227:323–350.
4. Waxman SG. Conduction in myelinated, unmyelinated, and demyelinated fibers. Arch Neurol 1977;34:585–589.
5. Bostock H, Sears TA. Continuous conduction in demyelinated mammalian nerve fibers. Nature 1976;263:786–787.
6. Bostock H, Sears TA. The internodal axon membrane: electrical excitability and continuous conduction in segmental demyelination. J Physiol 1978;280:273–301.
7. Smith KJ, Bostock H, Hall SM. Saltatory conduction precedes remyelination in axons demyelinated with lysophosphatidyl choline. J Neurol Sci 1982;54:13–31.
8. Felts PA, Baker TA, Smith KJ. Conduction in segmentally demyelinated mammalian central axons. J Neurosci 1997;17:7267–7277.
9. Redford EJ, Kapoor R, Smith KJ. Nitric oxide donors reversibly block axonal conduction: demyelinated axons are especially susceptible. Brain 1997;120:2149–2157.
10. Shrager P, Custer AW, Kazarinova K, Rasband MN, Mattson. Nerve conduction block by nitric oxide that is mediated by the axonal environment. J Neurophysiol 1998;79:529–536.
11. Smith KJ, Blakemore WF, McDonald WI. The restoration of conduction by central remyelination. Brain 1981;104:383–404.
12. Honmou O, Felts PA, Waxman SG, Kocsis, JD. Restoration of normal conduction properties in demyelinated spinal cord axons in the adult rat by transplantation of exogenous Schwann cells. J Neurosci 1996;16:3199–3208.

13. Baron-Van Evercooren A, Avellana-Adalid V, Lachapelle F, Liblau R. Schwann cell transplantation and myelin repair of the CNS. Mult Scler 1997;3:157–161.
14. Imaizumi T, Lankford KL, Waxman SG, Greer CA, Kocsis JD. Transplanted olfactory ensheathing cells remyelinate and enhance axonal conduction in the demyelinated dorsal columns of the rat spinal cord. J Neurosci 1998;18:6176–6185.
15. Trapp BD, Ransahoff R, Rudick R. Axonal pathology in multiple sclerosis: relationship to neurologic disability. Curr Opin Neurol 1999;12:295–302.
16. Kornek B, Lassmann H. Axonal pathology in multiple sclerosis. A historical note. Brain Pathol 1999;9:651–656.
17. Davie CA, Barker GJ, Webb S, Tofts PS, Thompson AJ, Harding AE, McDonald WI, Miller DH. Persistent functional deficit in multiple sclerosis and autosomal dominant cerebellar ataxia is associated with axon loss [published erratum appears in Brain 1996 Aug;119(Pt 4):1415]. Brain 1995;118:1583–1592.
18. Matthews PM, Pioro E, Narayanan S, De Stefano N, Fu L, Francis G, Antel J, Wolfson C, Arnold DL. Assessment of lesion pathology in multiple sclerosis using quantitative MRI morphometry and magnetic resonance spectroscopy. Brain 1996;119:715–722.
19. Narayanan S, Fu L, Pioro E, De Stefano N, Collins DL, Francis GS, Antel, JP, Matthews PM, Arnold DL. Imaging of axonal damage in multiple sclerosis: spatial distribution of magnetic resonance imaging lesions. Ann Neurol 1997;41:385–391.
20. De Stefano N, Matthews PM, Antel JP, Preul M, Francis G, Arnold DL. Chemical pathology of acute demyelinating lesions and its correlation with disability. Ann Neurol 1995;38:901–909.
21. De Stefano N, Matthews PM, Narayanan S, Francis GS, Antel JP, Arnold DL. Axonal dysfunction and disability in a relapse of multiple sclerosis: longitudinal study of a patient. Neurology 1997;49:1138–1141.
22. Fu L, Matthews PM, De SN, Worsley KJ, Narayanan S, Francis GS, Antel JP, Wolfson C, Arnold DL. Imaging axonal damage of normal-appearing white matter in multiple sclerosis. Brain 1998;121:103–113.
23. van Walderveen MA, Barkhof F, Hommes OR, Polman CH, Tobi H, Frequin ST, Valk J. Correlating MRI and clinical disease activity in multiple sclerosis: relevance of hypointense lesions on short-TR/short-TE (T1-weighted) spin-echo images. Neurology 1995;45:1684–1690.
24. van Walderveen MA, Kamphorst W, Scheltens, P, van Waesberghe JH, Ravid R, Valk J, Polman CH, Barkhof F. Histopathologic correlate of hypointense lesions on T1-weighted spin-echo MRI in multiple sclerosis. Neurology 1998;50:1282–1288.
25. van Walderveen MA, Barkhof F, Pouwels PJ, van Schijndel RA, Polman CH, Castelijns JA. Neuronal damage in T1-hypointense multiple sclerosis lesions demonstrated in vivo using proton magnetic resonance spectroscopy. Ann Neurol 1999;46:79–87.
26. Truyen L, van Waesberghe JH, van, Walderveen MA, van Oosten BW, Polman CH, Hommes OR, Ader HJ, Barkhof F. Accumulation of hypointense lesions ("black holes") on T1 spin-echo MRI correlates with disease progression in multiple sclerosis. Neurology 1996;47:1469–1476.
27. van Waesberghe JH, Kamphorst W, De Groot CJA, van Walderveen MA, Castelijns JA, Ravid R, Lycklama a Nijeholt G.J., van der Valk P, Polman CH, Thompson AJ, Barkhof F. Axonal loss in multiple sclerosis lesions: Magnetic resonance imaging insights into substrates of disability. Ann Neurol 1999;46:747–754.
28. Dal Canto MC, Lipton HL. Primary demyelination in Theiler's virus infection. An ultrastructural study. Lab Invest 1975;33:626–637.
29. Rodriguez M, Oleszak E, Leibowitz J. Theiler's murine encephalomyelitis: a model of demyelination and persistence of virus. [Review] [158 refs]. Crit Rev Immunol 1987;7:325–365.
30. McGavern DB, Murray PD, Rivera-Quinones C, Schmelzer JD, Low PA, Rodriguez M. Axonal loss results in spinal cord atrophy, electrophysiologic abnormalities, and neurologic deficits following demyelination in a chronic inflammatory model of multiple sclerosis. Brain 2000;123:519–531.

31. Rivera-Quinones C, McGavern D, Schmelzer JD, Hunter SF, Low PA, Rodriguez M. Absence of neurological deficits following extensive demyelination in a class I-deficient murine model of multiple sclerosis. Nat Med 1998;4:187–193.
32. De Stefano N, Narayanan SR, Matthews PM, Francis GS, Antel JP, Arnold DL. In vivo evidence for axonal dysfunction remote from focal cerebral demyelination of the type seen in multiple sclerosis. Brain 1999;122:1933–1939.
33. McGavern DB, Murray PD, Rodriguez M. Quantitation of spinal cord demyelination, remyelination, atrophy, and axonal loss in a model of progressive neurologic injury. J Neurosci Res 1999;58:492–504.
34. Ure DR, Rodriguez M. Extensive injury of descending neurons demonstrated by retrograde labeling in a virus-induced murine model of chronic inflammatory demyelination. J Neuropathol Exp Neurol 2000;59:664–678.
35. McGavern DB, Zoecklein L, Drescher KM, Rodriguez M. Quantitative assessment of neurologic deficits in a chronic progressive murine model of CNS demyelination. Exp Neurol 1999;158:171–181.
36. Losseff NA, Wang L, Lai HM, Yoo DS, Gawne-Cain ML, McDonald WI, Miller DH, Thompson AJ. Progressive cerebral atrophy in multiple sclerosis. A serial MRI study. Brain 1996;119:2009–2019.
37. Losseff NA, Webb SL, O'Riordan JI, Page R, Wang L, Barker GJ, Tofts PS, McDonald WI, Miller DH, Thompson AJ. Spinal cord atrophy and disability in multiple sclerosis. A new reproducible and sensitive MRI method with potential to monitor disease progression. Brain 1996;119:701–708.
38. Miller DJ, Rivera-Quinones C, Njenga MK, Leibowitz J, Rodriguez M. Spontaneous CNS remyelination in beta 2 microglobulin-deficient mice following virus-induced demyelination. J Neurosci 1995;15:8345–8352.
39. Ghatak NR, Hirano A, Lijtmaer H, Zimmerman HM. Asymptomatic demyelinated plaque in the spinal cord. Arch Neurol 1974;30:484–486.
40. Mews I, Bergmann M, Bunkowski S, Gullotta F, Bruck W. Oligodendrocyte and axon pathology in clinically silent multiple sclerosis lesions. Mult Scler 1998;4:55–62.
41. O'Riordan JI, Losseff NA, Phatouros C, Thompson AJ, Moseley IF, MacManus, G, McDonald WI, Miller DH. Asymptomatic spinal cord lesions in clinically isolated optic nerve, brain stem, and spinal cord syndromes suggestive of demyelination. J Neurol Neurosurg Psychiatry 1998;64:353–357.

The effect of virus-like infections on the course of multiple sclerosis

William A. Sibley*
Department of Neurology, University of Arizona, College of Medicine, 1501 N. Campbell Ave., Tucson, AZ 85724, USA

Introduction

Most research efforts in multiple sclerosis in recent years have focused on the natural history of the disease, as revealed by serial MRI scans, and elucidating possible immunologic *effector* mechanisms. The latter efforts have studied especially the role of T-lymphocytes, macrophages, and cytokines as the last links in a chain of events which might produce demyelinating lesions.

In contrast, there has been relatively little study of possible causative or *inducing* factors which might set these immunologic events into motion. Several years ago, a number of attempts to transmit MS to non-human primates produced negative results [1,2]. While these attempts do not entirely exclude direct infection as a cause of MS, these efforts and persistent failure to culture a causative organism, make this a less likely possiblility [3].

This review will concentrate on existing evidence that intercurrent virus-like infections influence the course of MS. This first seemed likely to us on the basis of preliminary results reported by Joseph Foley and I in 1965: among 39 patients studied over a period of 3 years, at 3-monthly intervals, there were 69 exacerbations of MS. Thirty-three of these (48%) occurred in a 4 week at-risk (AR) period: i.e. an infection occurred in a period extending from 3 weeks before to one week after the exacerbation [4].

An additional clue suggesting that infection might play a role has been the finding of a significant seasonal variation in the occurrence of MS exacerbations [4,5], and retrobulbar neuritis [5,6]. The latter is the easiest MS symptom to date accurately, because of its usual rapid and dramatic onset. A significant seasonal variation in retrobulbar neuritis in a number of northern latitudes has been found by in some studies, but not in all. Several positive reports have shown a tendency to a spring and late summer/early fall peak in exacerbation frequency, with significantly fewer new bouts in mid-winter [5–7].

Correspondence address: Department of Neurology, University of Arizona, College of Medicine, 1501 N. Campbell Avenue, Tucson, AZ 85724, USA. E-mail: was@u.arizona.edu

Table 1. Overall influence of 779 virus-like infections on MS exacerbations.

Period	No. pt.-yrs (%)	No. exacerbations (%)	Annual exac. rate
AR	105 (12%)	67 (27%)	0.64 (0.49–0.79)
NAR	791 (88%)	67 (27%)	0.23 (0.19–0.21)
Total	896 (100%)	246 (100%)	$X^2 = 56.3$, $p<0.001$

AR = At risk
NAR = Not at risk

Table 2. Influence of virus-like infections on annual exacerbation rates by mean disability.

Mean DSS	No. Pts.	Pt-yrs AR	Pt-yrs. NAR	Rate AR	Rate NAR
0–2	36	31.8	152.7	0.88*	0.31
2–4	32	19.3	123.7	0.99*	0.41
4–6	48	28.8	234.8	0.49*	0.20
6–8	44	21.9	222.5	0.27	0.14
8–10	10	3.1	57.7	0.00	0.06

* $p<0.001$ in comparison to NAR period.
Mean disability: average of entry and exit DSS scores.

Methods

Beginning in 1976, we began a prospective study of 170 MS patients and 134 sex- and age-matched controls, which lasted for 8 years. In this effort, patients were examined every three months, or whenever new symptoms suggested worsening of the disease. On a monthly basis, both patients and controls answered a series of questions about environmental incidents, with special emphasis on the occurrence of infections, traumatic episodes, and psychologically stressful life events. At the end of the investigation we had a dated list of all exacerbations, and were in a position to attempt to make time correlations between these environmental events and worsening of the disease. The many details of the methodology are available in the primary publications [8–10].

Results

Virus-like infections and the frequency of exacerbations

Table 1 shows an association between viral infections and increased exacerbation rate. Twelve percent of the time in the study, patients were at risk (AR) for infection, the AR period being defined as 5 weeks after the beginning of symptoms of infection, and two weeks prior to the beginning of these manifestations. In this 12% of the time, 27% of exacerbations occurred. The annual exacerbation rate AR was nearly three times the exacerbation rate in the same patients when not at risk (NAR), a highly significant difference.

Table 2 indicates that this same relationship between viral infections and exac-

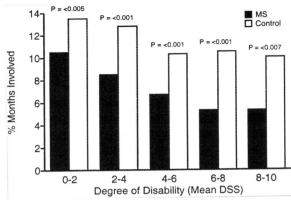

Fig. 1. Frequency of common viral infections in MS patients with varying degrees of disability and age-mathed controls. (Mean DSS = average of entry and exit DSS scores: mean follow-up 5.3 years.)

Table 3. Progression of MS disability and frequency of virus-like infections.

Entry DSS	Infections	Patients	Mean age	Mean inf./yr	Mean progression DSS/yr
0–4	<1/yr	37	43	0.53	0.43*
	>1/yr	46	33	1.68	0.19*
5–9	<1/yr	64	50	0.41	0.13
	>1/yr	23	44	1.55	0.15

*SE of difference in means = 0.08, p = 0.0027.

erbations occurred in patients with different levels of disability, although it was not seen in patients with advanced disability, a group who had few exacerbations.

Frequency of common viral infections in MS patients and controls

Fig. 1 illustrates a major finding of our study, i.e. significantly fewer viral infections occurred in MS patients than controls. There was an almost reciprocal proportionality between the number of infections and degree of disability.

An inverse relationship between the frequency of viral infection and the rate of disability development?

There are two points of evidence supporting this possibility. One, perusal of the figure indicates progressively fewer infections in patients with advancing disability. That this is not all due to sheltering is the fact that patients in the minimally disabled category (column one in the figure) had the same mean age, same number of young children at home, the same number working outside the home, and the same opportunity for infectious contacts as age-matched controls. Secondly, if one

Table 4. Seasonal variation in retrobulbar neuritis*.

Period	# cases			
	University Hospitals of Cleveland	Cleveland Clinic	Mayo Clinic	Stockholm*
Jan–Mar	12 (16%)	12 (17%)	9 (12%)	25 (17%)
Apr–Jun	23 (32%)	21 (30%)	31 (42%)	46 (31%)
Jul–Sep	20 (27%)	23 (33%)	19 (26%)	42 (29%)
Oct–Dec	18 (25%)	14 (20%)	15 (20%)	34 (23%)
	73 (100%)	70 (100%)	74 (100%)	147 (100%)
	p>0.05,<0.1	p>0.05,<0.1	p<0.01	

Cases from Cleveland Hospitals (Sibley and Foley [5]) and the Mayo Clinic (Taub and Rucker [6]) include young adults with and without an established diagnosis of multiple sclerosis.
*Cases from Stockholm (Jin, dePedro-Cuesta, Soderstrom, and Link [7]) are monosymptomatic optic neuritis. Stockholm data significant for difference between winter and spring (Dec–Feb vs Mar–May, $p < 0.001$).

studies the rate of disability increase in mildly affected patients (DSS <5), those with the fewest infections had a significantly higher rate of increase in DSS/yr (Table 3). This phenomenon was not seen in patients with a DSS of 5 or greater at entry into the study. There was a very slow advancement of DSS in patients with DSS of 6 or greater in this study: approximately 0.12 DSS points per year between DSS 6–8.

Discussion

The finding of an increase in MS exacerbations in association with common viral infections has been confirmed in Ottawa [11], Gothenberg [12], Baltimore [13], Nottingham [14], and Rotterdam [15], and there is now wide acceptance of this phenomenon.

If MS exacerbations are provoked by common virus-like infections, which viruses are the most important? Because a significant seasonal variation might provide some clues in this regard, the seasonal data should be examined carefully. As noted previously, our prospective studies in Cleveland found a seasonal difference in the rate of MS exacerbations [4], including a winter dip in the frequency of MS attacks and episodes of retrobulbar neuritis in young adults. The latter was first reported by Taub and Rucker in a small series [6], later confirmed by retrospective analysis of the dates of onset of retrobulbar neuritis in young adults in two Cleveland hospitals [5], and has recently been confirmed in Stockholm County for monosymptomatic optic neuritis (Table 4) [7]. In this table, representing a total of 364 patients, there is a striking similarity in the seasonal occurrence of optic neuritis in Cleveland hospitals 1959–1963, and the rates in Stockholm 1990–1995. The seasonal attack rate of MS in Cleveland also consisted of a spring and late summer peak, and a relatively quiescent winter. There was no substantial difference between seasonal occurrence of MS onsets and optic neuritis in Cleveland [5]. Also, a comparison of months of onset of MS in Arizona, and

London, Ontario, Canada showed significantly fewer MS cases beginning in the winter in Ontario, but not in Arizona [10]. Some have found different seasonal patterns including a spring increase in attacks [17–19] or no overall seasonality [20] In the Optic Neuritis Treatment Trial, which included 388 patients, cases were less common in the winter, and more frequent in spring and fall, but the differences were slight [21] It is not clear why this series should differ from those tabulated in Table 4, but the Optic Neuritis Treatment Trial was a 15-center study, and four study sites were in southern states; unfortunately the months of occurrence, by center, were not given.

Does the type of seasonal distribution seen in Table 4 provide any clues? The exact pattern of viral seasonality varies in different years, and also varies by locale; for example, in southern Arizona enteric viral infections occur year-round, while in northern climates they are more or less confined to summer months. The rhinoviruses as a class, on the other hand, produce infections throughout the year, but peak in the spring and fall, according to infection-monitoring viral isolation studies in Seattle and Chicago [22,23] Contrary to popular conception, they are less common in winter months. For example, in the Seattle Virus Watch, where viral agents were systematically cultured over a period of years, the mid-winter depression in rhinovirus isolations was consistent in each of four years.

There is thus some similarity between the seasonal occurrence that some have found for optic neuritis and MS attacks and the seasonal occurrence of rhinovirus infections recorded in some northern latitudes. This fact, and reports that in some studies rhinoviruses are the commonest isolate in URI's [24] adds to the weight of evidence suggesting a possible relationship. If a single viral group is responsible, rhinoviruses remain as a candidate which have not been adequately investigated. It is possible, of course, that different viruses are responsible in different individuals, which could help account for the wide variation in the course of MS in different patients.

Consistent with this idea, is the fact that commercially available viral antigens have been used to test MS sera before and after exacerbations by a number of authors [4,12–14,25] and while positive results (a four-fold increase in antibody titre between pre- and post exacerbation sera) were obtained in a few individual cases, these results have not been confined to a single viral class. None of these studies have included rhinovirus antigens, however.

If it is true that 27–48% of MS exacerbations are associated with virus-like infections, how does one explain exacerbation which shows no such association? Here the great frequency of inapparent infection can be invoked as a possible answer. Most primary viral infections are symptomatic. Re-infection by the same virus, however, is common, and in these cases inapparent or very mild infection is very frequent. This is especially true of the viruses which commonly produce upper respiratory symptoms [26].

Our interpretation of the finding of fewer virus-like infections in MS patients is that MS patients have fewer *symptomatic* infections, but probably the same number of totalinfections as controls. It would be unreasonable to believe that MS patients have fewer infections, since their sera contain a full complement of antiviral antibodies [4]. In fact, serum titers of measles and a number of other viruses are often higher in groups of MS patients [3].

Others have not compared the frequency of infections in MS patients and age- and sex-matched controls, and thus there has been no confirmation of our finding that patients have 25–60% fewer symptomatic viral infections than controls. While the mechanism whereby symptomatic common infections are reduced in MS patients is obscure, it is tempting to believe that it may be related to the same immune overactivity, genetically based, that is postulated to underlie the disease.

Other investigators have not compared the rate of disability advancement with the frequency of viral infection. Thus there has been no opportunity to confirm our suggestion of more rapid advancement of MS disability in patients who have the fewest such infections. This phenomenon was statistically significant in those patients with no disability, or mild disability on entry into the study; it could not be confirmed in patients with more advanced disability at entry, possibly due to the slow rates of disability progression in this group and the resultant loss of statistical power. If ultimately confirmed, we suspect that it will be found that the same immune mechanisms that determine a lesser frequency of symptomatic infection, also play a role in producing a more aggressive form of MS.

References

1. Gibbs CJ, Gajdusek DC, Alpers MP. Attempts to transmit subacute and chronic neurological diseases to animals. In: Burdzy K, Kallos P (eds) Pathogenesis and Etiology of Demyelinating Diseases. S. Karger, Basel, New York, 1969.
2. Sibley WA, Laguna JF, Kalter SS. Attempts to transmit multiple sclerosis to non-human primates. In: Bauer HJ, Poser S, Ritter G (eds) Progress in Multiple Sclerosis Research, Springer Verlag, Berlin, Heidelberg, New York, 1980.
3. Johnson RT. The virology of demyelinating diseases. Annals of Neurology 1994;36:S54–S60.
4. Sibley WA, Foley JM. Infection and immunity in multiple sclerosis. Annals New York Academy of Science 1965;122:457–468.
5. Sibley WA, Foley JM. Seasonal variation in multiple slcerosis and retrobulbar neuritis in Northeastern Ohio. Trans Am Neurol Assn 1965;90:295–297.
6. Taub RG, Rucker CW. The relationship of retrobulbar neuritis to multiple sclerosis. Am J Ophthal 1954;37:494–497.
7. Jin Y, de Pedro-Cuesta J, Soderstrom M, Link H. Incidence of optic neuritis in Stockholm, Sweden, 1990–1995. II. Time and space patterns. Arch Neurol 1999;56:975–980.
8. Sibley WA, Bamford CR, Clark K. Clinical viral infections and multiple sclerosis. Lancet 1985;1:1313–1315.
9. Sibley WA. Risk factors in multiple sclerosis-implication for pathogenesis. In: Serlupi Crescenzi G (ed) A Multidisciplinary Approach to Myelin Diseases,Plenum Press, New York, 1988; pp 227–232.
10. Sibley WA, Bamford CR, Clark KP, Smith MS, Laguna JF. A prospective study of physical trauma and multiple sclerosis. Journal of Neurol Neurosurg and Psychiat 1991;54:584–589.
11. Narod S, Johnson-Lussemburg CM, Zheng Q, Nelson R. Viral infections and MS (letter). Lancet 1985;2:165.
12. Andersen O, Lygner P, Bergstrom T, et al. Viral infections trigger multiple sclerosis relapses: a prospective seroepidemiological study. Journal of Neurology 1993;240:417–422.

13. Panitch HS. Influence of infection on exacerbations of multiple sclerosis. Annals of Neurology 1994;36:S25–S28.
14. Edwards S, Zvartau M, Clarke H, Irving, W, Blumhardt LD. Clinical relapses and disease activity on magnetic resonance imaging associated with viral upper respiratory tract infections in multiple sclerosis. J Neurol Neurosurg Psychiatry 1998;64:736–741.
15. Buljevac D, van Doorn PA, van der Meche FGA. Rotterdam study on infections and relapses in multiple sclerosis: a preliminary report on the occurrence of infections. Multiple Sclerosis 1999;5(Supp. 1):S35.
16. Sibley WA, Paty DW. A comparison of multiple sclerosis in Ontario, Canada and Arizona, USA. Acta Neurol Scand 1981;64(Supp. 87):60–68.
17. Hutchinson WM. Acute optic neuritis and the prognosis for multiple sclerosis. J Neurol Neurosurg Psychiatry 1976;39:283–289.
18. Bradley WG, Whitty CWM. Acute optic neuritis: its clinical features and their relation to prognosis for recovery of vision. J Neurol Neurosurg Psychiatry 1967;30:531–538.
19. Wuthrich R, Rieder HP. The seasonal incidence of multiple sclerosis in Switzerland. Eur Neurol 1970;3:257–264.
20. Compston DAS, Batchelor, Earl CJ, McDonald WI. Factors influencing the risk of multiple sclerosis developing in patients with optic neuritis. Brain 1978;101:495–511.
21. Optic Neuritis Study Group. The 5-year risk of MS after optic neuritis. Neurology 199749:1404–1412.
22. Fox JP, Cooney MK, Hall CE. The Seattle virus watch. V. Epidemiologic observations of rhinovirus infections, 1965–69, in families with young children. Am J Epidemiol 1975;101:122.
23. Levandowski RA. Rhinoviruses. In: Gorbach SL, Bartlett JG, Blacklow NR (eds) Infectious Diseases. Second Edition. W.B. Saunders Co., Philadelphia, London, 1998; p 2173.
24. Monto AS, Sullivan KM. Acute repiratory illness in the community. Frequency of illness and the agents involved. Epidemiol Infect 1993;110:145–159.
25. Compston DAS, Vakarelis BN, Paul E, McDonald WI, Batchelor JR, Mims CA. Viral infection in patients with multiple sclerosis and HLA-DR matched controls. Brain 1986;109:325–344.
26. Chang T. Recurrent viral infection (reinfection). New Eng. J. Med. 1971;284:765–773.

A clinical neuroradiological and virological longitudinal study to verify the role of virus in inducing relapses in multiple sclerosis patients: preliminary results

P. Ferrante[1*], R. Mancuso[2], R. Cavarretta[3], E. Pagani[2], M. Filippi[4], G. Comi[4], G. Iannucci[4], P. Melzi[3], S. Della Bella[5], D. Caputo[3]

[1]Chair of Virology, Dept. of Preclinical Sciences, University of Milan; [2]Laboratory of Biology and [3]Multiple Sclerosis Unit, Don C.Gnocchi Foundation ONLUS, IRCCS, Milan; [4]Neuroimaging Research Unit, Ospedale San Raffaele, IRCCS, Milan, 5Chair of Immunology, Dept. of Sciences and Biomedical Technology, University of Milan, Italy

Introduction

The possible role of one or more virus in the etiology of multiple sclerosis (MS) has been suspected from more than a century but, despite the very large amount of studies, clear, direct supports to the viral etiology of MS are still lacking [1,2]. In spite of this failure several evidences suggest that environmental factors, and in particular one or more viruses, can be implicated in the immunological pathogenic process that probably induces MS in genetically predisposed individuals [3–8].

Besides their possible etiologic role, based on clinical and laboratory observations, viruses have also been indicated as triggering factors in the clinical exacerbations that can be observed in relapsing remitting MS [3,9–12].

The hypothesis of a viral activity in the evolution of MS is in some way also supported from the recently demonstrated efficacy in MS of the therapy with beta interferon, a drug well known for its antiviral action, and from the report of the action of Acyclovir in reducing MS clinical relapses [13].

Recently our group performed a laboratory study to verify the possible role of different human herpesviruses in the triggering of clinical acute attack in relapsing remitting MS patients. We focused our attention on these neurotropic viruses because different members of this family have repeatedly suggested as possible etiologic agent in MS [14–17].

In that study we analysed longitudinally a group of relapsing remitting MS and of healthy subjects and searched, using specific nested Polymerase Chain Reaction (PCR) tests, the DNA belonging to herpes simplex virus type 1 and 2 (HSV-1 and HSV-2), human cytomegalovirus (HCMV), Epstein-Barr virus (EBV), human herpes virus six (HHV6), and the messenger RNA of HSV-1 in their peripheral blood. HSV-1 mRNA and DNA was detected in a significant number of acute MS patients but not in the control group and thus we suggested that HSV-1 is actively

---

Correspondence address: Pasquale Ferrante, MD, Laboratory of Biology, Don C. Gnocchi Foundation, IRCCS, Via Capecelatro, 66, I-20148, Milan, Italy. Tel: +39-02-40308211. Fax: +39-02-40308438. E-mail: pferrante@dongnocchi.it

replicating in the peripheral blood of MS patients during clinical acute attack and probably play a role in the triggering of MS relapses [18]

Based on this experience we have decided to verify these results performing a longitudinal survey on 30 clinically defined MS patients not treated with interferon or any other long term therapy. The patients will be followed for 18 months and, at the enrolment in the study and thereafter every three months and on the occasion of an intercurrent clinical acute attack, they are subjected to neurological evaluation and to magnetic resonance imaging with gadolinium enhancement. Every six months they are analysed using a complete battery of neuropsychological tests to asses their cognitive function. At the moment of the enrolment and at each neurological survey samples of peripheral blood, serum and urine are collected for the virological study. Using specific and sensitive nested PCR, DNA and or RNA belonging to HSV-1 and -2, varicella zoster virus (VZV), HCMV, EBV, HHV6 and JC virus (JCV) are searched in all the peripheral blood samples. A longitudinal analysis of antiviral antibodies toward the same viruses is also performed on the serum samples.

Since the study is still in progress, we here present the results concerning a first group of 10 MS patients that have been studied for the presence of HSV-1, HCMV and HHV6. Although very preliminary the data obtained seem to confirm our previous findings and suggest the possible involvement of HSV-1 in clinical acute attack of relapsing remitting MS.

Materials and methods

Patients

The study foresees the enrolment of 30 patients of both sex, of age between 18 and 50 years, affected by relapsing-remitting Multiple Sclerosis, diagnosed according the criteria of Poser et al. (1983) [19], with or without sequelae, not under immunosuppressing or immunomodulating therapy and free of other associated diseases involving the immune system.

A venous blood sample, a neurological examination (always executed by the same physician) and a NMR of the brain and spinal cord with gadolinium enhancement, are foreseen at the moment of the enrollment (T0), every 3 months (T1, T2, etc.), on the occasion of every neurological relapse and anyway before the steroid treatment. At T0 and every 6 months a basal neuropsychological evaluation is performed. The follow up duration is fixed at 18 months.

The patients must, beside signing an informed consensus, devote to write a monthly logbook on which are noted any known or new neurological symptoms, any concomitant diseases, the eventually administrated drugs, the occurrence of menstruation and any other relevant physical or psychological stress. This data acquisition improves the probability to find significant association between the relapse and events preceding the onset of the neurological symptoms.

Neuropsychological evaluation

All the patients enrolled in the study have undergone a short basal neuropsychological evaluation based on Dementia Rating Scale (DRS) [20] and Wechsler Memory Scale (WMS) [21].

The DRS is made of five sections that regard in order: attention, initiative and perseverance, constructive praxis, conceptualization and memory. The WMS is constituted by seven sections as follows: information, orientation, mental control, logic memory, repetition of numbers, visual reproduction and associations.

Biological sample collection and preparation

From all the patients samples of peripheral blood with EDTA, serum and urine were collected at each of the follow up points and on the first day of admission in the hospital during a clinical relapse.

For the viral DNA search, DNA is extracted from peripheral blood using Wizard genomic DNA purification kit (Promega, Madison, USA) and stored at $-20°C$. Briefly, after lysis of red blood cells, white blood cells and their nuclei are lysed in appropriate cells/nuclei lysis solution. Then cell proteins are removed by a salt-precipitation step that leaves the DNA in solution. Finally, it is concentrated and desalted by isopropanol precipitation. 700 ng of DNA are utilized in each nested PCR assay.

RNA is extracted from peripheral blood using the QIAamp RNA Blood Mini Kit (QIAGEN GmbH, Hilden, Germany) purification procedure; briefly, erythrocytes are selectively lysed and leukocytes are recovered by centrifugation. The leukocytes are then lysed using highly denaturation which immediately inactivate ribonucleases (RNAases) to allow the isolation of intact RNA; the sample is applied to a spin column provided with a silica-gel-based membrane; total RNA bound to this membrane, washed away from contaminants, eluted in RNAase free water and stored at $-80°C$. Prior to RT-PCR amplification , any possible DNA contamination is eliminated treating all the samples with Deoxiribonuclease I (DNAseI, GIBCO BRL- Life Technologies Italia Srl,, Milano, Italy) according to the protocol supplied by the producer. To perform each RT-nested PCR, approximately 500–1000 ng of RNA are used.

Urine surnatant and sediment are separated by centrifugation at 2500 RPM for 10 minutes and then stored at $-20°C$. Five microliters of these samples are utilized to detect viral genomes.

Serological methods

All the sera will be tested for the presence of antibodies to HSV-1, HSV-2, VZV utilizing commercial ELISA tests (Genzyme Virotech Gmbh, Russelsheim, Germany), and to HCMV, with a MEIA test. (Abbott Laboratories, Chicago, IL, USA)

Table 1. Nucleotide sequence, position on the genome and denomination of primers used in the specific nested PCR for the different viruses studied.

	Primers denomination	Position on genome	Sequence 5' → 3'
JCV	JC1	*LT Antigen, 4215-4232*	AAC ACA GCT TGA CTG AGG
	JC2	*LT Antigen, 4573-4592*	GCT TCA GAC AAT GGT TTG GG
	PEP1	*LT Antigen, 4255-4274*	AGT CTT TAG GGT CTT CTA CC
	PEP2	*LT Antigen, 4408-4427*	GGT GCC AAC CTA TGG AAC AG
HCMV	P2668	*DNA polymerase, 3396-3377*	TGT CCG TGT CCC CGT AGA TG
	P2675	*DNA polymerase, 2807-2826*	CGA CTT TGC CAG CCT GTA CC
	CIF	*DNA polymerase, 2873-2992*	GTG CTC TCG GAA CTG CTC AA
	CIR	*DNA polymerase, 3098-3079*	CGT TAC TTT GAG CGC CAT CT
HSV-1	P2668	*DNA polymerase, 65471-65452*	
HSV-2 (#)		*DNA polymerase, 2680-2661#*	TGT CCG TGT CCC CGT AGA TG
		DNA polymerase, 64953-64972	
	P2675	*DNA polymerase, 2163-2182#*	CGA CTT TGC CAG CCT GTA CC
		DNA polymerase, 2294-2314	
	HIF	*DNA polymerase, 65085-65105 #*	GAC GGC TGT TCT TCG TCA AGG
		DNA polymerase, 2458-2477	
	HIR	*DNA polymerase, 65268-65249 #*	GTG AAC CCG TAC ACC GAG TT
VZV	ZF	*ORF 4, 3271-3291*	CAGACTCCAACGCTTCAATCA
	ZR	*ORF 4, 3709-3690*	ACAACGCCTCCTGCAATGAC
	P2674	*ORF 4, 3377-3394*	ATGTCCGTACAACATCAA
	P2673	*ORF 4, 3643-3624*	CGCAGAGAGAACCTTTTAGC
EBV	EB3	*EBNA-1, 109332-109351*	AAGGAGGGTGGTTTGGAAAG
	EB4	*EBNA-1, 109609-109628*	AGACAATGGACTCCCTTAGC
	EB1	*EBNA-1, 109353-109372*	ATCGTGGTCAAGGAGGTTCC
	EB2	*EBNA-1, 109542-109561*	ACTCAATGGTGTAAGACGAC
HHV6	EX1	*MCP, 17072-17096*	GCGTTTTCAGTGTGTAGTTCGGCAG
	EX2	*MCP, 17592-17567*	TGGCCGCATTCGTACAGATACGGAGG
	IN-3	*MCP, 17267-17291*	GCTAGAACGTATTTGTCGAGAACG
	IN-4	*MCP, 17525-17501*	ATCCGAAACAACTGTCTGACTGGCA

Table 2. Nucleotide sequence, position on the genome and denomination of the primers adopted in the specific nested reverse transcriptase PCR used for the search of HSV-1 mRNA.

	Primers denomination	Position on genome	Sequence 5' → 3'
HSV-1	FHEX	DNA polymerase, 62990-63011	ACCCAGGCGCCATACGTACTATA
	RHEX	DNA polymerase, 63315-63334	TGTGATGGCGTCCATAAACC
	FHINT	DNA polymerase, 63021-63041	ATGAATTTCGATTCATCGCCC
	RHINT	DNA polymerase, 63302-63321	ATAAACCGCGCGTGGAACTG

Table 3. Clinical and neuroradiological evaluation, reported as EDSS score and MRI results, of the ten MS patients at the programmed times and during clinical relapses. The sign + indicates the presence of demyelinating lesions with gadolinium enhancement; nd = not done.

	T0	RELAPSE		T1		RELAPSE		T2	RELAPSE		T3	
	EDSS	EDSS	MRI	EDSS	MRI	EDSS	MRI	EDSS	EDSS	MRI	EDSS	
1(MR)	3.0	4.0	+	3.0	/	/	/	3.5	/	/	/	
2(SC)	1.0	/	/	2.0	/	3.0	+	1.0	2.0	+	4.0	
3(MG)	4.0	/	/	3.5	/	/	/	4.0	/	/	1.5	
4(WM)	1.5	/	/	1.5	+	/	/	1.5	/	/		
5(GG)	1.0	3.0	+	1.5	+	/	/	1.5	3.0	+	0.0	
6(FG)	0.0	2.0	+	1.5	+	/	/	0.0	/	/		
7(PF)	3.0	3.5	/	3.5	/	/	/	3.5	/	/	3.0	
8(RL)	3.0	/	/	3.5	/	/	/	3.0	/	/	2.0	
9(MMR)	3.5	4.0	nd	3.0	/	/	/	2.5	/	/		
10(IM)	0.0	/	/	0.0	/	/	/		/	/		

101

PCR analysis

To detect viral DNA, a region belonging to the DNA polymerase gene of HSV1/2 and HCMV, to the EBNA1 gene of EBV and to the LT region of JCV have been respectively searched by using specific nested PCR methods whose primers and protocols have been previously described in details [18]. In addition, the major capsid protein gene of HHV-6, using the primers and PCR protocol previously described by Secchiero *et al.* [22], and the ORF4 gene of VZV have been also searched.

The sequences of all the primers employed, their position on viral genome and denomination are reported in Table 1.

To evaluate the presence of HSV-1 mRNA, an indicator of viral replication, a reverse transcriptase PCR (RT-PCR) was performed on RNA samples utilizing oligonucleotide primers specific for HSV-1 DNA polymerase gene with the Titan One Tube RT-PCR System (Roche Diagnostics GmbH, Mannheim, Germany). The sequence of the primers employed, their position on the genome and their denomination are reported in Table 2. RNA from VERO cells infected with HSV-1 was included as positive control and to verify the sensibility of method. Likewise, RNA extracted from V0801 cell line infected respectively with HCMV and VZV was utilized as control to evaluate the specificity of RT-PCR.

Results

Neurological and neuroradiological evaluation

Up to now 22 MS patients have been enrolled in the study, however, due to the different follow up length, in this contest only 10 will be considered. They include 3 males and 7 females with the age ranging between 26 and 47 years, an history of disease between 2 and 22 years and an EDSS score at T0 between 0 and 4. Three of these patients present a relapsing remitting pattern with sequelae.

Four patients were followed up to the 9th month, five to the 6th and one to the 3rd month (Table 3). During the follow up, two patients presented 2 relapses, four patients had only 1 relapse and the other four remained stable. Before the relapse, one patient had an important psychological stress (mourning), one patient had an influenza syndrome, one patient had multiple teeth extraction and the remaining three patients did not have any association with relevant events. For five of these six patients it has been possible to execute the programmed NMR during the relapse. One or more enhanced lesions have been detected in all the NMR executed during the relapse. All the MS patients had an increase of the EDSS score varying from 0.5 to 2 points as a consequence of the relapse and the large majority of them returned at the previous EDSS score when tested later.

Neuropsychological evaluation

None of the ten MS patients considered in this report was depressed at the moment of enrollment in the study, moreover none of them had real pathological scores when tested with the adopted neuropsychological tests.

In this preliminary report we have considered only the variations of the scores observed in the patients who at least underwent the evaluation at the third and sixth month (T0 and at T2). The results obtained are summarized in Table 4. The patients marked with an asterisk are those that during the six months had a relapse. It is possible to see that the patients 1 and 9 maintained their performance substantially unchanged, the patients 2 and 7 showed similar scores at the global test but obtained markedly diminished scores in particular in subsection of logical memory. It is also possible to note that in patient 5 the scores remained unchanged during the follow up period, although he had, at T0, score values that were markedly lower than those obtained by the other MS patients. On the contrary patient 6 had a significant increase of the score values at T2 in comparison to those observed at T0.

Viral DNA search

The search of DNA belonging to HSV-1/2, HCMV, VZV, EBV, HHV-6 and JCV has been performed on all the peripheral blood samples so far collected.

Moreover it should be remembered that the sera and the urine have not yet tested and the serological analyses will be performed only when a significant number of sera will be collected.

All the ten MS patients considered had at least one sample positive for one or more virus. Among the 44 peripheral blood samples, 26 (63.7%) were DNA positive: only one virus has been detected in thirteen samples, a dual infection was found in eleven, three different viruses were detected in two samples and in the peripheral blood of one patient we amplified DNA belonging to four different viral agents.

The frequency of detection of each virus is reported in Table 5. EBV DNA was found in 29.5% and HHV-6 in 27.3%. samples, HSV-1 has been amplified in six samples (13.6%), HCMV and JCV in seven samples (15.9%), while VZV only in two (4.5%) out of 44 samples.

Viral DNA has been detected both in the samples collected at the programmed time of the follow up and in those collected during a clinical relapse (Table 6). In particular of the ten patients six were positive at T0, eight at T1, four at T2 and six at T3 for one or more virus. During the follow up, a total of six clinical relapses, involving five MS patients, have been observed, five in between T0 and T1, and one in between T2 and T3.

Patients 5 had two relapses during the observation period. Peripheral blood was collected only in four of the six observed relapses, and viral DNA, belonging respectively to EBV, HSV-1, and HHV6, has been detected in all three of the five

Table 4. Results of the neuropsychological evaluation of the ten MS patients at the moment of the enrolment (T0) and at the sixth month (T2).

	1 (MR)		2 (SC)		3 (MG)		4 (WM)		5 (GG)		6 (FG)		7 (PF)		8 (RL)		9 (MMR)		10 (IM)	
	T0	T2	T0	T2	T0	T2	T0	T2	T0	T2	T0	T2	T0	T2	T0	T2	T0	T2	T0	T2
DRS tot	143	143	142	142	143	144	136	134	133	133	144	144	139	136	136	136	138	136	144	
WMS (QM)	124	129	114	108	114	110	118	108	92	92	108	132	110	76	90	110	129	122	124	
Information	6	5	6	6	5	6	6	6	4	5	5	6	6	6	5	5	6	6	6	
Orientation	5	5	5	5	6	5	5	5	5	6	5	5	5	5	5	5	5	5	5	
Mental control	9	9	9	8	8	7	8	7	8	8	8	8	6	7	8	8	8	7	8	
Logic Memory	18	18.5	14	10	11.5	17	13	11.5	7	10	12.5	18.5	11.5	6.5	8	7.5	14	14	15.5	
Rep. Numbers	10	9	10	9	11	11	10	11	6	10	11	12	9	8	9	8	9	8	11	
Visual repr.	5	6	9	13	14	6	8	8	13	12	12	12	14	4	6	12	12	10	12	
Associations	16	19	15	14	15	16	16.5	13	13	7.5	11.5	14.5	17	8.5	13	11	19	20	20	

Table 6. Detection of DNA belonging to HSV-1, HSV-2, HCMV, HHV-6, VZV, EBV, JCV, HTLV-I/II, in the peripheral blood of ten multiple sclerosis patients at the different time of observation during the follow up and on the occasion of a clinical relapse (nd= not done).

Patients	T0	Clinical relapse	T1	T2	Clinical relapse	T3
1 (MR)	HHV-6, EBV	EBV	neg	neg	/	neg
2 (SC)	HHV6, HCMV	/	neg	neg	neg	neg
3 (MG)	EBV	/	HCMV, JCV	neg	/	JCV, HCMV, EBV, HHV-6
4 (WM)	EBV	/	HHV-6, EBV	neg	/	EBV
5 (GG)	HHV-6, EBV	/	HSV-1, HCMV, HHV6	HSV-1	HHV-6, EBV	neg
6 (FG)	EBV	/	HCMV, JCV	neg	/	JCV
7 (PF)	EBV	HSV-1	HCMV, VZV	neg	/	neg
8 (RL)	neg	/	HSV-1, HCMV, HHV-6	VZV, EBV	/	HSV-1, HHV-6
9 (MMR)	neg	HSV-1	HCMV, EBV	HHV-6, JCV	/	HHV6
10 (IM)	neg	/	neg	HHV-6	/	neg

Table 5. Detection of viral DNA, expressed as absolute number and percentage, in the 44 peripheral blood samples examined.

Virus	Positive number	Percentage
HSV-1	6	13.6 %
VZV	2	4.5 %
EBV	13	29.5 %
HCMV	7	15.9 %
HHV-6	12	27.3 %
JCV	7	15.9 %

samples collected.

Since our study, on the basis of previous experiences, was mostly focused on HSV-1 and HSV-2, a detailed analysis of the data obtained on the patients whose peripheral blood was positive for these two virus, is of particular interest. Patient 5 had a clinical relapse between T0 and T1, and a second one between T2 and T3. During the first relapse the peripheral blood was not collected but 15 days later, at T1, he was positive for HSV-1, HCMV and HHV6, while during the second relapse he resulted positive for HHV-6 and EBV.

Patient 6 was positive for HSV-1 when he had a clinical relapse between T0 and T1.

Patient 8 was apparently stable during the follow up and resulted positive for HSV-1 at T1 and also in T3.

Finally patient 9 was positive for HSV-1 during the unique relapse he developed between T0 and T1.

Moreover patient 1 were positive for EBV during relapse occurred in between T0 and T1 and patient 5 showed the presence of two viruses (EBV and HHV-6) in sample collected during a relapse between T2 and T3.

Discussion

The possible role of viruses in the triggering of clinical attack in relapsing remitting MS patients has been suggested from various authors [9,10], including our group [18]. However it is well known that during the clinical course of relapsing remitting MS, besides the clinically relevant relapses, it is also possible that the patient can develop new active demyelinating lesions, evidenced with magnetic resonance imaging, without detectable clinical symptoms. Moreover, it should be considered that also viral infections or reactivations of persistent viral infection may be asymptomatic. Thus the conventional studies performed, as we did, targeting the MS patients during a clinical acute attack and in a stable phase, and performing neuroradiological analysis only on the occasion of the relapse, have the problem that the subclinical demyelinating lesions could be considered as a period of stability.

In order to overcome these confounding factors, we designed the study presented here, in which the 30 MS patients are followed for eighteen months with neuroradiological analysis programmed every three months and performed also on the occasion of clinical relapses, and with a fully neuropsychological assessment every six months. A large set of viruses are taken in consideration, including six of the eight known human herpesviruses and JCVs, and their presence is investigated using updated molecular biology methods and more conventional serological assays.

Up to now EBV, HHV6 and HCMV have been detected more frequently than HSV-1 and HSV-2.

Both EBV and HHV-6 are already known for their capability of inducing neurological diseases [23,24] and they have also suggested as etiological agents of MS [17,25]. However in our previous study we did not observe significant differences in the frequency of EBV and HHV-6 DNA between MS patients and controls [18]. Moreover it should be pointed that the finding of EBV, HCMV and HHV6 DNA in the peripheral blood is, in some way, not completely unexpected, since these viruses are known for their capability to establish latency in peripheral blood cells of a large majority of normal population and of MS patients. To this regard, in order to verify the possible role of EBV and HHV6 in MS triggering, the search of DNA in the serum and a complete evaluation of the trend of the specific antibodies will be necessary and these studies are currently in progress.

As in our previous report also the present preliminary data seem to indicate a possible role of HSV-1 and HSV-2 in the clinical course and thus in the relapse of multiple sclerosis. First of all, in total HSV-1 and HSV-2 have been detected six times in the peripheral blood and in two occasion the viral presence was coincident with a clinical relapse (in patients 6 and 9). As regard to the HSV-1 DNA detection during the clinical stable phase on the occasion of the programmed controls, a careful evaluation of these results is useful. In the patient 5, HSV-1 DNA has been amplified at T1 only fifteen days after the clinical relapse. Patient 6 was already positive for HSV-2 at T0 when he was free of neurological symptoms, however the neuropsychological evaluation indicate that this patient had at T0 scores lower than at T2, with a consistent increase of the cognitive performances that has not been observe in any of the other MS patient. Moreover this patient had still HSV-2 during the relapse that he developed 30 days later.

In our opinion these results are interesting since HSV-1 and HSV-2 establish the latency in sensorial neuronal ganglia and during reactivation viral presence is restricted to the nervous system and to the innervated mucosal regions [26]. Thus the finding of HSV-1 and HSV-2 in the peripheral blood of MS patients is of particular relevance.

In conclusion, the results presented here, although preliminary, can be seen as a further support to the hypothesis, already suggested [27,28], that HSV-1 and HSV-2 could act as triggering factors of relapses in MS, and if they will be confirmed when this study will be completed new perspectives for MS treatment will be probably open.

Since the data on the possible involvement of other viruses and in particular of

EBV are still preliminary, we cannot rule out the hypothesis that, as already suggested [29], several different viruses could be involved in the etiology and clinical progression of MS.

Acknowledgement

This work was supported by a grant given from the Fondazione Cassa di Risparmio delle Provincie Lombarde (CARIPLO) to the Don C. Gnocchi Foundation, ONLUS, IRCCS; and by a grant Ricerca Corrente 2000 given from the Italian Ministry of Health to the Don C. Gnocchi Foundation, ONLUS, IRCCS.

References

1. Kurtzke JF. Epidemiologic evidence for multiple sclerosis as an infection. Clin Microbiol Rev 1993;6:382–427.
2. Dalgleish AG. Viruses and multiple sclerosis. Acta Neurol Scand 1997;S169:8–15.
3. Sibley WA, Bamford CR, Clark K. Clinical viral infections and multiple sclerosis. Lancet 1985;1:1313–1315.
4. Ferrante P, Mancuso R. Viral cerebrospinal fluid virological analysis in mulitpile sclerosis. In: Thompson EJ, Troiano M, Livrea P, editors. Cerebrospinal fluid analysis in multiple sclerosis. Milano: Springer-Verlag Italia, 1996;1–14.
5. Kirk J and Zhou AL. Viral infection at the blood-brain barrier in multiple sclerosis: -an ultrastructural study of tissues from a UK Regional Brain Bank. Multiple Sclerosis 1996;1:242–252.
6. Perron H, Garson JA, Bedin F, Beseme F, Paranhos-Baccala G, Komuria-Pradel F, Mallet F, Tuke PW, Voisset C, Blond JL, Lalande B, Seigneurin JM, Mandrand B and The Collaborative Research Group on Multiple Sclerosis. Molecular identification of a novel retrovirus repeatedly isolated from patients with multiple sclerosis. Proc Natl Acad Sci USA 1997;94;14:7583–7588.
7. Monteyne P, Bureau JF, Brahic M. Viruses and multiple sclerosis. Curr Opin Neurol 1998;4:287-291.
8. Ross RT, Cheang M, Landry G, Klassen L, Doerksen K. Herpes zoster and multiple sclerosis. Can J Neurol Sci 1999;26:29–32.
9. Andersen O, Lygner PE, Bergström T, Andersson M, Vahlne A. Viral infections trigger multiple sclerosis relapses: a prospective seroepidemiological study. J Neurol 1993;24:417–422.
10. Panitch HS. Influence of infection on exacerbations of multiple sclerosis. Ann Neurol 1994;36:S25–S28.
11. Gran B, Hemmer B, Vergelli M, McFarland , Martin R. Molecular mimicry and multiple sclerosis: degenerate T-cell recognition and the induction of autoimmunity. Ann Neurol 1999;45:559–567.
12. Wandinger KP, Jabs W, Siekhaus A, Bubel S, Trillenberg P, Wagner HJ, Wessel K, Kirkhner H, Hennig H. Association between clinical disease activity and Epstein-Barr virus reactivation in MS. Neurology 2000;55:178–184.
13. Bergstrom T. Herpesviruses: a rationale for antiviral treatment in multiple sclerosis. Antiviral Res 1999;41:1–15.
14. Martin JR. Herpes simplex virus type1 and 2 and multiple sclerosis. Lancet 1981;2:777–781.
15. Bray PF, Luka J. Antibodies against Epstein-Barr nuclear antigen (EBNA) in multiple sclerosis CSF and two pentapeptide sequence identities between EBNA and myelin basic protein. Neurology 1992;42:1798–1804.
16. Sanders VJ, Felisan S, Waddel A, Tourtellotte WW. Detection of herpesviridae in postmortem multiple sclerosis brain tissue and controls by polimerase chain reaction. J of Neurovirol

1996;2:249–258.
17. Soldan SS, Berti R, Salem N, Secchiero P, Flamand L, Calabresi PA, Brennan MB, Maloni HW, McFarland HF, Lin HC, Patnaik M, Jacobson S (). Association of human herpesvirus type 6 with multiple sclerosis: increased IgM response to HHV-6 early antigen and detection of serum HHV-6 DNA. Nature Med 1997;3:12:1394–1397.
18. Ferrante P, Mancuso R, Pagani E., Calvo MG, Saresella M, Speciale L, Caputo D. Molecular evidences for a role of HSV-1 in multiple sclerosis acute attack. J Neurovirol 2000;6:S109–114.
19. Poser CM, Paty DW, Scheinberg L, et al.. New diagnostic criteria for multiple sclerosis: guidelines for research protocols. Ann Neurol 1983;13:227–231.
20. Mattis S. Dementia rating scale: professional manual. Odessa, FL: Psychological assessment Resources; 1988.
21. Wechsler D. Wechsler memory scale. Revised Manual. The Psychological Corporation. San Antonio, TX. 1988.
22. Secchiero P, Carrigan DR, Asano Y, Benedetti L, Crowley RW, Komaroffet AL, Gallo RC, Lusso P. Detection of human herpesvirus 6 in plasma of children with primary infection and immunosuppressed patients by polymerase chain reaction. J Inf Dis 1995;171:273–280
23. Ishiguro N, Yamada S, Takahashi T, Takahashi Y, Togashi T, Okuno T, Yamanishi K. Meningoencephalitis associated with HHV-6 related exanthem subitum. Acta Pediatr Scand 1990;79:987–989.
24. Huang LM, Lee CY, Lee PI, Chen JM, Wang PJ. Meningitis caused by human herpesvirus-6. Arch Dis Child 1991;66:1443–1444.
25. Munch M, Riisom K, Christensen T, Moller-Larsen A, Haahr S. The significance of Epstein-Barr virus seropositivity in multiple sclerosis patients? Acta Neurol Scand 1998 ; 97:3:171–4.
26. Whitley RJ(1996). Herpes simplex viruses In: . Fields et al. editors. Virology. Philadelphia : Lippincott-Raven Publisher, II: 2297–2342.
27. Bergstrom T, Anderson O, Vahlne A. Isolation of herpes simplex virus type 1 during first attack of multiple sclerosis. Ann. Neurol 1989;26:283.
28. Lycke J, Svennerholm B, Hjelmquist E, Frisen L, Badr G, Andersson M, Vahlne A, Andersen O. Acyclovir treatment of relapsing-remitting multiple sclerosis. J Neurol 1996;243:214–224.
29. Hunter SF, Hafler DA. Ubiquitous pathogens. Link between infection and autoimmunity in MS? Neurology, 2000;55:164–165.

Viral and bacterial specificities of oligoclonal IgG bands in multiple sclerosis

S. Goffette and C.J.M. Sindic*
Laboratory of Neurochemistry, Faculty of Medicine, Université Catholique de Louvain, 53-59, Avenue Mounier, 1200 Brussels, Belgium

Introduction

The presence of oligoclonal IgG bands restricted to the cerebrospinal fluid (CSF) is the hallmark of the intrathecal humoral immune response observed in Multiple Sclerosis (MS). By means of very sensitive techniques, such as isoelectric focusing followed by immunoblotting, these IgG bands are present in about 95% of MS cases. We have previously defined 5 types of IgG pattern in an European Consensus Group [1]. Both type 2 and type 3 may be observed in MS and both types indicate an intrathecal synthesis.

Until now, all attempts at defining MS-specific antigens responsible for this oligoclonal immune response have been unsuccessful. It is still not clear whether these oligoclonal IgG are linked to a viral infection, an autoimmune reaction or some other manifestation of hypersensitivity unrelated to putative MS-specific antigens. Thus, the relevance of the CSF-restricted oligoclonal IgG to the pathogenesis of the MS process remains obscure.

In contrast, during acute infections of the central nervous system (CNS), oligoclonal IgG bands support the antibody activity against the infectious agent. The antigen-driven immunoblotting technique enables us to demonstrate the antibody specificities corresponding to oligoclonal IgG bands, and in addition, to detect oligoclonal antibodies from the polyclonal IgG background [2–4]. Whereas a strong monospecific reaction is observed in acute infections, a polyspecific immune reaction directed against various neurotropic viruses is present in MS and already in acute optic neuritis [5,6].

Patients

We studied five groups of patients. The first group (N=12) was used as a control and consisted of 11 non-neurological patients suffering from minor neurosis or tension headache and one patient with cervicarthrosis myelopathy. Group II (N=20) consisted of 20 patients with MS. All displayed oligoclonal IgG bands restricted to the CSF. Diagnosis was clinically definite in 18, and laboratory-supported probable in 2. Group III (N=27) consisted of 27 patients with acute monosymptomatic

---

Correspondence address: Prof. C.J.M. Sindic. Tel: 32-2-7645359. Fax: 32-2-7649337. E-mail: sindic@nchm.ucl.ac.be

optic neuritis (AMON) examined within the first 4 weeks of their disease by Dr. J. Fredericksen in Copenhagen. The final diagnosis of idiopathic AMON was based on clinical criteria (i.e. two or more of the following symptoms being required: blurred vision, decreased visual acuity, retrobulbar pain, dyschromatopsia and scotomas), other causes having been ruled out by ophthalmological, neurological and biological examinations. Patients with previous ON were excluded. Group IV (N=3) consisted of two patients with neurolupus and one with the Guillain-Barré syndrome (GBS). Group V (N=14) consisted of patients with various infectious diseases of the CNS: 4 with Varicella zoster (VZ) meningitis (isolated meningitis in one case, two cases associated with zoster ophthalmicus and one with zoster spinalis), 3 with herpetic encephalitis, one with rabies encephalitis, one with progressive multifocal leucoencephalopathy, two with neuro-AIDS (a patient with CNS toxoplasmosis and a patient with AIDS-related dementia), one each with neuroborreliosis, neurosyphilis and tuberculous meningitis.

Materials and methods

The antigen-driven immunoblots were performed with commercially available measles, VZ, cytomegalovirus (CMV), herpes virus hominis, mumps, and rubella antigens from Whittaker Bioproducts. Mycobacterial antigens were from Difco Laboratories. VP1 recombinant antigens of JC virus and herpesvirus 6 antigens were gifts from Profs. Luke and Linde, respectively. Rabbit anti-human IgG and alkaline phosphatase-conjugated rabbit anti-human IgG specific for gamma-chain were obtained from Dako (Copenhagen, Denmark) (code A 090 and D 336, respectively). Isoelectric focusing and capillary blotting were performed as previously published [7–8].

Results

Patients without inflammatory disorders of the CNS (control group; N=12)

In only one case (a patient suffering from tension headache), CSF displayed faint oligoclonal anti-VZ antibodies not detectable in the corresponding serum. An intrathecal synthesis of anti-VZ antibodies was thought to occur in this case.

Patients with MS (N=20)

Only two patients out 20 (10%) displayed either no immunostaining or a mirror pattern for the antigens from the four neurotropic viruses (measles, VZ, rubella and mumps) under study. CSF-restricted oligoclonal antibodies were detected against four antigens in three patients, against three antigens in one (measles, VZ and rubella), against two antigens in 11 (measles and rubella in six, measles and VZ in two, measles and mumps in one, rubella and VZ in one, VZ and mumps in one) (Fig. 1), and against one antigen in three (measles, rubella and mumps in one case

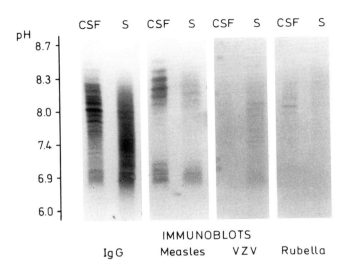

Fig. 1. Representative capillary blotting of IgG, and anti-measles, VZ and rubella antibodies in CSF and serum (S) from a MS patient. The presence of CSF-restricted oligoclonal IgG and antibodies against measles and rubella antigens demonstrated intrathecal synthesis. In contrast, the absence of oligoclonal anti-VZ antibodies in the CSF in spite of the presence of such antibodies in the serum indicated the intrathecal synthesis of antibodies with other specificities.

each). The local production of anti-measles IgG antibodies was thus the most frequent, occurring in 14 of the 20 cases (70%). Intrathecal production of anti-rubella, anti-VZ and anti-mumps antibodies was present in 12 (60%), 8 (40%) and 6 (30%) cases, respectively. Oligoclonal anti-herpesvirus 6 antibodies were present in three out of a series of 10 other MS patients (30%) (Fig. 2). In contrast, only one MS CSF out of 20 displayed a slight intrathecal production of anti-CMV antibodies and only one out of 16 displayed oligoclonal antibodies against the recombinant VP1 antigen of the JC virus. Antibody reactivity against mycobacterial antigens was never detected.

Patients with AMON (N=27)

Sixteen patients (59%) showed a polyspecific intrathecal synthesis of oligoclonal IgG antibodies against one or more viruses (12 measles, nine VZ, six rubella, six mumps). The presence of virus-specific oligoclonal IgG was significantly related to the results of oligoclonal IgG (P=0.003), free kappa chain bands (P=0.002), and brain MRI abnormalities (P=0.040).

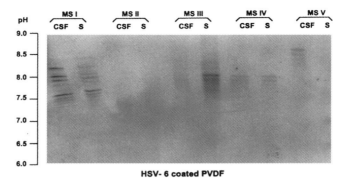

Fig. 2. Representative capillary blotting of anti- herpesvirus 6 (HSV-6) antibodies in CSF and serum (S) from five MS patients. The presence of CSF-restricted oligoclonal antibodies (patient V) or the the more marked immunostaining of such antibodies in the CSF than in the serum (patient I) demonstrated intrathecal synthesis. No reaction was observed in patient II, but a mirror pattern was present in patient IV (passive transfer of antibodies from the blood). Absence of oligoclonal antibodies in the CSF of patient III, in spite of their presence in the serum, indicated the intrathecal synthesis of antibodies with other specificities.

Patients with other inflammatory diseases of the nervous system (N=3)

The two patients with neurolupus also displayed, in addition to the presence of CSF-restricted oligoclonal IgG bands, a polyspecific intrathecal immune response against measles, VZ and rubella antigens in one case, against VZ antigens in the other. In contrast, the GBS patient had numerous oligoclonal IgG bands and antibodies in both CSF and serum ("mirror pattern") without detectable intrathecal production of antibodies against measles, rubella, VZ, mumps and CMV antigens.

Patients with infectious diseases of the CNS (N=14)

Antibody production unrelated to the causal infectious agent was observed in five cases : against measles antigens in a case of VZ meningitis, against VZ antigens in a case each of neuroborreliosis and of neuro-AIDS complicated by brain toxoplasmosis, against VP1 in one case of neuroborreliosis, and against mumps antigens in a case of herpetic encephalitis. It should be noted that in the first two cases, the samples under study were collected three months after onset of the disease and specific treatment, and that in the last case, the CSF sample was collected 10 years after the acute disease at the occasion of an epileptic fit (an intrathecal synthesis of oligoclonal anti-herpes simplex virus antibodies was still present in this sample).

Discussion

The main result of this investigation is to demonstrate that most MS patients (18 out of 20; 90%) and AMON patients (16 out of 27; 59%) displayed an intrathecal production of antibodies against one or more of the neurotropic viruses under study. Our prevalence rates of intrathecal synthesis in MS are very close to the ones reported by Felgenhauer and Reiber in a study of the antibody index calculated from enzyme immunoassays [9]. They found an intrathecal synthesis in 79% of cases for measles, in 70% for rubella and in 62% for VZ, whereas our figures are 70, 60 and 40% respectively.

These anti-viral antibodies displayed an oligoclonal pattern which did not correspond to the major oligoclonal IgG bands detected in the same CSF sample after the same run of isoelectric focusing. Such a polyspecific reaction was not restricted to CSF samples from MS patients but was also present in two cases of neurolupus and, at a lesser degree, in one case of neuro-AIDS and in four cases of infectious diseases of the CNS after clinical recovery. In the latter cases however, the bulk of the CSF oligoclonal IgG bands was directed against the causal infectious agent (VZ, herpes simplex and *Borrelia burgdorferi*).

The long-term persistence within the CNS of memory B-cells after a viral brain infection has been reported in a murine model [10]. Measles, rubella, VZ, mumps, and herpesvirus 6 infections are very common during childhood. These neurotropic viruses probably affect the brain more frequently than clinically observed [11] and could induce the recruitment of specific B-cells within the CNS and their persistence as memory cells. The influx of activated T-helper cells across the blood-brain and the blood-CSF barriers seems to play a key role in initiating the MS process and coud lead to the re-activation of memory B cells present in the CNS. Such a re-activation could result in the production of antibodies against various neurotropic infectious agents and this production has been called "polyspecific" [9].

In conclusion, this polyspecific intrathecal immune response has to be considered as a side-product of a chronic immune stimulation within the CNS. Its relationship to the etiopathogenesis of MS seems to be weak. These anti-viral antibodies are likely mere "bystanders" of the ongoing immune response and could be produced without de novo replication of the corresponding viral genome [12–13]. The use of this polyspecific reaction for diagnostic purposes must take into account its occurrence not only in MS but also in chronic CNS infections [14] and in other inflammatory CNS disorders. However, the continuous synthesis of these antibodies in absence of their antigenic target may be linked to the same mechanisms which are responsible for the very chronic immune activation observed in the MS process.

Acknowledgement

The authors are thankful to Mrs. M.P. Van Antwerpen for skilful technical assistance and to J.L. Frederiksen, M.D., for clinical data and CSF samples from patients with optic neuritis. This work was supported by a grant from the "Groupe belge d' Etude de la Sclérose en plaques".

References

1. Andersson M et al. Cerebrospinal fluid in the diagnosis of multiple sclerosis: a consensus report. J Neurol Neurosurg Psychiatry 1994;57;897–902.
2. Bukasa K et al. Anti-HIV antibodies in the CSF of AIDS patients. A serological and immunoblotting study. J Neurol Neurosurg Psychiatry 1988;51;1063–1068.
3. Boucquey D et al. Herpes simplex virus type 2 meningitis without genital lesions: an immunoblot study. J Neurol 1990;237;285–289.
4. Sindic CJM et al. Intrathecal synthesis of anti-mycobacterial antibodies in patients with tuberculous meningitis. An immunoblotting study. J Neurol Neurosurg Psychiatry 1990;53;662–666.
5. Sindic CJM, Monteyne Ph, Laterre EC. The intrathecal synthesis of virus-specific oligoclonal IgG in multiple sclerosis. J Neuroimmunol 1994;54;75–80.
6. Fredericksen JL, Sindic CJM. Intrathecal synthesis of virus-specific oligoclonal IgG, and of free kappa and free lambda oligoclonal bands in acute monosymptomatic optic neuritis. Comparison with brain MRI. Multiple Sclerosis 1998;4;22–26.
7. Sindic CJM, Laterre EC. Oligoclonal free kappa and lambda bands in the cerebrospinal fluid of patients with multiple sclerosis and other neurological diseases. An immmunoaffinity- mediated capillary blot study. J Neuroimmunol 1991;33;63–72.
8. Sindic CJM et al. Detection of specific oligoclonal antibodies to recombinant JC virus VP1 in patients with progressive multifocal leukoencephalopathy. J Neuroimmunol 1997;76;100–104.
9. Felgenhauer K, Reiber H. The diagnostic significance of antibody specificity indices in multiple sclerosis and herpes virus induced diseases of the nervous system. Clin Investig 1992;70;28–37.
10. Gerhard W, Koprowski H. Persistance of virus-specific memory B cells in mice CNS. Nature 1977;266;360–361.
11. Gibbs F, Gibbs E, Carpenter P. Electroencephalographic abnormality in "uncomplicated" childhood diseases. JAMA 1959;171;1050–1055.
12. Godec MS et al. Absence of measles, Mumps, and Rubella Viral Genomic sequences from Multiple Sclerosis brain tissue by polymerase chain reaction. Ann Neurol 1992;32;401–404.
13. Nicoll JAR, Kinrade E, Love S. PCR-mediated search for herpes simplex virus DNA in sections of brain from patients with multiple sclerosis and other neurological disorders. J Neurol Sci 1992;113;144–151.
14. Vandvik B, Nilsen RE, Vartdal F, Norrby E. Mumps meningitis : specific and non-specific antibody responses in the central nervous system. Acta neurol scandinav 1982;65;468–487.

Hepatitis B and multiple sclerosis. Hepatitis B vaccination in France*

Michel Clanet[1], David Brassat

Federation de Neurologie, CHU Toulouse Purpan, 31059 Toulouse cedex, France

Introduction

Active or passive immunisation with vaccines may lead to acute lesions of immunomediated pathogenesis involving the central or peripheral nervous systems. These adverse events raise the problem of safety of these vaccines in some individuals. Some reports have suggested that MS-like demyelinating episodes might be observed following vaccinations. Because the lack of clear scientific data, many neurologists decide to avoid vaccinations in MS patients especially vaccines prepared from attenuated alive agents. However some vaccinations are clearly safe in MS patients as it was confirmed for vaccination against influenzae virus [1].
In France, there is a major controversy amplified by media on the possible severe consequences of vaccination against hepatitis B virus (HBV): it should induce acute episodes of central demyelination suggesting relapses of multiple sclerosis (MS).

Hepatitis B and vaccination against hepatitis B in France

Concern about the vaccine appeared in France during 1994–1995 after a vaccination campaign was launched (1993).

A large number of individuals who received the vaccine shots complained adverse effects, some of them serious, usually in the spectrum of autoimmune or nervous system disorders. Some groups of people are seeking compensation from manufacturers or government and militate for the suppression of the mandatory vaccine. Recently a magistrate court sentenced the manufacturer to pay a financial compensation for a patient affected by a CNS demyelinating event following such a vaccination. An associative network linking people concerned by adverse events post HBV vaccination was created some years ago (REVAHB).

---

*The main part of the data presented in this article come from official reports of the Agence Française de Sécurité Sanitaire des Produits de Santé (AFSSPS) and are published under authorization of the Direction de l'Evaluation des Médicaments et des Produits Biologiques. The Pharmacovigilance of HBV vaccination is collected by the Centre Régional de Pharmacovigilance de Strasbourg (Pr JL IMBS).
[1]*Corresponding author*: Tel: 33-5-61772067. Fax: 33-5-61779443. E-mail: clanet@cict.fr

Hepatitis B and vaccination against HBV in France: some epidemiological data.

Among the countries with low endemicity, France occupies an intermediate level of endemicity. Although some uncertainty exists on these figures, the estimation is about 3000 hepatitis B acute symptomatic in France every year. According to the most confident hypothesis by the experts, 20 to 40% are symptomatic; 2 to 5% become chronic. It may be considered there is every year in France between 200 and 500 chronic hepatitis which lead finally to 50 / 120 cirrhosis and cancer. Other estimations suggest that there are 1500 death related to VHB in France every year.

In March 1998 near to 78 millions of doses had been sold. More than 26 millions of individuals have received this vaccine in France since 1991 suggesting that between 35 to 45% of the French population is vaccinated against the virus. The ratio adults/children is about 70% for adults and 30% for children. The vaccination was first mandatory for workers at risk (1991) then recommended in the general program of vaccination, new born and pre-adolescents (school-campaign vaccination). Over 20% of people aged from 20 to 44 years were vaccinated in 1993–1994.

Pharmacovigilance reports from the National Committee of Pharmacovigilance (Pr IMBS Strasbourg)

The first evaluation (1989–1998) collected 249 CNS demyelinating events, 71% of which reported in women. In the majority of cases (70%) it happened in the two months following the spot. These events correspond to: 131 MS cases with 74 «first relapse», 46 optic neuritis, 23 «myelitis», 49 miscellaneous symptoms. Twenty-seven (27) cases concerned children (7–15 years), with 7 MS. The profile was similar for the different vaccines used. In the same time 63 PNS demyelinating disorders were reported (29 GB syndrome, 5 chronic demyelinating polyneuropathy, 25 brachial plexus neuritis, 5 « neuropathies »). The conclusion of the Committee was that the comparison between the number of spontaneous report of demyelinating events in the vaccinated population with the number of incidental cases of MS expected in this population did not provide any evidence for a causal relationship between hepatitis B vaccination and the occurrence of CNS demyelination.

A comparison expected cases/observed cases was performed on the spontaneous notified cases between 1/94–12/96, under 3 different prevalences figures (45, 50, 60/100 000) coming from MS epidemiological studies performed in France. No significant difference was observed between observed and expected cases, but an undernotification of cases is probable and a slight increase of observed cases should reach the significance threshold (A. Fourrier et al. Oral communication). These neurological complications occurred especially in the patients more susceptible to develop MS: women in the third decade, with an overrepresentation of the HLA DR2 antigen [2]. Warning and precautions were added in the medical information of this vaccine for patients with multiple sclerosis and more epidemiological studies were implemented.

These figures were updated until 12/98: 450 CNS demyelinating events and 76 PNS disorders were finally notified which means an increase of 44.6% and 17.1% respectively. This increase was mainly linked to notification of retrospective cases following the warning effect of the first announcement. It appeared that the notified cases could represented 50% of the total of cases (Capture-recapture study: preliminary results, D. Costagliola et al. Oral communication). However the conclusions of the Pharmacovigilance National Committee were that the hypothesis of a pure coincidence between a spontaneous occurrence of neurological events and the HBV cannot be ruled out. Inversely, a weak risk (triggering event?) cannot be completely rejected.

Prospective epidemiological studies

A pilot case-control survey conducted in 1996 (A. Alperovitch, O.Lyon-Caen oral communication) showed that there was a non significant increase (odds-ratio 1.7; CI 95% = 0.8–3.7) in the proportion of demyelinating events in people vaccinated against hepatitis B compared to the proportion observed in the controls.

A new, larger French multicentric case-control study included 236 cases and 355 controls [3]. Case was defined as a first episode of CNS demyelination within the 6 months preceding the examination. Controls were patients presenting headaches, non-inflammatory rheumatological disorders and vascular or acute neurological diseases matched to cases on sex, age, referral date. A subject (case/control) was classified as exposed to HBV vaccine if he received at least one injection within the 2 months preceding the index date. Use of vaccination certificates was requested. No significant difference was found between patients and controls: odds ratio 1.8 (CI 95%: 0.7–4.6), (for all subjects: exposed cases = 13/236, controls 12/355). This study was considered to have a limited power to detect a significant association because the low frequency of exposure in the two months at-risk time window. However, this risk estimate accorded with a study conducted in UK on the GPRD database (L. Abenhaim et al. Oral Communication)

Taken together the results available do not confirm a significant correlation between vaccination against hepatitis B and MS. However the weak, non significant, increase of risk cannot exclude a slight risk for this vaccine to trigger an episode of MS. It must be noted that a few cases of a second demyelinating event after a rechallenge have been reported. Obviously more research in this field must be conducted and three different epidemiological studies are in progress [4].

Since the writing of the first version of this review three important reports have been published which confirms there is no association between hepatitis B vaccination and the development of MS:
- *Zipp et al. did not find any increase in demyelinating disease after hepatitis B vaccination in a retrospective cohort study among subjects included in a US health care data base [5].*

- *Confavreux and the Vaccines in MS study group (Vaccimus) in a case-crossover study with EDMUS data base (screening the patients with relapses between 1993–1997) confirms that vaccinations does not appear to increase the short term risk of relapse in MS [6].*
- *Ascherio et al. conducted a nested case-control study in two large cohorts of nurses in the US. (Nurses Health Study since 1976, 121700 women, and the Nurses' Health study II since 1989, 116 671 women). The results indicate no association between hepatitis B vaccination and the development of MS [7].*

Conclusions

Following the discussion of all these studies at the AFSSPS the French health minister (B. Kouchner) decided to stop the school vaccination campaign, decision taken under the principle of precaution, which led to a loss of confidence of French people in this vaccine. It must be outlined that HBV currently infects 350 millions people world-wide and through the chronic liver failure or liver cancer leads to 1500 deaths in France or 4000 deaths in US each year. Blaming the vaccine can conduct to major deleterious effects in the world vaccination campaign particularly in countries in which hepatitis B is a major public health problem and MS absent or at a very low prevalence. This is the first vaccine which can prevent cancer [4]. A risk /benefit study of HBV vaccination performed by D. Lévy-Bruhl et al (oral communication), under epidemiological figures of 1994, in a scenario of a 30 years follow-up of 800 000 adolescents vaccinated, suggested that in the worst hypothesis, 29 cirrhosis / carcinoma and 7 fulminans hepatitis are avoided for a risk of 2 acute CNS demyelinating event.

In waiting new epidemiological data French neurologists follow the health ministry recommendations: in hepatitis B seronegative patients with MS the risk–benefit ratio of antihepatitis B vaccination must be accurately assessed. In our current neurological practice this risk benefit is also strongly discussed for vaccination of children or relatives of MS patients. Finally it must be precised that no case of neurological adverse event has been reported after vaccinations of children under 2 years confirming that the vaccination is undoubtedly safe under this age. There are strong efforts among the health professionals to clarify the confusion which exist around this vaccination and to limit the risk of hepatitis B in France by a global approach of a general prevention of risks [8].

References

1. Miller AE, Morgante LA, Buchwald LY. A multicenter, randomized, double blind, placebo controlled trial of influenzae immunization in multiple sclerosis. Neurology 1997;48:312–314.
2. Gout O, Lyon-Caen O. Sclérose en plaques et vaccination contre le virus de l'hépatite B Rev Neurol (Paris) 1998;154:3:205–207.
3. Fourrier A, Bégaud B, Touzé E, Alpérovitch A. Association between hepatitis B vaccine and central nervous system demyelinating disorders: a case-control study. Pharmacoepidemiol. Drug Safety 1999;8(Suppl:S140–S141),abstract.

4. Marshall E. A shadow falls on hepatitis B vaccination effort. Science 1998;281:630–631.
5. Zipp F, Weil JG, Einhäulp KM. No increase in demyelinating diseases after hepatitis B vaccination. Nat. Med. 1999;5:964–965.
6. Confavreux C, Suissa S, Saddier P, Bourdes V, Vukusic S and the Vaccines in MS Study Group. N Engl J Med 2001;344:319–326.
7. Ascherio A, Zhang SM, Hernan MA, Olek MJ, Coplan PM, Brodovicz K, Walker AM. Hepatitis B Vaccination and the risk of MS N Engl J Med 2001;344:327–332
8. Limiter le risque d'Hépatite B Vaccinations individuelles et vaccinations systématiques, agir en professionnels Rev Prescr 1999;19(201):854–858.

Hepatitis B vaccination and central nervous system demyelination: an immunological approach

E Corcuff[1], J Reboul[2], A Tourbah[1,3], G Edan[4], O Gout[3], O Lyon-Caen[3] and R Liblau[1,2,3,*]

[1]*INSERM U546,* [2]*Cellular Immunology Laboratory,* [3]*Fédération de Neurologie, Hôpital de la Salpêtrière, 75651 Paris Cedex 13, France;* [4] *Neurology Department University Hospital, Rennes, France*

Introduction

Several cases of neurological impairment consistent with central nervous system (CNS) demyelination or multiple sclerosis (MS) have been reported following injections of recombinant hepatitis B vaccine. The initial observations involved two female patients: one exhibited severe neurological exacerbation of an already diagnosed relapsing-remitting MS and the other presented with *de novo* neurological symptoms consistent with a diagnosis of MS [1]. Since then, several other cases have been reported [2–8].

Although a meta-analysis of this small number of cases is of limited value, it is striking that most cases involved women, the onset was in young adults, some patients had familial history of MS and a high prevalence of HLA-DR2 was noted [1,2]. Cases were reported following immunization with hepatitis B surface antigen (HBs) produced by different manufacturers: the recombinant antigen may or may not contain the pre-S2 region, and is either produced in mammalian cells (Chinese Hamster Ovary cells) or in *Saccharomyces cerevisiae*. No clear association was found between the onset of neurological symptoms and the number of vaccine injections. The post-vaccinal period during which neurological symptoms might be considered to be associated with vaccination is clearly arbitrary. A period of 2 to 3 months is usually considered relevant [9]. CSF analysis showed the presence of intrathecal IgG synthesis as revealed by isoelectrofocussing. MRI findings were initially compatible with the diagnosis of acute disseminated encephalomyelitis. However, repeated MRI showed temporal evolution of lesions with appearance of new lesions both on T2-weighted and post-gadolinium T1-weighted sequences [8]. In a few cases, multiple recurrences of neurological signs followed repetition of the vaccinal injection [8].

These observations clearly raise the question of whether these bouts of neurological impairment represent a purely fortuitous chronological association with no causal link or are genuine complications of the hepatitis B vaccination. In favor of a fortuitous association is the fact that both events (hepatitis B vaccination

---

Correspondence address: Hôpital de la Salpêtrière, Fédération de Neurologie, Paris Cedex 13, 75651, France. E-mail: 106063.1005@compuserve.com

and CNS demyelination) are relatively frequent occurrences in the young population. In addition, of the few epidemiological studies performed to date, none has detected a statistically significant increase in the frequency of CNS demyelination in the vaccinated population [10]. Moreover, the demographical characteristics of the affected population match those of MS patient [2].

However it remains possible that the immune stimulation provided by the hepatitis B vaccine acts as a triggering event in patients with subclinical MS. The predominance of female cases and the high frequency of the HLA-DR2 haplotype and of a first-degree relative with MS lends additional support to this hypothesis. This hypothesis is not inconsistent with the epidemiological results as they did not have the power to detect an increased incidence of CNS demyelination in the potentially at risk population. Not all immunological challenges would necessarily result in an exacerbation of previously silent MS, due to the heterogeneity of the intensity and quality of the immune response to different antigens. Indeed, Miller et al. showed no statistically significant increase in MS relapses of following *influenza* virus vaccination in MS patients [11].

Case report

We have recently observed a case of CNS demyelination following HBs vaccination in a previously healthy twin of an MS patient. A 45-year-old man with no medical history received three monthly injections of recombinant HB vaccine (GenHevac, Pasteur). Six weeks after the last injection the patient experienced horizontal diplopia. He also complained of transient but unusual tingling in both hands. The clinical examination was normal except for the sixth nerve palsy. He was treated with intravenous methylprednisolone and the diplopia disappeared after several weeks.

Auditory evoked potentials recorded a bilateral delay indicative of a supra-bulbar lesion. Somatosensory evoked potentials showed prolonged supra-medullar latencies. The cerebrospinal fluid (CSF) contained 11 cells per microliter, consisting of lymphocytes and a few macrophages, and showed evidence of intrathecal IgG synthesis. MRI of the brain revealed four significant lesions on T2-weighted sequences (2 periventricular, 1 in the subcortical white matter and 1 in the pons) consistent with demyelination. Anti-HBs antibodies were > 1000 mlU/ml. The HLA typing was A1/30, B18/37, DRB1*03/04, DQB1*0201/0301.

The following paraclinical tests were normal or negative: blood cell count, ESR, blood glucose, complement levels, circulating immune complexes, rheumatoid factors, anti-cardiolipin, anti-nuclear and anti-DNA antibody, serum and CSF angiotensin-converting enzyme, anti-HBc antibody, Lyme, HTLV and HIV serology, acetylcholine receptor autoantibody, electromyography, visual evoked potentials.

The monozygotic twin brother has clinically definite, relapsing-remitting and benign MS. At 32 years, he suffered from a left optic neuritis that lasted three weeks and was spontaneously regressive. Four years later, optic neuritis occurred

Table 1. Genetic microsatellites tested in twin brothers.

Microsatellite (number of alleles)	Alleles of twin 1	Alleles of twin 2
D1S213 (n=11)	107/115	107/115
D1S2860 (n=8)	179/181	179/181
D3S3727 (n=9)	119/123	119/123
D4S2971 (n=8)	147/149	147/149
D8S532 (n=9)	249/253	249/253
D8S1791 (n=8)	240/247	240/247
D9S176 (n=9)	132/136	132/136
D9S1786 (n=9)	199/203	199/203
D10S1483 (n=14)	138/152	138/152
D12S318 (n=9)	253/255	253/255
D14S999 (n=10)	251/258	251/258
D15S130 (n=7)	227/233	227/233
D15S1014 (n=7)	196/196	196/196
D19S900 (n=11)	161/175	161/175
HLA-DRB1* (n=16)	DRB1*03/04	DRB1*03/04
HLA-DQB1* (n=13)	DQB1*0201/0301	DQB1*0201/0301

on the right side. At 39 years, he complained of weakness in the right hand and experienced a recurrence of left optic neuritis. The neurological symptoms improved following intravenous methylprednisolone therapy. We confirmed zygosity of the twins by DNA genotyping using 16 highly polymorphic microsatellite probes on 11 different chromosomes (Table 1).

One could interpret this case as indicating that, in a small population of subjects genetically prone to developing MS (in whom subclinical disease may already exist), the HBs vaccine might trigger clinical expression of the disease. The immunological mechanisms behind this observation are not yet understood but several hypotheses have been put forward to link exogenous antigens to MS [12].

Immunological links between autoimmunity, infections, and vaccination

A large body of epidemiological and clinical evidence in various autoimmune diseases indicates that infection may trigger autoimmunity. Several mechanisms have been proposed to account for induction/exacerbation of an autoimmune process by infections [13–15]. One is molecular mimicry, implying some level of structural homology between a self antigen and a microbial antigen [16,17]. A second mechanism relies on activation of a subset of T cells containing self-reactive lymphocytes by a viral or bacterial superantigen [18,19]. Experimental evidence for such a mechanism exists and whether it could come into play in a human autoimmune disease has recently been debated [20–23]. An additional possibility, recently put forward, is based on the fact that allelic exclusion at the T-cell receptor (TCR) loci is incomplete. Therefore, some T cells can express dual TCR specificities. If one TCR is specific for a microbial antigen and the other is self-reactive there is potential for activation of autoreactive T cells with microbial antigens. Recent experimental evidence for the plausibility of such a hypothesis was given in TCR transgenic mouse models [24,25]. Whether this mechanism applies only to reductionist transgenic model systems or could also be relevant for polyclonal T-cell populations is still unknown. Importantly, these three mechanisms do not require the immune response to the foreign antigen to take place in the tissue that will be the target of the autoimmune disease. Epitope spreading has been proposed as a fourth mechanism linking infection to autoimmunity. An infection occurring in a tissue will indeed promote both inflammation and release of tissue-specific autoantigens that can be processed and presented to self-reactive T cells leading to perpetuation of the disease process. This mechanism has recently been neatly demonstrated in the Theiler's virus encephalomyelitis model [26,27]. Finally, immune responses against a microbial antigen may activate self-reactive T cells in a bystander manner through the potentialisation of the antigen presenting properties of APCs or through the release of cytokines [28–30]. Of these different mechanisms, at least two (molecular mimicry and bystander activation) might operate in the vaccine situation.

Molecular mimicry

The molecular mimicry concept stipulates that cross-reactivity at the T-cell or B-cell level between an exogenous antigen and a self antigen can promote or exacerbate autoimmunity [16,17]. Advances in basic immunology have underlined the potential of immunological cross-reactivity in the T-cell compartment. Indeed, the degeneracy of TCR recognition is wider than previously thought and a given TCR is in fact able to recognize a variety of antigenic peptides bound to the same MHC molecule [31–33]. This plasticity of the T-cell recognition has been understood at the molecular level through the X-ray resolution of the crystal structure of a MHC-peptide:TCR trimolecular complex [34]. Only a few amino acids in an antigenic peptide are exposed to and can contact the TCR, while the others are buried within the MHC groove. At the cellular level, the demonstration that a given T-cell clone can recognize multiple epitopes derived from self and non self antigens in the same MHC context gave further credence to the molecular mimicry hypothesis [33]. In particular it has become apparent that an myelin basic protein (MBP)-reactive TCR can recognize unrelated microbial mimic peptides [31,33,35,36].

Providing the first experimental evidence *in vivo* for molecular mimicry, Fujinami and Oldstone showed that inflammatory infiltration of the CNS can be induced in rabbits through immunization with a viral peptide bearing homology with an MBP peptide [16]. While this was a remarkable finding, it was not clear whether the same TCR recognized both the MBP peptide and the viral peptide. In H-2^u mice, Gautam et al. have shown that only five native MBP amino acids in a polyalanine peptide (**AAAQKRP**AAAAA) could stimulate MBP-reactive T cells and induce experimental autoimmune encephalomyelitis (EAE) with the same severity and incidence as the wild-type MBP Ac1-11 peptide, ASQKRPSQRHG [37]. These experiments have led us to show that a peptide derived from herpes virus *saimiri* (HSV), **AAQRRPS**RPFA, with homology to the MBP peptide at five crucial MHC-binding and TCR-contact residues, is able to induce EAE in susceptible H-2^u mice [38]. Similarly, after having identified a peptide from the human papillomavirus that can be recognized by an MBP-specific T-cell clone, Brocke et al. showed that this viral peptide can generate T cells that are not only able to recognize the MBP peptide but can also transfer EAE [36].

Molecular mimicry has even been proposed as the initial cause of human inflammatory diseases such as rheumatic fever following streptoccocal infections and Guillain-Barré syndrome following *Campylobacter jejuni* infections [39]. Cross-reactive T cells between MBP and human coronavirus 229 E have been demonstrated in MS [40]. Furthermore, human MBP-reactive CD4$^+$ T-cell clones have been shown to respond in vitro to a number of viral and bacterial peptides [31,35,41]. Interestingly, in one instance it was possible to stimulate MBP-reactive T-cell clones with Epstein-Barr virus-transformed B-cell lines without addition of exogenous peptide [35]. This indicates that, for this antigen at least, a naturally processed viral peptide can be presented by MHC class II molecules to cross reactive T cells.

However, one has to explain the relative rarity of autoimmunity given the widespread T cell cross-reactivity. Several mechanisms might contribute to this apparent paradox: control mechanisms preventing autoreactive T cells from becoming pathogenic might develop in the majority of the individuals [42]. On the other hand, cross-reactivity between self and foreign antigen might also result in silencing of the autoreactive T cells due to activation-induced cell death or due to antagonistic activity of the foreign peptide on autoreactive T cells. The latter has been demonstrated by Steinman et al. in the EAE setting [43].

Bystander activation

Bystander activation of autoreactive T cells during an immune response to an unrelated antigen has been suggested as the cause of CNS inflammation occurring in a transgenic model developed by Oldstone and colleagues [29]. In this model, a chronic CNS autoimmune disease mediated by CD8$^+$ T cells is induced through a molecular mimicry mechanism after infection with a virus sharing epitopes with a protein expressed by oligodendrocytes. A second infection with the same virus leads to an enhancement of the inflammatory demyelinating lesions. Interestingly, a subsequent infection with an unrelated virus also results in enhanced pathology associated with activation of the autoreactive CD8$^+$ T cells. Therefore, bystander activation of clinically inapparent CNS inflammation might result from a potent immune activation in the periphery. The molecular basis of this bystander effect has not yet been explored but cytokines are likely candidates.

In vitro observations have also demonstrated that MBP-specific autoreactive T cells incubated in the presence of lipopolysaccharide or bacterial DNA containing CpG dinucleotides secrete a large amount of interferon-γ and, in contrast to unstimulated cells, are able to transfer EAE [44]. This conversion to a deleterious phenotype is dependent on IL12 production by APCs stimulated by the microbial

Table 2. Inclusion and exclusion criteria for patients presenting CNS manifestations following hepatitis B vaccination.

Inclusion criteria	Exclusion criteria
Neurological symptoms confirmed by a neurologist within 8 weeks of any HBs vaccine injection	Pregnant women
Age: between 18 and 50 years	Patients with pre-existing neurological disease
Blood samples taken within 3 months following the onset of the clinical manifestations	Patients treated with immunosuppressive treatments or corticosteroids between vaccination and the date of sampling
Patient's written consent to participate in the study	

products. Similar results were obtained in vivo after injection of bacterial oligonucleotides in transgenic mice expressing an MBP-specific TCR [45]. In addition to this cytokine-mediated effect, bystander inflammation can also promote autoimmunity through enhanced presentation of autoantigens to autoreactive T cells in predisposed individuals, as recently shown by Flavell et al. in nonobese diabetic mice [46]. Another theoretical consideration is the fact that increased B7.1 and B7.2 expression by APCs during infection or immunization can favor transcostimulation, i.e., delivering of the costimulatory signal 2 by a cell type different from the one triggering the antigen-specific signal 1 [30]. Therefore, instead of undergoing anergy or death due to lack of costimulatory signals, the autoreactive T cells could undergo a full-blown activation.

Towards the testing of these hypotheses

We decided to address these hypotheses in the situation of CNS demyelination following hepatitis B vaccination. We therefore collected peripheral blood mononuclear cells (PBMC) from 10 patients fulfilling predefined criteria (Table 2). Vaccinated healthy individuals matched for age and sex served as controls. The characteristics of the patients are: mean age (31 years), sex ratio men/women (6/4). Six cases occurred following the second vaccinal injection and four cases following the third injection. The mean delay between the last vaccinal injection and onset of neurological signs was 1.7 months. Sixty percent of the HLA-typed patients are HLA-DR15.

The following parameters will be studied and compared between groups: T-cell proliferation to recombinant HBs and tetanus toxoid in a six-day proliferation assay, production of Th1 (interferon-γ) and Th2 (IL5) cytokines in response to a mitogenic signal, to a recall antigen (tetanus toxoid), and to HBs.

If the precursor frequency of HBs reactive T cells is sufficient, T-cell lines against HBs will be derived from both vaccinated patients and controls. These lines will be characterized phenotypically and functionally and their cross-reactivity towards a panel of myelin antigens will be assessed. Indeed, the fact that sequence analyses did not reveal a striking homology between the vaccinal antigen and selected myelin proteins does not eliminate the possibility of cross-reactivity at the T-cell level, as discussed above.

To conclude, we would like to stress that, in terms of public health, the benefits of the vaccine in adolescents/adults at risk for hepatitis B virus far outweigh the potential neurological risk. However, the systematic reporting and analysis of cases with CNS demyelination following HBs vaccination may allow the identification of a small group potentially at risk. Moreover, a thorough immunological analysis in patients developing MS-like disease following hepatitis B vaccination might help unravel the immunopathogenesis of MS and help define, at the cellular and molecular level, the mechanisms responsible for MS relapse after infection.

Acknowledgement

We are grateful to C. Chrétien-Dumont for typing of the manuscript. E. Corcuff was supported by a fellowship from Smithkline Beecham. This work is supported by INSERM, ARSEP, and the French Health Ministry (AOM96038).

References

1. Herroelen L, De Keyser J, Ebinger G. Central nervous system demyelination after immunization with recombinant hepatitis B vaccine. Lancet 1995;338:1174–1175.
2. Gout O, Theodorou I, Liblau R, Lyon-Caen O. Central nervous system demyelination after recombinant hepatitis B vaccination: report of 25 cases. Neurology 1997;48:A424.
3. Kaplanski G, Retornaz F, Durand JM, Soubeyrand J. Central nervous system demyelination after vaccination against hepatitis B and HLA haplotype. J Neurol Neurosurg Psychiatry 1995;58:758–759.
4. Salvetti M, Pisani A, Bastianello S, Millefiorini E, Buttinelli C, Pozzilli C. Clinical and MRI assessment of disease activity in patients with multiple sclerosis after influenza vaccination. J Neurol 1995;242:143–146.
5. Nadler JP. Multiple sclerosis and hepatitis B vaccination. Clin Infect Dis 1993;17:928–929.
6. Deisenhammer F, Pohl P, Bösch S, Schmidauer C. Acute cerebellar ataxia after immunization with recombinant hepatitis B vaccine. Acta Neurol Scand 1994;89:462–463.
7. Tartaglino LM, Heiman-Patterson T, Friedman DP, Flanders AE. MR imaging in a case of post-vaccination myelitis. Am J Neuroradiol 1993;16:581–582.
8. Tourbah A, Gout O, Liblau R, Lyon-Caen O, Bougniot C, Iba-Zizen MT, Cabanis EA. Encephalitis after hepatitis B vaccination. Recurrent disseminated encephalitis or MS? Neurology 1999;53:396–401.
9. Hall A, van Damme P. A technical consultation on the safety of hepatitis B vaccines. Report of a meeting organized by the Viral Hepatitis Prevention Board. Antwerpen, Belgium; 1998.
10. Zipp F, Weil JG, EinhaUpl KM. No increase in demyelinating diseases after hepatitis B vaccination. Nature Medicine 1999;5:964–965.
11. Miller AE, Morgante LA, Buchwald LY, Nutile SM, Coyle PK, Krupp LB, Doscher CA, Lublin FD, Knobler RL, Trantas F, Kelley L, Smith CR, La Rocca N, Lopez S. A multicenter, randomized, double-blind, placebo-controlled trial of influenza immunization in multiple sclerosis. Neurology 1997;48:312–314.
12. Liblau RS, Fontaine B. Recent advances in immunology in multiple sclerosis. Curr Op Neurol 1998;11:293–298.
13. Steinman L, Conlon P. Viral damage and the breakdown of self tolerance. Nature Medicine

1997;3:1085–1087.
14. Benoist C, Mathis D. The pathogen connection. Nature 1998;394: 227–228.
15. Wekerle H. The viral triggering of autoimmune disease. Nature Medicine 1998;4:770–771.
16. Fujanimi RS, Oldstone M. Amino acid homology between the encephalitogenic site of myelin basic protein and virus: a mechanism for autoimmunity. Science 1985;230:1043–1046.
17. Oldstone M. Molecular mimicry and autoimmune disease. Cell 1987;50:819–821.
18. Brocke S, Gaur A, Piercy C, Gautam A, Gijbels K, Fathman CG, Steinman L. Induction of relapsing paralysis in experimental autoimmune encephalomyelitis by bacterial superantigen. Nature 1993;365:642–644.
19. Schiffenbauer J, Soos J, Johnson H. The possible role of bacterial superantigens in the pathogenesis of autoimmune disorder. Immunol Today 1998;19:117–120.
20. Conrad B, Weissmahr RN, Böni J, Arcari R, Schüpbach J, Mach B. A human endogenous retroviral superantigen as candidate autoimmune gene in type I diabetes. Cell 1997;90:303–313.
21. Murphy VJ, Harrison LC, Rudert WA, Luppi P, Trucco M, Fierabracci A, Biro PA, Bottazzo GF. Retroviral superantigens and type 1 diabetes mellitus. Cell 1998;95:9–11.
22. Lower R, Tonjes RR, Boller K et al. Development of insulin-dependent diabetes mellitus does not depend on specific expression of HERV-K. Cell 1998;95:11–13.
23. Lan MS, Mason A, Coutant R, Chen Q-Y, Vargas A, Rao J, Gomez R, Chalew S, Garry R, Maclaren NK. HERV-K10s and immune-mediated (type1) diabetes. Cell 1998;90:14–16.
24. Sarukhan A, Garcia C, Lanoue A, von Boehmer H. Allelic inclusion of T cell receptor α genes poses an autoimmune hazard due to low-level expression of autospecific receptors. Immunity 1998;8:563–570.
25. Fossati G, Cooke A, Quartey Papafio R, Haskins K, Stokinger B. Triggering a second T cell receptor on diabetogenic T cells can prevent induction of diabetes. J Exp Med 1999;190:577–583.
26. Miller SD, Vanderlugt CL, Begolka WS, Pao W, Yauch RL, Neville KL, Katz-Levy Y, Carrizosa A, Kim BS. Persistent infection with Theiler's virus leads to CNS autoimmunity via epitope spreading. Nature Medicine 1997;3:1133–1136.
27. Katz-Levy Y, Neville KL, Girvin AM, Vanderlugt CL, Pope JG, Tan LJ, Miller SD. Endogenous presentation of self myelin epitopes by CNS-resident APCs in Theiler's virus-infected mice. J Clin Invest 1999;104:599–610.
28. Ehl S, Hombach J, Aichele P, Hengartner H, Zinkernagel RM. Bystander activation of cytotoxic T cells: studies on the mechanism and evaluation of in vivo significance in a transgenic mouse model. J Exp Med 1997;187:1241–1251.
29. Evans CF, Horwitz MS, Hobbs MV, Oldstone MBA. Viral infection of transgenic mice expressing a viral protein in oligodendrocytes leads to chronic central nervous system autoimmune diseases. J Exp Med 1996;184:2371–2384.
30. Ding L, Shevach EM. Activation of CD4+ T cells by delivery of the B7 costimulatory signal on bystander antigen-presenting cells (trans-costimulation). Eur J Immunol 1994;24:859–866.
31. Hemmer B, Fleckenstein BT, Vergelli M, Jung G, McFarland H, Martin R, Wiesmüller KH. Identification of high potency microbial and self ligands for a human autoreactive class II-restricted T cell clone. J Exp Med 1997;185:1651–1659.
32. Mason D. A very high level of crossreactivity is an essential feature of the T-cell receptor. Immunol Today 1998;19:395–404.
33. Gran B, Hemmer B, Vergelli M, McFarland H, Martin R. Molecular mimicry and multiple sclerosis degenerate T-cell recognition and the induction of autoimmunity. Ann Neurol 1999;45:559–567.
34. Garcia KC, Degano M, Pease LR, Huang M, Peterson PA, Teyton L, Wilson IA. Structural basis of plasticity in T cell receptor recognition of a self peptide-MHC antigen. Science 1998;279:1166–1172.
35. Wucherpfenning KW, Strominger JL. Molecular mimicry in T cell-mediated autoimmunity: viral peptides activate human T cell clones specific for myelin basic protein. Cell 1995;80:695–705.

36. Ufret-Vincentry RL, Qingkey L, Tresser N, Pak SH, Gado A, Hausmann S, Wucherpfennig W, Brocke S. In vivo survival of viral antigen-specific T cells that induce experimental autoimmune encephalomyelitis. J Exp Med 1998;188:1725–1738.
37. Gautam AM, Pearson CI, Smilek DE, Steinman L, McDevitt HO. A polyalanine peptide with only five native myelin basic protein residues induces autoimmune encephalomyelitis. J Exp Med 1992;176:605–609.
38. Gautam AM, Liblau R, Chelvanayagam G, Steinman L, Boston T. A viral peptide with limited homology to a self peptide can induce clinical signs of experimental autoimmune encephalomyelitis. J Immunol 1998;161:60–64.
39. Yuki N, Taki F et al. A bacterium lipopolysaccharide that elicits Guillain-Barré syndrome has a GM1 ganglioside-like structure. J Exp Med 1993;178:1771–1776.
40. Talbot PJ, Paquette JS, Ciurli C, Antel JP, Ouellet F. Myelin basic protein and human coronavirus 229E cross-reactive T-cells in multiple sclerosis. Ann Neurol 1996;39:233–240.
41. Wucherpfenning KW, Catz I, Hausmann S, Strominger JL, Steinman L, Warren KG. Recognition of the immunodominant myelin basic protein peptide by autoantibodies and HLA-DR2-restricted T cell clones from multiple sclerosis patients. J Clin Invest 1997;100:1114–1122.
42. Olivares-Villagomez D, Wang Y, Lafaille JJ. Regulatory CD4$^+$ T cells expressing endogenous T cell receptor chains protect myelin basic protein-specific transgenic mice from spontaneous autoimmune encephalomyelitis. J Exp Med 1998;188:1883–1894.
43. Ruiz PJ, Garren H, Hirschberg DL, Langer-Gould AM, Levite M, Karpuj MV, Southwood S, Sette A, Conlon P, Steinman L. Microbial epitopes act as altered peptide ligands to prevent experimental autoimmune encephalomyelitis. J Exp Med 1999;189:1275–1283.
44. Segal BM, Klinman DM, Shevach EM. Microbial products induce autoimmune disease by an IL12-dependent pathway. J Immunol 1997;159:5087–5090.
45. Brabb T, Goldrath AW, von Dassow P, Paez A, Liggit HD, Goverman J. Triggers of autoimmune disease in a murine TCR-transgenic model for multiple sclerosis. J Immunol 1997;159:497–507.
46. Green EA, Eynon EE, Flavell RA. Local expression of TNFα in neonatal NOD mice promotes diabetes by enhancing presentation of islet antigens. Immunity 1998;9:733–743.

Is there a causal link between hepatitis B vaccination and multiple sclerosis?

Philippe Monteyne*

GlaxoSmithKline Biologicals, rue de de l'Institut, 89, B-1330 Rixensart, Belgium

Introduction

Today, after a mass vaccination programme that started in 1994, some 27 million French people have been vaccinated against hepatitis B, a large proportion of whom are aged between 20 and 40, the age group in which most MS cases are encountered. In this context, it was important to know if administration of the hepatitis B vaccine and onset or relapse of MS followed each other purely by coincidence, or if these two events were causally linked. A group of international experts that met by end 1998 to discuss these questions concluded that all available data show no demonstration of a causal link between hepatitis B vaccination and MS, and that vaccination policies in place around the world should be maintained [1].

Hepatitis B

Health authorities regard hepatitis B as one of the most serious diseases on a world-wide basis on account of the frequency with which it occurs, its often chronic nature and the associated complications that can lead to liver cancer. It is estimated that two billion people, or more than one in three people in the world, have at some time been infected by the hepatitis B virus [2].

In 30 per cent of cases, it is not known how the virus was acquired [3]. Chronic carriers can transmit the hepatitis B virus without knowing that they are infected. This group is also at risk of eventually developing serious complications, such as cirrhosis and cancer of the liver. Indeed, the hepatitis B virus is second only to tobacco as the most frequent known cause of cancer [4–5].

The WHO estimates that there are currently 350 million chronic carriers of the virus world-wide. Around one quarter of these chronic carriers will die of cirrhosis or cancer of the liver within thirty years [6], bringing the number of people who die from the disease every year world-wide to at least one million.

The Centers for Disease Control (CDC) in the US regards hepatitis B vaccine as the first vaccine that helps to prevent cancer [7]. In terms of public health, widespread vaccination of all children (long before they are exposed to the risks of sexual transmission) is an investment for the future. To date, more than 500 million people around the world have been vaccinated against hepatitis B.

---

*Tel: (32 2) 656 72 79. Fax: (32 2) 656 8113. E-mail: philippe.monteyne@sbbio.be

Multiple sclerosis

The cause of the disease is still unknown. Interestingly, the clinical symptoms of MS were first described by Jean Martin Charcot at the Salpêtrière hospital in 1868 and, less than twenty years later, one of his famous successors to the Chair of Neurology, Pierre Marie, suspected that the disease might by caused by an infection.

For 100 years, many different infections and many viruses, in particular, were suggested as the possible cause of MS. However, none of these hypotheses has stood up to the test of time and scientific investigation [8]. Today, experts agree that there are multiple factors, including a complex genetic background (i.e. it is probably caused by many different genes that have not yet been identified) and environmental factors (one or perhaps many viruses or other external agents).

The only link that has been established is that certain viral infections appear to trigger the relapses of MS. Researchers have recently demonstrated the benefits of vaccinating MS patients against influenza in order to prevent the real risk that they will develop a relapse if infected by the virus. Influenza vaccination presents no danger to these patients [9]. Finally, it should be noted that no causal link has ever been identified between the disease caused by the hepatitis B virus and MS.

Analysis of the scientific data on the possibility of a link between hepatitis B vaccination and multiple sclerosis

The possibility of a link between hepatitis B vaccination and multiple sclerosis has been suggested, in particular, because with the French experience it is the first time that a population within the age group that develops MS has been vaccinated on a very large scale. In addition, neurology manuals conventionally state that any stimulation of the immune system, such as a vaccination, may theoretically cause flare-ups in MS patients (which does not mean to say that it can cause MS in a healthy subject) [10]. However, hepatitis B vaccine does not contain the whole virus and merely stimulates antibodies against the surface antigen of hepatitis B. Furthermore, the only vaccine for which a prospective study has provided data enabling the effects of vaccination to be compared with the effects of a simple placebo is the influenza vaccine. We now know that this vaccine represents no risk to MS patients whatsoever [11].

The theory of molecular mimicry

Some years ago, researchers demonstrated that part of rabbit myelin is closely related to the "polymerase" protein of the hepatitis B virus [12]. There is no such close relationship between hepatitis B polymerase and human myelin, however, and the hepatitis B vaccine does not contain the polymerase protein from the hepatitis B virus.

Pharmacovigilance

Up to December 1998, the number of new MS cases reported after vaccination with Engerix B (the hepatitis B vaccine produced by SmithKline Beecham) was 165 for the whole world, with 127 of these cases being reported in France. This disproportion between France and the rest of the world can probably be explained by the heightened awareness of the causal link hypothesis in France.

From these figures, we can calculate an average frequency in France of 1.04 "new" cases of MS per 100,000 people vaccinated, compared to the frequency that would normally be expected of 1 to 3 cases per 100,000 people per year [13].

If these cases of demyelination after hepatitis B vaccination are categorised by age and by sex, they correspond exactly to the usual distribution of MS in the rest of the population, and have thus not been affected by the higher vaccination of the younger age groups (large-scale vaccination of children in France would have lowered the age at which MS occurred in the whole of the vaccinated population if there were a causal link).

Similarly, the time at which demyelination occured after hepatitis B vaccination was totally random (cases of MS should have appeared within a specific time interval after the vaccination if there were a causal link).

Epidemiology.

A retrospective study was recently carried out in the US using data provided by health care insurance organisations. 27,000 vaccinated people were compared with 107,000 people not vaccinated against hepatitis B. No significant difference between vaccinated and non-vaccinated individuals was found for demyelinating episodes at any time point analysed from 2 months to three years follow-up [14]. This is the first population-based controlled study published about this question. The main criticism that could be made of this study concerns the source of data which, by nature of coming from a health care database, was not designed for such a study, and does not have all the details that would be available from a medical record.

More recently, a retrospective study in Canada showed no rise in new cases of multiple sclerosis in adolescents following the start of a school-based hepatitis B vaccination programme in 1992 in British Columbia [15] (period 1986–1992, 288, 657 students, compared with period 1992–1998, 267, 412 vaccinated students).

Two other retrospective epidemiological studies were recently carried out at the request of the French health authorities. Preliminary results were communicated to the press by Bernard Kouchner, Secretary of State for Health, on October 1, 1998. In both studies, the relative risk was 1.4 with respect to the link between the hepatitis B vaccine and the occurrence of demyelination within a period of two months after vaccination. The figure has a deviation from statistical significance including 1 ("confidence interval of the results at 95%"). Once again, taking into account this statistically non significant result, it was concluded that no causal link had been demonstrated between hepatitis B vaccination and MS.

Conclusion

The conclusion reached on the basis of available data is that the most plausible explanation for the observed temporal association between vaccination and MS is that it is a coincidental association. Prof. S. Katz, representing the Infectious Diseases Society of America and the American Academy of Pediatrics, testified as follows in May 1999 before the Subcommittee on Criminal Justice, House Committee on Government Resources. "I can think of no group who would like to find a cause for multiple sclerosis more than the National MS Society. It is therefore noteworthy that their Scientific Advisory Board has reviewed the hypothesis of a causal association between hepatitis B vaccine and multiple sclerosis and has rejected it" [16].

Acknowledgement

I am grateful to my colleagues and my management, particularly to F. André, VP, senior medical director, for critical reading of the manuscript.

References

1. Halsey NA, Duclos P, Van Damme P, Margolis H. Hepatitis B vaccine and central nervous system demyelinating diseases. Pediatr Infect Dis J 1999;18:23–24.
2. Grob P et al. Hepatitis B: a serious public health threat. Vaccine 1998;16:S1–S2.
3. Smedile A, Wands J. Hepatitis B: basic science and epidemiology/Hepatitis D. In: Rizetto M et al. (eds.) Viral Hepatitis and Liver Diseases, Minerva Medica, Turin, 1997;906–910.
4. Anonymous. Prevention of liver cancer. WHO TRS 691, 1983;1–30.
5. Kane M. Global programme for control of hepatitis B infection. Vaccine 1995;13:S47–S49.
6. Anonymous. Hepatitis B Vaccine – making global progress. WHO, EPI, Update October 1996.
7. Anonymous. Hepatitis B Vaccine Fact Sheet. http://www.cdc.gov/ncidod/diseases/hepatitis/b/factvax.htm.
8. Monteynie P, Bureau JF, Brahic M. Viruses and multiple sclerosis. Curr Opin Neurol 1998;11:287–291.
9. De Keyser J, Zwanikken C, Boon M. Effects of influenza vaccination and influenza illness on exacerbations in multiple sclerosis. J Neurol Sci 1998;159:51–53.
10. Cambier J, Masson M, Dehen H. Neurologie, Ed Masson; 5th edition, p. 300.
11. Miller AE et al. A multicenter, randomised, double-blind, placebo-controlled trial of influenza immunization in multiple sclerosis. Neurology 1997;48:312–314.
12. Fujinami RA, Oldstone MBA. Amino acid homology between the encephalitogenic site of myelin basic protein and virus: mechanism for autoimmunity. Science 1995;230:1043–1045.
13. Kane M. Absence d'arguments en faveur d'une relation entre la sclérose en plaques et la vaccination contre l'hépatite B. Virologie 1997;1:363–364.
14. Zipp F, Weil JG, Einhäupl KM. No increase in demyelinating diseases after hepatitis B vaccination. Nat Med, 1999;5 (9):964–965.
15. Sadovnick AD, Scheifele DW. School-based hepatitis B vaccination programme and adolescent multiple sclerosis. Lancet 2000;355:549–550.
16. Anonymous, http://www.idsociety.org/vaccine/testify_hepb.htm.

Public health aspects of hepatitis B vaccination and multiple sclerosis

Dr. Philippe Duclos[1],*, Dr. Andy Hall[2] and Dr. Pierre Van Damme[3]

[1]*Department of Vaccines and Biologicals, World Health Organization, 20 Via Appia, CH 1211, Geneva-27, Switzerland;* [2]*London School of Hygiene & Tropical Medicine, Keppelstreet, London WCIE 7HT, UK;* [3]*University of Antwerp, Department of Epidemiology and Community Medicine, Universiteitsplein 1, B-2610 Antwerp, Belgium*

Background

Hepatitis B is one of the major diseases of mankind and is preventable with safe and effective vaccines. More than 2,000 million persons worldwide have serologic evidence of past or current HB virus infection and there are more than 350 million chronic carriers of the virus at high risk of death from cirrhosis and liver cancer, diseases which kill almost 1 million persons a year [1]. Hepatitis B vaccines have been available since 1982 and more than 1 billion doses have been used. Hepatitis B vaccines are considered one of the safest and least reactogenic vaccines. Hepatitis B vaccines are composed of highly purified preparations of HBs antigen (HBsAg). HBsAg is a glycoprotein that makes up the outer envelope of hepatitis B virus, and is also found as 22-nm spheres and tubular forms in the serum of people with acute and chronic infection. Vaccines are prepared by harvesting HBs Ag from the plasma of people with chronic infection (plasma derived vaccine) or by inserting plasmids containing the viral gene in yeast or mammalian cells (recombinant DNA vaccine). An adjuvant, aluminum phosphate or aluminum hydroxide is added to the vaccines which are sometimes preserved with thimerosal. The concentration of HBs Ag varies from 2.5 to 40 µg per dose, according to target population. The vaccine is approximately 95% effective in preventing the HBV chronic carrier state, and direct reduction of liver cancer has already been documented in immunized children.

Considering the major public health burden of hepatitis B infections and the availability of safe and effective vaccines that could prevent most of this burden, in 1992, the World Health Assembly, recommended that hepatitis B vaccines should be integrated into national immunization programmes in all countries with a hepatitis B carrier prevalence (HBsAg) of 8% or greater by 1995 and in all countries by 1997 [2]. It was acknowledged that target groups and strategies may vary with the local epidemiology. When carrier prevalence is >= 2%, the most effective strategy is incorporation into routine infant immunization schedule. Countries with lower prevalence may consider immunization of all adolescents as an addition or alterna-

---

*Correspondence address: World Health Organization, Department of Vaccines and Biologicals, 20 Via Appia, Geneva-27, CH 1211, Switzerland. E-mail: duclos@who.ch

tive to infant immunization.

As a result over 100 countries successfully introduced hepatitis B vaccination into their routine vaccination schedule [3]. The use of hepatitis B vaccine has resulted in dramatic reductions in the prevalence of the carrier state in many areas [4,5]. In areas without significant perinatal transmission the carrier rate has been reduced to less than 1% from levels of 15–20%. In areas where perinatal transmission is important (e.g. China) the reduction has been to less than 2% of the vaccinated population [5]. There is now clear data from Taiwan showing that hepatitis B vaccination reduces the incidence of liver cancer in children [6,7].

In recent years, following intensive use of the vaccine in France several case reports raised concerns that hepatitis B immunization may be linked to new cases or reactivation of multiple sclerosis (MS) or other demyelinating diseases within a period of 2 to 3 months after vaccination [8]. Articles published in the media stating a link between hepatitis B immunization and MS have further fuelled the worry over the safety of the vaccine.

Viral Hepatitis Prevention Board (VHPB) Technical consultation on the safety of hepatitis B vaccines

On September 28–30, 1998, the VHPB, whose activities are incorporated into the WHO collaborating center on prevention and control of viral hepatitis from the University of Antwerp, Belgium, as a result of concern generated in France on the safety of hepatitis B vaccine and potential link between hepatitis B and MS, called a technical consultation on the safety of hepatitis B vaccines to review accumulated data and policy implications [9–11]. This consultation took place in Geneva from September 28 through September 30, 1998 and brought together representatives from national public health and regulatory authorities, academia, the hospital sector, the pharmaceutical industry and the World Health Organization. Participants included experts in the fields of public health, epidemiology, immunology, neurology and pharmacology.

Participants were provided with information on the epidemiology of hepatitis B, hepatitis B vaccination, MS, the possible biological plausibility in particular molecular mimicry that could explain an association between administration of the hepatitis B vaccine and MS as well as information on adverse events following the use of hepatitis B vaccine (data from literature and clinical trials, with surveillance data from the United States, Canada, Italy and from industry). They were also presented with a review preliminary data and summary of study protocols of epidemiological studies in progress in France, the UK and the US.

The French studies included a comparison of expected number of cases of central nervous system demyelinating disease (CDD) that would have occurred by chance in association with vaccination and that was calculated based on the underlying rate of disease and the number of doses of hepatitis B vaccine distributed in France. This comparison indicated that the observed number of cases was less than the expected. The other French study was a case control study conducted in neurological wards and which did not show a statistically significant associations between HB vaccination and CDD. The UK study was an analysis of a computerized, UK-based General Practitioner data base covering some 4 million people was carried out. Further refinements to the analysis are needed but preliminary results did show no association with multiple sclerosis but a marginal association with acute demyelination not diagnosed as multiple sclerosis. Finally the study from the USA involved the analysis of a research database from the Diversified Pharmaceutical Services, a pharmacy benefits management company. The research database was formed from six primary healthcare programmes and records reimbursement claims for both hepatitis B vaccination and medical treatment for demyelinating disorders. A cohort analysis was performed defining a group of 27,229 who had claimed for hepatitis B vaccination and a group of 107,469 who had not claimed for hepatitis B vaccination. These groups were matched on age and sex. Analysis of rates showed rate ratios of less than one for all outcomes indicating a potential protective effect of hepatitis B vaccination. This was interpreted as potentially due to a 'healthy vaccinee' effect.

The presentation of the epidemiologic data generated a discussion in which following points were made:

- The presentations highlighted the range of variables that were potential confounders in the study of a possible association between hepatitis B vaccine and multiple sclerosis. In particular the need to control for adolescent infections and socio-economic group.
- The interval from vaccination to onset of demyelination in the cases reported from France was of the order of two months. It was unclear how this fitted with data suggesting that the aetiology of MS acted some decade before the peak incidence.
- In the light of this, the possibility was raised that vaccination might only act as a trigger precipitating MS in individuals who would inevitably develop the disease in time.
- No studies appeared to have addressed an association between natural hepatitis B infection and MS. Such a study could possibly be performed using stored sera from completed case control studies.
- The geographical pattern of multiple sclerosis seemed almost to be a mirror image of persistent hepatitis B infection and this might deserve formal analysis.
- The putative association raised the question of the incidence of multiple sclerosis in the highly vaccinated, and high latitude, Eskimo population in Alaska.
- The question was raised whether it is biologically plausible that no association exists between natural hepatitis B carriage (where there are massive amounts of HBsAg in the circulation) and MS whereas there is a putative association with the injection of relatively small quantities of surface antigen as vaccine. The immunologists felt that this was plausible given the potential for large dose tolerance induction. In addition, natural replicative infection is quite different from vaccination with only part of the non–replicative antigenic repertoire of

the virus.

There are three hypotheses that could explain the observed cases of demyelinating disorders following hepatitis B vaccine:
1. Coincidence, due to the large number of hepatitis B vaccine doses administered, many of them in age groups where symptoms of MS first occur.
2. "Triggering": An increased risk of symptomatic demyelination following hepatitis B vaccine which would act as a "trigger" in individuals predisposed to develop MS or central nervous system (CNS) demyelinating diseases. These individuals would have developed demyelination with or without an altered natural history following some immunologic or other precipitating factor.
3. A true causal relationship between hepatitis B vaccination and MS or other CNS demyelinating disease.

Evidence to support the first hypothesis includes the fact that no statistically significant association was found between hepatitis B vaccine and MS in the limited studies conducted to date. Further, the age and sex distributions of MS cases reported through spontaneous reporting systems match the recognised age and sex distribution of MS cases that preceded the use of the vaccine and are not correlated with vaccine administration.

In support of the second hypothesis of an increased risk of MS following hepatitis B vaccination seen as a precipitating factor is that some studies have shown slightly elevated odds ratios although these were not statistically significant. Evidence inconsistent with this second hypothesis is the observation that no increased risk was found in another study. No tangible evidence was presented for the biologic plausibility of any association.

There is no evidence whatsoever of a causal link between hepatitis B virus infection and CNS demyelinating disorders including multiple sclerosis. Additional epidemiological and immunological research is ongoing or planned to further examine any association between vaccination, including hepatitis B, and CNS demyelinating disease. Altogether, evidence in favour of an increased risk following vaccination is weak and does not meet the criteria for causality.

The group reached the following consensus:
- Epidemiological studies have identified that transmission of hepatitis B virus occurs through contact with contaminated blood and other clinical materials, sexual transmission, child to child transmission and through household contacts. Yet in almost all countries studied, approximately 1/3 of hepatitis cases occurs in people who are not included in these identified high risk groups. Strategies that target only groups at high risk of hepatitis B virus infection have failed, therefore strategies that protect the population at large are encouraged.
- Those who have already been vaccinated should be reassured that no studies demonstrate that they are at increased risk of developing MS.
- MS does not occur in children <1 year of age. There is no reason to believe that infant immunization with hepatitis B vaccine will change this fact.

- MS is a very rare disease in children and adolescents. Amongst the cases of MS that have been reported to surveillance systems, there is no evidence that MS was caused by the hepatitis B vaccine in these cases.
- By better understanding the pathogenesis and aetiology of MS and related demyelinating disorders one would be in a position to identify factors that influence its initiation and outcome. Therefore, research into the causation and immunopathogenesis of CNS demyelinating disorders including MS should be encouraged and supported.
- Similarly, the influence of HB vaccination on immune functions, including immunological memory, should be further investigated. An important aspect of immunological research concerns the availability of biological material including T and B lymphocytes, monocytes, serum and plasma which has been stored properly. Therefore a close collaboration between epidemiologists and clinical researchers is essential. This also implies close liaison with regulatory authorities.
- Any future or ongoing study of demyelinating disorder and HB vaccine should take advantage of the large experience gathered to date on conducting MS aetiologic studies and the Guidelines developed. The vaccine and MS communities would benefit from jointly organizing the best possible case-control study on aetiology of MS. Such a study would examine a wide range of hypotheses (though each being as specific as possible); including different viral, bacterial and parasitic infections, environmental factors, and vaccinations, and control for various potential confounders.
- Principal investigators of existing studies should standardise methodology to facilitate future meta-analysis. In addition, validation of vaccination exposure should be improved by obtaining a copy of the vaccination certificate and/or serology for hepatitis B surface antibody. Researchers should also collect better information on potential confounders.
- The data available to date, although limited, does not demonstrate a causal association between HB immunization and CNS demyelinating diseases, including MS. No evidence presented at this meeting indicates a need to change public health policies with respect to HB immunization. Therefore, based on demonstrated important benefits – including the prevention of cirrhosis and cancer, and a hypothetical risk, the group supports the WHO recommendations that all countries should have universal infant and/or adolescent immunization programs and continue to immunize adults at increased risk of HB infection as appropriate.

Discussion

Since this meeting no new evidence has been provided to support a potential link between hepatitis B vaccination and MS. On the contrary several recent publications have been reassuring while results of other studies are still pending [12,13].

Yet, as a result of the public and professional concern of a potential link between hepatitis B vaccination and MS generated following reports of case series, on October 1, 1998, the French Authorities temporarily suspended the school-based adolescent hepatitis B vaccine program [14]. These authorities, however, maintained the recommendation of universal infant immunization as well as the recommendation to administer the vaccine to adults at increased risk and reiterated continued support for adolescent vaccination through primary care physician. The French decision was misquoted in the media and interpreted as a ban of hepatitis B immunization in France. It generated lots of concern in both developed and developing countries and could have had a real disastrous effect on hepatitis B immunization way beyond France if it was not all the efforts spend to maintain confidence in the program. Several countries considering stopping immunization with hepatitis B vaccination while several countries others were contemplating withholding introduction of hepatitis B vaccination. The final impact on vaccine coverage globally was limited but is hard to measure and has not been quantified.

This situation highlights the fact that although spontaneous surveillance, case report, case series, and ecologic studies are used and useful to build hypothesis about adverse events, they do present an incomplete picture and have to be treated with caution as they may end up being unduly damaging to health programs when taken out of context i.e. one does not take into account the differential risk of a given medical outcome in vaccinated versus unvaccinated individuals and one fails to acknowledge the large number of vaccine doses administered which will result in many health outcome following vaccination as mere coincidental events although they would be by no means causally related. This is particularly critical for vaccines administered to healthy individuals for their long term or life long protection through an injection which is likely to lead to selective recall bias. Indeed case reports and case series have to be considered as merely hypothesis generating and one has to use epidemiologic studies to further investigate the situation.

Acknowledgement

Many thanks to the VHPB which gave kind permission to reproduce part of the meeting's report which is heavily abstracted in this document.

References

1. Van Damme P, Kane M, Meheus A. Integration of hepatitis B into national immunization programmes. BMJ 1997;314:1033–36.
2. Mahoney FJ, Kane M. Hepatitis B vaccine. In: Plotkin SA and Orenstien WA (eds) Vaccines Third Edition, WB Saunders Company, Philadelphia; 1999.
3. Kane M. Status of hepatitis B immunisation programmes in 1998. Vaccine 1998;16 (suppl.):104–8.
4. Kane M. Status of hepatitis B immunisation programmes in 1998. Vaccine 1998;16 (suppl.):104–8.
5. Kane M. Hepatitis B control through immunisation. In: Rizzetto M, Purcell R, Gerin J, and Verme G (eds) Viral Hepatitis and Liver Disease. Turin: Edizioni Minerva Medica; 1997:638–42.

6. Chang MH, Chen CJ, Lai MS, Hsu HM et al. Universal hepatitis B vaccination in Taiwan and the incidence of hepatocellular carcinoma in children. N Engl J Med 1997;336:1855–9.
7. Hsu HM, Lu CP, Lee SC et al. Seroepidemiologic survey for hepatitis B virus infection in taiwan: the effect of hepatitis B mass immunization. J Infect Dis 1999;179:367–70.
8. Gout O, Lyon-Caen O. Sclérose en plaques et vaccination contre le virus de l'hépatite B. Rev Neurol 1998;154:205–7.
9. Hall A, Van Damme P. on behalf of the Viral Hepatitis Prevention Board. A Technical Consultation on the Safety of Hepatitis B Vaccines. Report of a meeting organized by the Viral Hepatitis Prevention Board on 28–30 September 1998. Viral Hepatitis Prevention Board, University of Antwerp, Antwerp, 1999.
10. Hall A, Kane M, Roure C, Mehus A. Multiple sclerosis and hepatitis B vaccine? Vaccine 1999;17:2473–5.
11. Halsey NA, Duclos Ph, Van Damme P, Margolis H on behalf of the Viral Hepatitis Prevention Board. Pediatr Infect Dis J 1999;18:23–4.
12. Zipp F, Weil JG, Einhäupl KM. No increase in demyelinating diseases after hepatitis B vaccination. Nature Med 1999;5:964–5.
13. Levy-Bruhl D, Ribiere I, Desenclos JC, Drucker J. Comparaison entre les risques de premières atteintes démyélinisantes centrales aigües et les bénéfices de la vaccination contre l'hépatite B. Bulletin Epidémiologique Hebdomadaire 1999;9:33–5.
14. Dossier de Presse, Conférence de presse du Secrétaire d'Etat à la Santé, 01/10/98.

Influenza vaccination in multiple sclerosis

Jacques De Keyser*, Geeta Ramsaransing and Cornelis Zwanikken

Department of Neurology, Academisch Ziekenhuis Groningen, Groningen, Postbus 30.001, 9700 RB Groningen, The Netherlands

Influenza and influenza vaccine

Influenza is a highly contagious acute respiratory disease occurring in epidemics and occasionally in pandemics. Typical symptoms include sudden onset of fever, myalgia especially in the back, sore throat, non-productive cough, a clear nasal discharge, malaise and headache. The most frequent complication is pneumonia, which can be fatal. Epidemics are associated not only be significant morbidity but also with increased mortality [1].

Vaccination is generally recommended for people with factors that predispose them to severe morbidity and mortality. These include all persons 65 years of age or older, patients with chronic pulmonary or cardiovascular disease, diabetes mellitus, renal disease, haemoglobinopathies, immunosuppression, and children receiving chronic aspirin therapy [2]. The vaccine, which is given on a yearly basis, currently consists of three inactivated formulations that contain 15 µg each of the haemagglutinin of influenza A (H1N1), influenza A (H3N2), and influenza B strains. The vaccine is well tolerated, safe and induces immunity in 60 to 90%. Efficacy of the vaccine varies by the similarity of the vaccine strains to the circulating strains, and the age and underlying disease of the recipient [1–2].

Traditionally neurologists have been very cautious in recommending vaccinations in patients with multiple sclerosis (MS) because of concerns that immunization would precipitate relapse. This concern is still reflected in recent textbooks [3]. What is the body of evidence available on influenza vaccination in MS and what recommendations can be based on this evidence?

Influenza vaccine and relapse in MS patients

In a retrospective analysis Sibley and colleagues studied 93 patients with MS who had received a total of 209 doses of influenza vaccine between 1962 and 1975 [4]. Only one patient developed retrobulbar neuritis 24 hours after her second dose of influenza vaccine. In none of the 92 other patients did any new symptoms suggestive of a new lesion develop during the month following vaccination.

Bamford and colleagues conducted a prospective but uncontrolled and unblinded study with swine influenza vaccine in 65 MS patients. The MS patients had no

---

Correspondence address: Prof. J. De Keyser, Department of Neurology, Academisch Ziekenhuis Groningen, P.O. Box 30.001, 9700 RB Groningen, The Netherlands. Tel: (+) 31-50-3612430. Fax: (+) 31-50-3611707. E-mail: j.h.a.de.keyser@neuro.azg.nl

increase in relapse rate compared with the 62 non-immunized control subjects during the 4 weeks follow-up period [5].

In 1977 Myers and coworkers reported a double blind placebo-controlled study in 66 MS patients [6]. The vaccine used contained 200 chick cell-agglutinating units of influenza A/New Jersey/76 and 200 chick cell-agglutinating units of influenza A/Victoria/75 whole viruses. The frequency of clinical relapses of MS was the same in the vaccine-treated (4 of 33 patients) and placebo-treated (4 of 33) groups. An untreated control group had a slightly higher rate of relapses (4 of 22). The efficacy of the vaccination as measured by titers of hemagglutination-inhibiting antibody was comparable to that reported for the general population.

Miller et al. reported on a multicenter, prospective, randomized, double-blind trial of influenza vaccination in patients with relapsing-remitting MS [7]. In the autumn of 1993, 104 patients at 5 MS centers received either standard influenza vaccine or placebo. Patients were examined at 4 weeks and 6 months after inoculation and were contacted by telephone at 1 week and 3 months. They were also examined at times of possible attacks but not when they were sick with flu-like illness. Exacerbation rates in the first month for both groups were not statistically different between the vaccine and placebo groups, and equal to or less than that which would have been expected in the natural course of the disease. The two groups also showed no difference in attack rate or disease progression over 6 months. Surprisingly, 7 vaccinated patients developed clinical manifestations consistent with influenza, compared with 3 in the placebo group (no significant difference).

Effect of influenza vaccine on MRI

Michielsens and coworkers studied 11 patients with relapsing-remitting MS with magnetic resonance imaging (MRI) scans 3 weeks before, at the day of vaccination with killed influenza virus and 3 weeks afterwards [8]. No exacerbations were noted in the pre- or postvaccination period. Eight contrast-enhanced or active lesions were present at the onset of the study. Three new active lesions appeared at the end of the prevaccination period while only 1 new active lesion was found at the end of the postvaccination period. They concluded that vaccination with killed influenza virus had no short-term effect on the activity of MS.

Salvetti and colleagues performed gadolinium-enhanced MRI one day before and at days 15 and 45 after vaccination in 6 MS patients [9]. They did not observe MRI disease activity following vaccination, except in one patient with the highest clinical disease activity during the year preceding the vaccination.

Influenza vaccine versus influenza illness

De Keyser et al. compared the effects of influenza illness (1995–1996 season) and influenza vaccination (autumn of 1996) on neurological symptoms in 233 patients with MS [10]. The vaccine contained the A/Singapore/6/86(H_1N_1), A/Wuhan/

359/95(H_3N_2), and B/Beijing/184/93 strains. No clinically relevant effects were reported in 53 patients with primary progressive multiple sclerosis, either following vaccination or the illness. In a group of 180 patients with relapsing multiple sclerosis, an exacerbation occurred within the following 6 weeks in 32% after influenza illness, whereas it occurred in only 5% after vaccination. The exacerbation rate following influenza illness was significantly higher regardless of whether patients were essentially restricted to wheelchair or not. Thus, the risk of experiencing a relapse was 6 times greater after influenza illness (which may have included other viruses than influenza viruses) than after influenza vaccination. These results are in agreement with previous reports showing that viral upper respiratory infections are a trigger for exacerbations in MS [11,12].

Recommendations

Current evidence indicates that influenza immunization in MS patients is safe, as it is neither associated with an increased exacerbation rate nor worsening of the disease in the postvaccination period. Therefore, vaccination may be a prudent precaution for all MS patients with moderate to severe disability because of an increased risk of complications of influenza. Annual vaccination is also recommended for all patients with a relapsing form of MS because the substantial risk of relapse following influenza illness. At present, what we do not know is whether influenza vaccination in MS patients is as effective as in healthy individuals.

Acknowledgement

This research was supported in part by a grant from Multiple Sclerosis International (Amsterdam, The Netherlands).

References

1. Cox NJ, Subbarao K. Influenza. Lancet 1999;354:1277–1282.
2. Nicholson KG, Snacken R, Palache AM. Influenza immunization policies in Europe and the United States. Vaccine 1995;13:365–369.
3. Schapiro RT, Baumhefner RW, Tourtelotte WW. Multiple sclerosis: a clinical viewpoint to management. In: Raine CS, McFarland HF, Tourtellotte WW, editors. Multiple Sclerosis. Clinical and pathogenetic basis. London: Chapman & Hal, 1997:391–420.
4. Sibley WA, Bamford CR, Laguna JF. Influenza vaccination in patients with multiple sclerosis. JAMA 1976;236:1965–1966.
5. Bamford CR, Sibley WA, Laguna JF. Swine influenza vaccination in patients with multiple sclerosis. Arch Neurol 1978;35:242–243.
6. Myers LW et al. Swine influenza virus vaccination in multiple sclerosis. J Infect Dis 1977;136:S546–S554.
7. Miller AE et al. A multicenter, randomized, double-blind, placebo-controlled trial of influenza immunization in multiple sclerosis. Neurology 1997;48:312–314.
8. Michielsens B et al. Serial magnetic resonance imaging studies with paramagnetic contrast medium: assessment of disease activity in patients with multiple sclerosis before and after influenza vaccination. Eur Neurol 1990;30:258–259.

9. Salvetti M et al. Clinical and MRI assessment of disease activity in patients with multiple sclerosis after influenza vaccination. J Neurol 1995;242:243–246.
10. De Keyser J, Zwanikken C, Boon M. Effects of influenza vaccination and infuenza illness on exacerbations in multiple sclerosis. J Neurol Sci 1998;159:51–53.
11. Sibley WA, Bamford CR, Clark K. Clinical viral infections and multiple sclerosis. Lancet 1985;1:1313–1315.
12. Andersen O et al. Viral infections trigger multiple sclerosis relapses: a prospective seroepidemiological study. J Neurol 1993;240:417–422.

Infection, immunization and immunomodulation; what's the difference?

N. Moriabadi[1], S. Niewiesk[2], N. Kruse[1], K.V. Toyka[1], V. ter Meulen[2] and P. Rieckmann[1,*]

[1]Clinical Research unit for multiple sclerosis and neuroimmunology, [2]Departments of Neurology and Virology*, Julius-Maximilians-University Würzburg, Germany

The role of infections in multiple sclerosis

The chronic inflammatory nature of multiple sclerosis is undoubtable, but an eliciting factor for this devastating disease is still unknown. Several infectious agents have been claimed to cause a pathophysiological cascade of mononuclear cell infiltrates in perivascular white matter of the central nervous system with consecutive demyelination and even axonal loss. Some of the latest candidates for infections as a cause of MS are chlamydia pneumoniae, herpes hominis virus type-6 (HHV-6) or endogenous retrovirvuses, like the multiple sclerosis associated retrovirus (MSRV) [1]. Until now, no single infectious agent could either explain the variable clinical course or the heterogeneous histopathological pattern of multiple sclerosis lesions.

On the other hand, several studies have provided evidence that disease exacerbations are related to recent upper respiratory infections. Therefore, complex interactions between infectious agents and the immune system are anticipated to induce activation of autoreactive T cells (Fig.1). Several direct and indirect mechanisms may be involved in this process:
1) Direct infection of the antigen presenting cells can induce upregulation of co-stimulatory molecules and pro-inflammatory cytokines
2) Virus infected cells may induce bystander activation in secondary lymphatic organs. This could be of particular importance for the role of upper respiratory infections, as the eliciting agents (e.g. adenovirus, rhinovirus, influenza virus) affect the mucosal epithelium and lymphatic system of the nose and throat where autoreactive T cells against components of the central nervous system myelin have been detected.
3) Structural similarities between infectious particles and auto-antigens may lead to cross-reaction of epitope sequences with T cells specific for self peptides. This process is called molecular mimikry [2,3]. This concept was recently challenged by the finding that the T cell receptor antigen recognition is more het-

---

Correspondence address: PD Dr. Peter Rieckmann, Clinical Research unit for multiple sclerosis and neuroimmunology, Department of Neurology, Julius-Maximilians-University, Josef-Schneider-Str. 11, D-97080 Würzburg, Germany. Tel: 49-931-2015766. Fax: 49-931-2013488. E-mail: p.rieckmann@mail.uni-wuerzburg.de

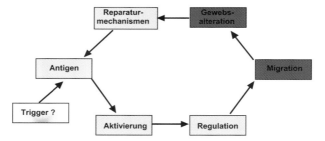

Fig.1. Evolution of MS lesions. Auto-antigen recognition may be triggered by an infectious agent which will then lead the activation and regulation of auto-reactive T cells. Cytokine activated auto-reactive T cells can penetrate the blood brain barrier and secrete toxic factors which will lead to demyelination or even axonal damage. At later stages active downmodulation of the inflammation can be induced by apoptosis of invaded T cells or active remyelination.

erogeneous when initially recognized. Even structurally unrelated peptides may activate the same T cell receptor [4].
None of these mechanisms has been proven to be associated with the disease process in MS, but there is increasing evidence that the complex interactions between various infectious agents and the immune system is a key factor in T cell-mediated auto-immunity [5].

Vaccination in multiple sclerosis

A recent report of the therapeutics and technology assessment subcommittee of the American Academy of Neurology (AAN) stated that "no vaccine currently licensed for use in the United States is known to cause or exacerbate a demyelinating disorder of the CNS" [6]. As immunisation programmes are among the most cost-effective public health measures they should be the first line of defense against infectious diseases. The overall incidence of neurological complications after immunisation is very low and there is up to now no scientific evidence for a causal relation between the manifestation of MS and any preceding vaccination. Published case reports of neurological complications usually do not account for the overall frequency of vaccine application, which is an absolute requirement for any risk calculation [7].

Even more recent concerns about a possible relationship between hepatitis B virus immunisation and MS in adults does not provide any epidemiological evidence [8,9]. On the other hand, the only available class I study on immunisation in MS has demonstrated that influenza vaccination is safe and should be recommended to patients at risk [10]. Similar results have been published recently, concerning the safety of tetanus vaccination [11].

Therefore, several questions concerning the different effects of infections versus immunisations in MS may be addressed:
1. Route of contact: Is the anatomical site of first encounter with the antigen important (mucosal epithelium in most infections versus intramuscular site in

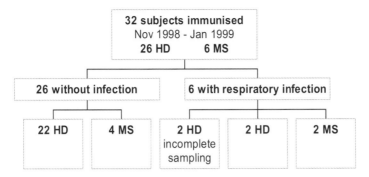

Fig. 2. Flow chart of the VIMS study (HD = healthy donors).

vaccination)?.
2. Bystander activation. Do natural infections provide stronger adjuvant effects than immunisations to elicit auto-immune reactions?
3. Circulating autoreactive immune cells. Are activated autoreactive T cells increased in the peripheral blood during infections versus immunisation?

Vaccination in multiple sclerosis (VIMS study)

During the flu season from november 1998 to january 1999 we performed a prospective, controlled, blinded, immunological follow-up study of healthy individuals (n = 26) and stable MS patients (n = 6) who underwent influenza vaccination for medical reasons. Six study participants had an upper respiratory infection during the observation period (Fig. 2). The study was approved by the local ethics committee and a written informed consent was obtained from each participant. Influvac® was used as the standard vaccine. Venous blood was drawn before, 2, 4, and 16 weeks after immunisation. During follow-up a febrile upper respiratory infection was diagnosed in six individuals. Additional blood sampling was performed at the time of infection as well as 2, 4, and 16 weeks after infection (Fig. 3). None of the participants received any immuno-modulatory treatment. No relapse occured during the observation period. Short term cultures of peripheral blood mononuclear cells were stimulated with either medium alone, influvac® split vaccine, UV-inactivated influenza virus strain APR 8.34, tetanus toxoid, and human myelin basic protein (MBP) or recombinant myelin oligodendrocyte glycoprotein (MOG). Antigen induced IFN-γ production was quantified by the ELISPOT assay and reverse transcriptase polymerase chain reaction (RT PCR). Both methods were proven to be very sensitive, reproducible and yielded comparable results (Fig. 3).

The primary focus of this study was to analyse IFN-γ producing autoreactive T cells against MBP or MOG, as an indicator for a Th1-type auto-immune response after influenza vaccination. Our initial results revealed no significant induction of autoreactive T cells at any time point after immunisation in MS patients and healthy individuals (Fig. 4). However, a significant increase in influenza-induced,

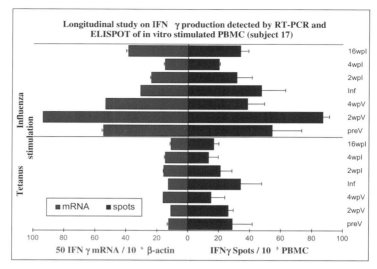

Fig. 3. Induction of IFN-γ mRNA (white bars) and IFN-γ producing blood cells (black bars) during influenza vaccination and natural occuring upper respiratory infection in an MS patient. WpV: weeks post vaccination; wpI: weeks post infection.

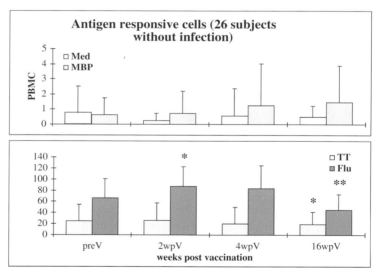

Fig. 4. No induction of MBP specific T cells after influenza vaccination. Peripheral blood mononuclear cells (PBMCs) were obtained from 26 subjects prior to influenza vaccination (preV), 2 weeks (2wpV), 4 weeks (4wpV) and 16 weeks (16wpV) after vaccination. Cells were cultured in the presence of myelin basic protein (MBP), tetanus toxoid (TT), influvac (Flu) or medium alone and IFN-γ specific T cells were determined by the Elispot technique. *Significant increase over baseline ($p<0.05$).

in vitro IFN-γ producing cells could be detected in 80% of study participants 2 weeks after influvac® injection. These results indicate a protective cytokine response against influenza virus without induction of autoreactive T cells. A similar approach was used by Mokhtarian et al. who did not observe any significant changes of MBP induced T cell proliferation after influenza vaccination [12].

Despite the fact that our results represent only initial data from an ongoing study, there is now increasing evidence from epidemiology and immunology that influenza vaccination is a safe medical procedure in MS patients at risk for infection induced relapses. Certain limitations should be applied for patients during a relapse. They should not be vaccinated until the symptoms have resolved after pulse steroid treatment [13].

References

1. Monteyne P, Bureau JF, Brahic M. Viruses and multiple sclerosis. Curr Opin Neurol 1998;11:287–91.
2. Fujinami RS, Oldstone MBA. Amino acid homology between the encephalitogenic site of myelin basic protein and virus: mechanism for autoimmunity. Science 1985;230:1043–5.
3. Wucherpfennig KW, Strominger JL. Molecular mimicry in T cell-mediated autoimmunity: Viral peptides activate human T cell clones specific for myelin basic protein. Cell 1995;80:695–705.
4. Hemmer B, Fleckenstein BT, Vergelli M, Jung G, McFarland H, Martin R, et al. Identification of high potency microbial and self ligands for a human autoreactive class II-restricted T cell clone. J. Exp. Med. 1997;185:1651–9.
5. Wekerle H. The viral triggering of autoimmune disease. Nat Med 1998;4:770–1.
6. Fenichel GM. Assessment: Neurologic risk of immunization. Neurology 1999;52:1546–52.
7. Medicine Io. Adverse events associated with childhood vaccines: evidence bearing on causality. National Academic Press 1993;Washington, DC.
8. Tourbah A, Gout O, Liblau R, Lyon-Caen O, Bougniot C, Iba-Zizen MT, et al. Encephalitis after hepatitis B vaccination. Recurrent disseminated encephalitis or MS? Neurology 1999;53:396–401.
9. Zipp F, Weil JG, Einhäupl KM. No increase in demyelinating diseases after hepatitis B vaccination. Nat Med 1999;5:964–5.
10. Miller AE, Morgante LA, Buchwald LY, Nutile SM, Coyle PK, Krupp LB, et al. A multicenter, randomized, double-blind, placebo-controlled trial of influenza immunization in multiple sclerosis. Neurology 1997;48:312–4.
11. de Keyser J. Safety of tetanus vaccination in relapsing-remitting multiple sclerosis. Infection 1998;26:319.
12. Mokhtarian F, Shirazian D, Morgante L, Miller A, Grob D, Lichstein E. Influenza virus vaccination of patients with multiple sclerosis. Multiple Sclerosis 1997;3:243–7.
13. Kollegger H, Schmied M, Zebenholzer K, Zeiler K, Hittmair K, Mallek R. Vaccination and steroids in MS: effects on disease progression and mood. Acta Neurol Scand 1999;100:69–73.

Detection of human herpesvirus 6 by quantitative real-time PCR in serum and cerebrospinal fluid of patients with multiple sclerosis

Giuseppe Locatelli[1], Mauro S. Malnati[1], Diego Franciotta[2], Roberto Furlan[3], Giancarlo Comi[3], Gianvito Martino[3] and Paolo Lusso[1,*]

[1]*Unit of Human Virology, and* [3]*Unit of Experimental Neuroimmunotherapy, DIBIT, San Raffaele Scientific Institute, Milano, Italy;* [2]*Laboratory of Neuroimmunology, Fondazione Mondino, University of Pavia*

HHV-6 and its role in human disease

Human herpesvirus 6 (HHV-6) is a β-herpesvirus originally isolated in 1986 from the peripheral blood of immunocompromised patients with lymphoproliferative disorders [1]. Although it was initially described as a B-lymphotropic virus, subsequent studies demonstrated that HHV-6 replicates primarily in T-lymphocytes both *in vitro* [2] and *in vivo* [3]. However, several other cell types of different lineage origin can be infected, either productively or nonproductively [4,5]. Consistent with this broad cellular tropism, we recently identified the CD46 molecule, a complement regulatory protein expressed on virtually all nucleated human cells, as a cellular receptor for HHV-6 [6]. Two major viral subgroups, A and B, have been identified, which form two segregated clusters with unique genetic, biologic and immunologic characteristics [7–9]. While HHV-6 B is highly prevalent in the human population in all geographic areas [4,5], the distribution of HHV-6 A appears to be more restricted. Primary infection with HHV-6 B occurs almost universally in early childhood and is linked to the etiology of *exanthema subitum*, a usually benign febrile disease [10]. Conversely, little is known about the time and pathological consequences of primary infection with HHV-6 A. In the adult life, HHV-6 reactivation and/or reinfection have been associated with a wide array of human diseases [4], although most of these associations have not been substantiated by rigorous criteria. Due to the ubiquitous nature of HHV-6, most conventional diagnostic methodologies, such as IgG serology or qualitative DNA PCR, do not allow to establish an etiologic link, as they fail to discriminate between latent and active infection. Using markers of active HHV-6 infection, including immunohistochemistry, RNA *in situ* hybridization and plasma DNAemia, convincing evidence has been provided that HHV-6 can cause severe, often life-threatening, opportunistic infections in immunocompromised patients [4,5]. Moreover, HHV-6 may have a direct immunosuppressive role *in vivo* and has been implicated as a

---

Correspondence address: San Raffaele Scientific Institute, Unit of Human Virology, Via Olgettina 60, Milan, 20132, Italy. E-mail: paolo.lusso@hsr.it

potential cofactor in the progression of HIV infection toward full-blown AIDS [11].

HHV-6 in neurological disorders

Multiple lines of evidence indicate that both HHV-6 subgroups possess a distinctive neurotropism. *In vitro,* HHV-6 can infect microglia, oligodendrocytes and astrocytes [12,13], although only in the latter was productive infection documented. Neurological complications, such as febrile seizures, meningitis and meningo-encephalitis, are relatively frequent during primary HHV-6 infection [14–16], and may be a consequence of direct viral invasion of the central nervous system [17]. Persistence of HHV-6 in brain tissue during the adult life has been documented [18,19]. Moreover, active HHV-6 infection was reported in several acute and chronic neurologic disorders, particularly in immunocompromised subjects; these include encephalitis [20,21], fulminant demyelinating encephalomyelitis [22,23], subacute leukoencephalitis [24], necrotizing encephalitis during Griscelli's syndrome [25], chronic myelopathy [26], AIDS-related demyelinization [27] and progressive multifocal leukoencephalopathy [28]. Altogether, these data suggest that HHV-6 may have an important neuropathogenic role *in vivo.*

HHV-6 in multiple sclerosis

Over the past few years, a possible implication of HHV-6 in the pathogenesis of multiple sclerosis (MS) has been investigated, although the evidence is still controversial [19,29–42]. The first strong association between HHV-6 infection and MS was established in a study conducted in 1995 by representational difference analysis (RDA), a molecular subtraction technique that identifies genetic sequences that differ between two genomes [29]. Using a nested PCR assay and immunohistochemistry, the same study documented the presence and active expression of HHV-6 in the brain tissue of more than 75% of patients with MS. Expression of viral antigens was observed in the nucleus of the oligodendrocytes surrounding the plaques that constitute the hallmark of the disease. Somewhat surprisingly, however, both HHV-6 DNA and HHV-6 antigen expression were detected in a similar proportion of non-MS brains, including brains of people who died of non-neurological causes, although viral antigen expression was not specifically associated with oligodendrocytes.

Conflicting results have emerged from a series of subsequent studies that investigated the role of HHV-6 in MS using different experimental approaches. Unfortunately, the majority of such studies was conducted on small numbers of patients and lacked adequate controls. Three reports [35,38,40] documented the presence of specific IgM antibodies against HHV-6 in MS patients, whereas no correlation was seen in a fourth report [41]; it is important to emphasize that, unlike the anti-HHV-6 IgG titers that have little diagnostic significance in the adult life, IgM responses may be specific indicators of HHV-6 reactivation or reinfection.

Similarly, the detection of plasma/serum HHV-6 DNAemia represents an accurate marker of active HHV-6 infection [43]. Nevertheless, conflicting results were obtained using this approach: while in most studies the rates of HHV-6 positivity were very low (ranging from 0 to 4%) [30,33,39,40], one study reported the presence of HHV-6 DNA in the serum of 15 of 50 (30%) MS patients, 14 of whom had the relapsing-remitting (RR) form of the disease [35]. Although these discrepancies are apparently difficult to reconcile, they could be related to the different sensitivity of the assays used in different laboratories.

Use of a calibrated real-time PCR assay for the diagnosis of active HHV-6 infection in MS patients

To investigate the role of HHV-6 in MS, we have evaluated the presence of cell-free HHV-6 DNA in paired serum and CSF samples obtained from 47 MS patients, 31 with RR disease, 8 with primary progressive (PP) disease and 8 with secondary progressive (SP) disease. Paired serum and CSF samples from 10 patients with non-MS neurological disorders, as well as sera from 22 patients with myastenia gravis and from 7 patients with Guillain-Barre' sindrome, were tested in parallel as controls. The CSF samples were spun at 1500 rpm for 15 min. to remove the cellular fraction. All the samples were stored frozen at $-80°C$ until use. A newly developed quantitative assay for HHV-6 DNA, based on the TaqMan real-time PCR technology, was used, which was shown to provide a sensitive and accurate measurement of the HHV-6 viral load in tissues and biological fluids (44). To further improve the sensitivity and accuracy of the assay, we introduced a virion-concentration step, by subjecting the samples to high-speed centrifugation, and we used a synthetic calibrator DNA, which allows to control for the efficiency of each step of the procedure (Locatelli G. *et al.*, manuscript in preparation). This calibrator molecule, which anneals with the same primers as the target DNA but is revealed with a different probe, was spiked into each sample before DNA extraction (500 copies/sample). DNA was extracted from serum or CSF by proteinase K digestion and purified by the phenol-chloroform method following standard protocols. Briefly, a 0.1 ml aliquot of each sample was incubated at 56°C for 1 hr in 0.5 ml lysis buffer (10 mM Tris-HCl pH 8.0, 5 mM EDTA, 0.5% V/V SDS, 0.1 µg/ml proteinase K). The extraction was followed by repeated purification with phenol-chloroform-isoamyl alcohol (24:24:1), followed by chloroform-isoamyl alcohol (24:1); DNA was subsequently isolated by ethanol/high-salt precipitation and finally resuspended in 0.1 ml of 5 mM Tris-HCl, 0.5 mM EDTA pH 8.0 buffer. Due to the generally poor DNA recovery from CSF, a carrier molecule (glycogen, 2 µg/ml) was added to all the samples prior to DNA precipitation. To reveal potential cross-contamination during DNA extraction or PCR preparation, 1 negative control (human serum that tested repeatedly negative for the presence of HHV-6 DNA) was included every 2 samples; moreover, all the different steps (DNA extraction, PCR mix preparation, addition of DNA to the PCR mix, amplification) were performed in separated and dedicated rooms.

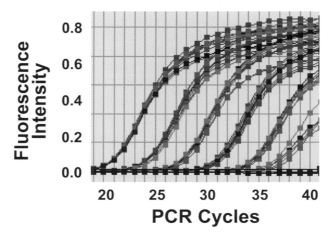

Fig. 1. Graphic representation of the quantitative real-time PCR assay amplification plots elaborated by the Sequence Detection Software. Multiple replicates of serial ten-fold dilutions of the standard plasmid DNA were amplified to generate a reference curve ranging (right to left) from 10^1 to 10^6 HHV-6 genome equivalents.

The HHV-6 viral load in serum or CSF was quantitated using the TaqMan fluorogenic detection system on the ABI PRISM 7700 Detector (PE Applied Biosystems, Foster City, CA). The primers and the probe were selected within open reading frame U67, which is highly conserved in both subgroups (A and B) of HHV-6. The primers were 5'-CAAAGCCAAATTATCCAGAGCG-3' and 5'-CGCTAGGTTGAGGATGATCGA-3'; the probe was 5'-CACCAGACGT-CACACCCGAAGGAAT-3'. The probe was covalently linked with the reporter dye 6-carboxy-fluorescein (FAM) in 5' position and the quencher dye 6-carboxy-tetramethyl-rhodamine (TAMRA) in 3' position. The amplified product was 133 bp in length. The reaction was performed in a total volume of 25 µl, with dATP, dCTP and dGTP at 100 µM each and dUTP at 200 µM, TaqMan buffer A 1x, magnesium chloride 5.5 mM, primers 300 nM each, probe 50 nM, 0.625 units of AmpliTaq Gold, 0.25 units of Uracil-N-glycosylase, 40% V/V of sample DNA. The TaqMan PCR protocol included 2 min degradation of preamplified templates at 50°C, followed by 15 min denaturation at 95°C and 40 cycles of denaturation at 95°C for 15 sec and annealing/extension at 58°C for 60 sec. Data analysis was performed using the Sequence Detection Software (PE Applied Biosystems). As a reference for HHV-6 DNA quantitation, the HHV-6 fragment amplified by the above mentioned primers was cloned into the pCRII plasmid (Invitrogen). A standard curve was obtained by serially diluting the plasmid stock quantitated by spectrophotometry (Fig. 1). The data were normalized based on the rate of recovery of the calibrator DNA.

To verify the sensitivity of the optimized TaqMan assay, we directly compared it with a standard nested PCR system, performed as previously described [43], on a selected group of paired patient samples (n=7). For nested PCR, DNA was

Table 1. Comparison of nested PCR and optimized real–time PCR for the detection of cell–free HHV–6 DNA in serum and CSF of patients with MS.

Patient no. (disease course)*	HHV–6 DNA Detection (viral genome equivalents/ml)			
	Nested PCR Assay		Real–Time PCR Assay	
	Serum	CSF	Serum	CSF
1 (PP)	–	–	– (<31)	+ (39)
2 (SP)	–	–	– (<48)	– (<57)
3 (RR)	–	–	– (<26)	– (<43)
4 (PP)	–	–	– (<31)	+ (181)
5 (RR)	–	–	– (<83)	– (<1000)
6 (RR)	–	–	+ (7454)	+ (12000)
7 (PP)	–	–	– (<29)	– (<20)

*RR = relapsing–remitting; PP = primary progressive; SP = secondary progressive.

extracted from 0.2 ml of the biological samples using the QIAblood kit (Qiagen). As illustrated in Table 1, none of the serum or CSF samples was found to contain HHV-6 DNA by standard nested PCR. The absence of Taq-polymerase inhibitors was demostrated by the correct amplification of end-point diluted HHV-6 DNA spiked into the samples (data not shown). By contrast, real-time PCR was able to identify the viral nucleic acid in 1 of the MS sera (14%) and in 3 of the CSF samples (43%). Quantitation of the calibrator DNA recovery permitted to determine, for each negative sample, a cut-off value of HHV-6 detectability (Table 1). All the HHV-6-negative samples, with one exception, had cut-off values lower than 100 viral genome equivalents/ml. One of the HHV-6 DNA-positive patients exhibited a remarkably high viral load both in serum and in CSF (7,545 and 12,000 viral genome equivalents/ml, respectively), whereas the other 2, with positivity only in CSF, had a markedly lower viral load (39 and 121 viral genome equivalents/ml). The presence of one negative control sample after each pair of patient samples allowed to exclude false positivity due to PCR cross contamination. The nested PCR test was also used to screen the CSF of additional MS patients, some of whom were followed longitudinally, but it consistently failed to detect the presence of HHV-6 DNA (data not shown). These data demonstrate that the optimized real-time PCR system has a greater sensitivity, compared to a standard nested PCR assay. Since the lower detection limit of the two systems is similar (1 viral genome equivalents/test), the improved sensitivity of the real-time PCR assay is likely to be related to the introduction of the virion concentration step, as well as to the more efficient DNA-extraction method used.

The analysis by real-time PCR was then extended to 40 additional MS patients from whom serum and CSF obtained at the same time point were available. Table 2 summarizes all the results obtained with the optimized real-time PCR assay. HHV-6 DNA was detected in 10 of 46 sera (17.9%) and in 12 of 47 CSF samples (25.5%). Among positive samples, the mean viral load was 968 in serum and 3,492 in CSF. HHV-6-positive cases were found among patients with all three types of disease course. In cases with PP disease, none of the sera was found to

Table 2. Detection of HHV-6 DNA by quantitative calibrated real-time PCR in serum and CSF of MS patients.

Disease		No. of patients	HHV-6 DNA-positive samples (%)	
			Serum	CSF
Multiple Sclerosis*	RR	31	8/30 (26.6%)	8/31 (25.8%)
	PP	8	0/8 (0%)	2/8 (25.0%)
	SP	8	2/8 (25.0%)	2/8 (25.0%)
	Total	47	10/46 (17.9%)	12/47 (25.5%)
Other neurological disorders		10	1/10 (10%)	1/10 (10%)

*RR = relapsing-remitting; PP = primary progressive; SP = secondary progressive.

harbor HHV-6 DNA, although the total number was low (n=8); by contrast, 25% of such cases had detectable HHV-6 DNA in CSF. In RR and SP cases, HHV-6 DNA was detected with similar frequency in CSF and in serum. In 5 patients, all with RR disease, HHV-6 DNA was detected both in serum and in CSF, whereas in all the other positive cases HHV-6 was harboured in only one of the two biological fluids. The negative results obtained with all the negative controls confirmed the lack of artifactual positivity due to PCR cross-contamination.

To evaluate the specificity of the results obtained in MS patients, we screened 20 paired samples (10 sera and 10 CSF) derived from patients affected by other neurological disorders. In two instances (1 CSF and 1 serum sample from different patients) a low viral burden was detected (73 and 263 genome equivalents/ml, respectively). As additional controls, we tested sera from 22 patients with myastenia gravis and 7 patients with Guillain-Barre' sindrome using the same optimized PCR method. With two exceptions, in which the viral load was low (11 and 236 genome equivalents/ml), all the samples were negative for the presence of HHV-6 DNA (data not shown). Altogether, these results indicate that active HHV-6 infection does occur in a significant proportion of patients with different forms of MS.

Concluding remarks

Despite intensive investigation, the causative mechanism of MS remains unknown. Several factors have been proposed to be involved in pathogenesis, including viral, genetic and environmental causes. In particular, different viruses have been implicated as etiologic agents, which would be responsible for triggering the diverse autoimmune phenomena documented in the course of MS [45]. However, no definitive evidence has been hitherto provided for the etiologic role of any such agents. We have documented active HHV-6 infection in a significant proportion of MS patients. The results of our study emphasize the remarkable differences in sensitivity among different assay systems used for the detection of active HHV-6 infection. While a conventional nested PCR was unable to detect cell-free HHV-6 DNA in any of our samples, a recently developed optimized real-time PCR system showed the presence of HHV-6 DNA in 18% of the sera and 25% of the CSF of MS patients. The variable sensitivity among different PCR assays may also explain the conflicting results obtained in previous reports, with rates of positivity in CSF ranging from 0% [33] to 6% [40], 11% [31], 14% [30] or 17% [38].

Although these results are preliminary, they are compatible with a possible involvement of HHV-6 in at least some cases of MS. A potential mechanism whereby HHV-6 could be involved in MS pathogenesis is suggested by our recent identification of CD46, a critical regulator of complement activation, as a cellular receptor for HHV-6 [6]. Engagement of CD46 may result in the abnormal activation of the complement cascade on autologous cells, leading to widespread tissue damage. In light of the relatively low expression of CD46 in the brain tissue, the threshold for spontaneous complement activation is likely to be reduced, compared to other anatomical sites, which might enhance the pathogenic potential of CD46 downmodulating agents, such as HHV-6 and measles virus. Further studies are needed to clarify the relation between HHV-6 reactivation (or reinfection) and the clinical stage of MS, in particular with the disease flare-ups that are commonly observed in patients with RR disease course. Carefully controlled longitudinal studies, taking advantage of optimized diagnostic tools for the detection of active HHV-6 infection, will be fundamental in elucidating the role of HHV-6 in the etiology of MS.

References

1. Salahuddin SZ et al. Isolation of a new virus, HBLV, in patients with lymphoproliferative disorders. Science 1986;234:596–601.
2. Lusso P et al. In vitro cellular tropism of human B-lymphotropic virus (human herpesvirus 6). J Exp Med 1988;167:1659–1670.
3. Takahashi K et al. Predominant CD4 T-lymphocyte tropism of human herpesvirus 6-related virus. J Virol 1989;63:3161–3163.
4. Lusso P. Human herpesvirus 6 (HHV-6). Antivir Res 1996;31:1–21.
5. Braun DK, Dominguez G, and Pellett PE. Human herpesvirus 6. Clin Microbiol Rev 1997;10:521–567.
6. Santoro F et al. CD46 is a cellular receptor for human herpesvirus 6. Cell 1999;99:817–827.
7. Aubin JT et al. Several groups among human herpesvirus 6 strains can be distinguished by southern blotting and polymerase chain reaction. J Clin Microbiol 1991;29:367–372.
8. Schirmer EC et al. Differentiation between two distinct classes of viruses now classified as human herpesvirus 6. Proc Natl Acad Sci USA 1991;88:5922–5926.
9. Ablashi et al. Genomic polymorphism, growth properties and immunologic variations in human herpesvirus-6 isolates. Virology 1991;184:545–552.
10. Yamanishi K, Okuna T, Shiraki K. Identification of human herpesvirus 6 as a causal agent for exhantem subitum. Lancet 1988;1:1065–7.
11. Lusso P, Gallo RC. Human herpesvirus 6 in AIDS. Immunol Today 1995;16:67–71.
12. He J et al. Infection of primary human fetal astrocytes by human herpesvirus 6. J Virol 1996;70:1296–1300.
13. Albright AV et al. The effect of human herpesvirus-6 (HHV-6) on cultured human neural cells: oligodendrocytes and microglia. J Neurovirol 1998;4:486–494.
14. Ishiguro N et al. Meningo-encephalitis associated with HHV-6 related exanthem subitum. Acta

Paediatr Scand 1990;79:987–989.
15. Asano Y et al. Fatal encephalitis/encephalopathy in primary human herpesvirus-6 infection. Arch Dis Child 1992;67:1484–1485.
16. Breese-Hall C et al. Human herpesvirus 6 infection in children. N Engl J Med 1994;331:432–438.
17. Kondo K et al. Association of human herpesvirus 6 infection of the central nervous system with recurrence of febrile convulsions. J Infect Dis 1993;167:1197–1200.
18. Hall C et al. Persistence of human herpesvirus 6 according to site and variant: possible greater neurotropism of variant A. Clin Infect Dis 1998;26:132–137.
19. Sanders VJ, Felisan S, Waddell A, Tourtellotte WW. Detection of herpesviridae in postmortem multiple sclerosis brain tissue and controls by polymerase chain reaction. J Neurovirol 1996;2:249–258.
20. Drobyski WR, Knox KK, Majewski D, and Carrigan DR. Fatal encephalitis due to variant B human herpesvirus-6 infection in a bone marrow-transplant recipient. N Engl J Med 1994;330:1356–1360.
21. Knox KK, Harrington DP, and Carrigan DR. Fulminant human herpesvirus 6 encephalitis in a human immunodeficiency virus-infected infant. J Med Virol 1995;45:288–292.
22. Kamei A et al. Acute disseminated demyelination due to primary human herpesvirus-6 infection. Eur J Pediatr 1997;156:709–712.
23. Novoa LJ et al. Fulminant demyelinating encephalomyelitis associated with productive HHV-6 infection in an immunocompetent adult. J Med Virol 1997;52:301–308.
24. Carrigan DR, Harrington D, Knox KK. Subacute leukoencephalitis caused by CNS infection with human herpesvirus-6 manifesting as acute multiple sclerosis. Neurology 1996;47:145–148.
25. Wagner M et al. Human herpesvirus-6 (HHV-6)-associated necrotizing encephalitis in Griscelli's syndrome. J Med Virol 1997;53:306–312.
26. Mackenzie IR, Carrigan DR, Wiley CA. Chronic myelopathy associated with human herpesvirus-6. Neurology 1995;45:2015–2017.
27. Saito Y et al. Cellular localization of human herpesvirus-6 in the brains of children with AIDS encephalopathy. J Neurovirol 1995;1:30–39.
28. Mock DJ et al. Association of human herpesvirus 6 with the demyelinative lesions of progressive multifocal leukoencephalopathy. J Neurovirol 1999;5:363–373.
29. Challoner PB et al. Plaque-associated expression of human herpesvirus 6 in multiple sclerosis. Proc Natl Acad Sci USA 1995;92:7440–7444.
30. Wilborn F et al. Detection of herpesvirus type 6 by polymerase chain reaction in blood donors: random tests and prospective longitudinal studies. Br J Haematol 1994;88:197–192.
31. Liedtke W, Malessa R, Faustmann PM, Eis-Hubinger AM. Human herpesvirus 6 polymerase chain reaction findings in human immunodeficiency virus associated neurological disease and multiple sclerosis. J Neurovirol 1995;1:253–258.
32. Carrigan DR, Knox KK. Human herpesvirus six and multiple sclerosis. Mult Scler 1997;3:390–394.
33. Martin C et al. Absence of seven human herpesviruses, including HHV-6, by polymerase chain reaction in CSF and blood from patients with multiple sclerosis and optic neuritis. Acta Neurol Scand 1997;95:280–283.
34. Merelli E et al. Human herpes virus 6 and human herpes virus 8 DNA sequences in brains of multiple sclerosis patients, normal adults and children. J Neurol 1997;244:450–454.
35. Soldan SS et al. Association of human herpes virus 6 (HHV-6) with multiple sclerosis: increased IgM response to HHV-6 early antigen and detection of serum HHV-6 DNA. Nat Med 1997;3:1394–1397.
36. Kimberlin DW, Whitley RJ. Human herpesvirus-6: neurologic implications of a newly-described viral pathogen. J Neurovirol 1998;4:474–485.
37. Mayne M et al. Infrequent detection of human herpesvirus 6 DNA in peripheral blood

mononuclear cells from multiple sclerosis patients. Ann Neurol 1998;44:391–394.
38. Ablashi et al. Human Herpesvirus-6 (HHV-6) infection in multiple sclerosis: a preliminary report. Mult Scler 1998;4:490–496.
39. Goldberg SH, Albright AV, Lisak RP, Gonzalez-Scarano F. Polymerase chain reaction analysis of human herpesvirus-6 sequences in the sera and cerebrospinal fluid of patients with multiple sclerosis. J Neurovirol 1999;5:134–139.
40. Enbom M et al. Similar humoral and cellular immunological reactivities to human herpesvirus 6 in patients with multiple sclerosis and controls. Clin Diagn Lab Immunol 1999;6:545–549.
41. Friedman JE et al. The association of the human herpesvirus-6 and MS. Mult Scler 1999;5:355–362.
42. N. Akhyani et al. Tissue distribution and variant characterization of human herpesvirus (HHV-6): increased prelevance of HHV-6A in patients with mulyiple sclerosis. J Inf Dis 2000;182:1321–1325.
43. Secchiero P et al. Detection of human herpesvirus 6 in plasma of children with primary infection and immunosuppressed patients by polymerase chain reaction. J Infect Dis 1995;171:273–280.
44. Locatelli G, Santoro F, Veglia F, Gobbi A, Lusso P, Malnati MS.: Real-time quantitative PCR for human herpesvirus 6 DNA. J Clin Microbiol 2000;38:4042–4048.
45. Johnson, R.T. The virology of demyelinating diseases. Ann Neurol 1994;36:S54–60.

The ability of candidate viruses to explain epidemiological findings in multiple sclerosis

Sven Haahr* and Per Höllsberg

Institute of Medical Microbiology and Immunology, The Bartholin Building, Aarhus University, DK-8000 Aarhus, Denmark

Introduction

The hypothesis of an infectious cause of multiple sclerosis (MS) dates back to Pierre Marie's proposal in 1884 [1], only 16 years after Jean M. Charcot's initial scientific description of the disease [2]. Since then, only epidemiological data have lend significant support to the idea that an environmental agent may trigger MS in genetically predisposed individuals. Over the years, a growing list of agents that were promoted as the cause of MS – only to vanish when examined in greater details – has unfairly impeached the concept of an infectious etiology of MS [3].

The tremendous progress in immunology research over the last decades has shed light on potential pathogenic processes in MS. In particular, research in experimental autoimmune encephalomyelitis (EAE), which in many ways have similarities to MS [4], has nourished the belief that MS is an autoimmune disease, in which an immune reaction is directed against self antigen(s). Although the autoimmune hypothesis essentially has ignored the triggering event, the "mimicry hypothesis" has gained in interest recently providing a bridge between an infectious etiology and an autoimmune pathogenesis [5,6]. The premise for this concept is the structural analogy between even distantly related protein sequences that allow infectious agents to trigger autoreactive T cells [5].

Whereas the mimicry hypothesis implicates a large variety of infectious agents, solely based on the structural relationship between the microorganism and self proteins, it is the premise of this paper that epidemiological data on MS are in conflict with this idea. Here we will analyse whether the viruses, which have attracted interest as candidate agents in MS during the last decade, may explain existing epidemiological findings of the disease. We will argue that such an analysis restricts the number of interesting viral candidate agents in MS.

Epidemiology of MS

Paradoxically, the virologic research efforts in MS seem to have ignored more or less the epidemiology of the disease, although the major argument for an infectious agent in MS originates from epidemiological reports. In this section, we will

---

*Correspondence address: Aarhus University, Institute of Medical Microbiology and Immunology, The Bartholin Building, Aarhus, DK-8000, Denmark. E-mail: haahr@microbiology.au.dk

review the essence of the epidemiology of MS as it may apply to the evaluation of viruses.

Prevalence of MS

The prevalence studies clearly indicate that MS occurs with the highest frequency in Caucasians with British or Scandinavian ancestors. On the northern hemisphere, a north-to-south gradient with the highest prevalence of MS in the north (reverse on the southern hemisphere) is documented in North America, Europe and Australia [7,8].

Within some countries, areas with significant variation in MS prevalence have been found, in spite of a homogenous population [7,8,9]. A convincingly lower prevalence of MS is present among non-Caucasians whether or not they live in their area of origin. Thus, the susceptibility to MS differ among races and ethnic groups, but this alone can hardly explain the differences in the national and global distributions of the disease [7,8]. While genetic factor contribute to these differences, an infectious agent may also be involved. Such an agent may explain clusters, epidemics, local variations in incidence, and even gradients of MS prevalence. In addition, transmission of an infectious agent may be influenced by cultural factors, including pre-chewing of meals for babies as used in Southern Europe. Indeed, the infectious agent hypothesis predicts that early transmission of the agent, e.g. from mother to child, may have a prophylactic effect on the development of MS.

Migration studies in MS

The migration studies provide the most substantiated epidemiological data in MS. They clearly demonstrate that the risk of acquiring MS may change after migration between places with varying prevalence of the disease [7,8,10]. This strongly suggests that environmental factors influence development of the disease. Moreover migration studies may give further information about the ages at which such factors are at play.

The majority of the studies have dealt with migrations from high to low prevalence areas. Taken together, these studies have shown that both migration within the same country and between countries lead to a lower risk of developing MS [7,8,10]. The exact age at which migration influences the risk for developing MS has not been established. Several studies have shown that persons migrating from high to low prevalence areas before the age of 15 reduce their risk of developing MS. A few studies have found that the reduction in risk happens if migration to a low prevalence area occurs through the first two decades of life [10]. An agent that may initiate MS when infecting after puberty, but protecting against MS when infecting before puberty would be able to explain the migration studies in MS [11].

Individuals migrating from a low to a high prevalence area seem to retain the same risk of acquiring MS as those in the low prevalence area [10]. Studies on migration inside USA showed an increase in risk when moving from low to high prevalence areas, although the increase was less than the decrease found when migration goes from high to low prevalence areas [7,10]. Thus, genetically susceptible individuals may lower their risk of acquiring the disease by avoiding an environmental agent, whereas genetically "protected" individuals may tolerate the environmental agent. It follows that such an environmental agent could be ubiquitous.

Familial occurrence of MS

A higher frequency of MS in families is a well-known phenomenon. In a Caucasian population with a prevalence of 100 MS patients per 100,000 inhabitants, the mother-to-daughter occurrence is 2%, whereas the father-to-son occurrence is 1%. Among siblings, the occurrence is 2–5%, depending on gender [12]. In monozygotic (MZ) twins, the concordance rates for MS varies between 6% and 50%, depending on the methods of recruitment. The average concordance rate in nine studies encompassing 153 MZ twins was 26%, which is in agreement with the most reliable of the studies, where twins were recruited from twin registers and MS clinics [13,14]. The MZ twin studies clearly indicate that MS is caused more by nurture than by nature. Nonetheless, the family studies can hardly be used for evaluating the nature of the possible factors involved in the initiation and the pathogenesis of MS. Interestingly, the same or even higher concordance rate in MZ twins has been demonstrated for the development of paralytic polio (36% versus 26% in MS) [15,16]. This emphasizes the role of genetic susceptibility to infections.

Geographical clustering and "epidemics" of MS

A cluster is defined as a statistical significant increase in the occurrence of a trait in a geographically bounded group [17]. Several clusters of MS cases, outside families, have been reported, although scepticism concerning the significance of these phenomenons exists [8]. A Norwegian study showed that MS patients within the same birth cohort had lived closer to each other between their 13th to 20th year of age than would be expected by chance, with a peak clustering at age 18 years. The authors found the results compatible with the involvement of a common infectious agent, acquired in adolescence [18], although investigations of potential etiologic agents in clusters have never been done [19].

MS "epidemics" are geographical clusters of MS cases that suddenly appear and subsequently disappear several years later [8]. The Faroe Island "epidemic" is the most extensively studied and the most famous of these "epidemics" [20,21]. It has been considered to be caused by an infectious agent introduced by the 8000 British servicemen stationed on the islands during the World War II (1940–45). The "epidemic" includes 24 persons developing MS among 30.000 inhabitants in the period 1943–60, with an exceptional high incidence in 1944–1945 [20]. MS "epidemics" may also have occurred in Iceland [22], on the Orkney and Shetland Islands [23,24], and on Sardinia [25]. In all these "epidemics" an increased incidence of MS has been recorded in association with servicemen stationed in these isolated communities during [20,22,24] or after [25] World War II.

Socio-economic status of MS patients

High socio-economic status has in several studies been associated with a higher prevalence of MS [8], which constitute a parallel to the higher prevalence of paralytic polio [26], clinical symptoms following hepatitis A infection [27] and infectious mononucleosis [28] in these settings. Thus, the association of socio-economic status with the development of MS may be related to acquiring an infectious agent late in life.

Antibody studies

Serological studies in MS patients are based on the following postulate: if a significant number of MS cases is caused by one infectious agent, a higher number of MS patients than healthy controls displays antibodies toward this agent. In addition MS patients may have higher titres and potentially antibodies of both the IgG and IgM class. For certain members of the herpes virus family it is a well-known phenomenon that reactivations of the virus may take place. If it can be documented that such reactivations are correlated and especially if they are preceding MS attacks (presence of IgM antibodies), it may support a pathogenic role of the agent in MS. Such antibody studies are done best on MS patients during longitudinal MRI surveillance. Intrathecal synthesis of antibodies with a higher level of IgG in the cerebrospinal fluid (CSF) from MS patients than from healthy controls has been found frequently [11]. The significance of this, if any, is not clear.

Studies on virus in blood, CSF and CNS

Several reports have attempted to demonstrate and quantify candidate viruses in CNS, CSF, and blood [29,30,31]. Previously, *in situ* hybridization and immunohistochemical methods were preferentially used [32], but in recent years PCR has been the preferred method [29,30,31]. Quantifying virus load and expression in mononuclear cells, serum, plasma, and the CNS may be of help in establishing an association with MS. Caution should be taken, however, in the interpretation of such data, as non-specific activation of viruses may occur. Thus, positive results do not necessarily indicate a disease association.

Treatment studies

The definitive indication that a specific agent is involved in the pathogenesis of MS is the ability to ameliorate clinical disease by specific chemotherapy against the agent. These studies are impeded by the lack of specific agents towards most viruses. Considering its general antiviral properties, treatment of MS patients with interferon-alpha was performed nearly twenty years ago [33], but significant efficacy on the clinical course of MS was not reached. In more recent studies, using interferon-beta and MRI evaluation, significant efficacy has been obtained. The mechanism behind the benefit of interferon-beta in MS is unknown, but has been suggested to be inhibition of the immune processes in the disease [34]. In a Swedish treatment study with aciclovir, which inhibits multiplication of certain members of the herpes virus family, a non-significant reduction of the attack rate was obtained [35].

Previous candidate virus in MS

As alluded to earlier, the general assumption has been that MS is initiated and triggered by infections with one of a number of common unspecified viruses [36,37]. A late infection was considered a prerequisite for explaining the migration studies [38]. However, observations from a case-control study indicate that MS patients recall the same number of several common infectious diseases as do healthy controls [39]. In addition, a recent study demonstrates that MS patients and their age, gender, and milieu-matched controls had their child diseases at the same age (Haahr et al., unpublished observations). Moreover, prevention of a number of diseases in childhood during the last 25–30 years, through vaccination against these diseases early in life, has apparently not reduced the number of new cases of MS in young people. Age of infection with common diseases does not seem to influence the tendency for development of MS.

An association of herpes simplex virus (HSV), varicellae-zoster virus (VZV) and cytomegalovirus (CMV) to MS, which has been suggested, is unlikely because the epidemiology of these viruses cannot explain the epidemiology of MS. Moreover, the frequency of seropositive MS patients and healthy controls to HSV and VZV is the same [40,41]. For CMV, a significantly lower number of MS patients are seropositive when compared with healthy controls [41]. The number of MS patients and healthy controls that suffer from cold sores and shingles do not differ, most likely indicating a comparable level of reactivation of HSV and VZV in patients and healthy controls (Haahr et al., unpublished observations).

The association of measles with MS, which was hypothesized several years ago, was discredited, when some MS patients got measles [8], indicating that they had no previous infection with measles virus. The suggestion of an association has now been abandoned after the observation that the number of persons developing MS has been unchanged despite widespread vaccination against measles throughout the last 25–30 years. The supposition that MS may follow vaccination with live measles vaccine can be excluded, because such a vaccination is given at the same age in various countries, therefore it cannot explain the migration studies. Furthermore, the epidemiology of natural measles virus infections deviates from the epidemiology of MS [42].

Association of MS with canine distemper virus, which is antigenically related to measles virus, has also been suggested [8]. As infections in humans with canine distemper virus are dependent on dogs infected with this virus, such an association is contradicted by the migration studies. Because of the high homology between canine distemper virus and measles virus proteins, it has been difficult to distinguish antibodies to distemper virus from those against measles virus [43].

Candidate viruses

Retrovirus

Retrovirus in animals are associated with demyelinating diseases. Probably the most important example for MS is the Visna-Maedi lentivirus infection in sheep in Iceland. This virus induces a strong cell-mediated immune response after a long period of incubation, followed by a chronic perivascular inflammation, and subsequent demyelination [44]. In humans, HTLV-I gives rise to a chronic inflammatory demyelinating disease (HAM/TSP), that in many ways resembles MS [45]. Nevertheless, a retrovirus causing MS has not been identified.

In 1989, a French research group described the production of extracellular virions associated with reverse transcriptase activity in a culture of leptomeningeal cells obtained from the cerebrospinal fluid of a patient with MS [46]. The same retrovirus was later demonstrated in macrophage cultures from patients with MS, and subsequently also in Epstein-Barr virus (EBV) immortalized B cells from MS patients. The virus was not observed in these cell types from non-MS persons [47]. Partial molecular characterization revealed that the retrovirus was related (about 75% homology) to the endogenous retrovirus ERV9. It is therefore unclear whether the virus is endogenous or exogenous [48]. Virus RNA was detected in circulating virions in serum from MS patients (9 of 17), but less frequently in non-MS controls (3 of 44) [48].

Almost simultaneously, a Danish research group observed retrovirus-like particles (RVLP) in T-cell lines from peripheral blood of an MS patient [49], and later from spontaneously developing B-lymphoblastoid cell lines from MS patients. These cell-lines expressed EBV together with RVLP [50,51]. Several *gag* and *env* fragments with homology to the human endogenous retrovirus RGH-2 have been identified in four MS cell lines. Expression of RGH-2 sequences at particle levels was detected in plasma from 22 of 31 MS patients, among these in plasma from all nine MS patients with active MS at the time of sampling. All healthy controls (n=18) and all patients with non-neurological diseases (n=29) were negative [52].

The significance of the expression of these endogenous retroviruses in MS is still unknown (e.g. cause or consequence of the disease) as is the relationship between the two reported viruses. The observations from the migration studies seem difficult to explain by endogenous retroviruses without involving other viruses as well.

Coronavirus

Coronaviruses are ubiquitous, respiratory pathogens involved in up to 30% of common colds in both children and adults. The virus may also be associated with diarrhoea and enterocolitis. All human isolates of coronavirus (HCV) can be grouped into two serotypes. However, certain animal coronaviruses cross-react serologically with the human strains [53]. In rodents, the neurotropic JHM strain of mouse hepatitis virus induces a chronic demyelinating encephalomyelitis, which in many ways resembles MS. This animal model has been used to study viral pathogenesis for a demyelinating disease [54,55].

The possible association of coronaviruses with MS evolved after isolation of coronavirus from two patients with MS [56]. Both isolates later turned out to be murine coronaviruses [57,58]. Sera from MS patients and matched controls had similar levels of antibody to both serotypes of HCV. In contrast, CSF from MS patients contained HCV antibodies more frequently and in higher titers than matched controls [59]. Two studies have demonstrated HCV in CNS from a few MS patients, especially in active demyelinating plaques, but not in CNS from neurological and healthy controls [60,61]. These observations have yet to be confirmed by others.

In order to evaluate a possible pathogenic mechanism of coronavirus infection in MS patients, myelin basic protein- and HCV-reactive T-cell lines were established from 16 MS patients and 14 healthy donors. A significantly higher number of cell lines from MS patients than controls showed cross-reactivity between myelin and HCV antigens. From this study it was suggested that molecular mimicry between HCV and myelin basic protein was a possible immunopathological mechanism in MS [62].

Infections with HCV occur at all ages with similar clinical features. Antibody studies have indicated 100% seropositivity for some HCV strains in an adult healthy population, but the serological data depends on the time that has elapsed since infection. Reinfection may occur with the same HCV strain [53]. Epidemiological knowledge on HCV infections is scanty and it is difficult to evaluate, whether the epidemiology of HCV may explain part of the epidemiology of MS.

Human herpes virus 6

Human herpes virus 6A (HHV-6A) was isolated from persons with lymphoproliferative syndromes in 1986 [63]. When HHV-6A is isolated, it is most often from chronically ill adults. Subsequently another variant named HHV-6B was isolated from children with the common childhood disease: exanthema subitum. HHV-6B has also been detected in serious diseases in adults. Like other herpesviruses,

HHV-6 establishes latency after primary infection, and reactivation of viral replication may occur in both immunocompromised and immunocompetent adults and children. The cell tropism for HHV-6 *in vivo* includes CD4+ T-cells, macrophages, endothelial and epithelial cells, neurones, oligodendrocytes, and glial cells of the brain and spinal cord. HHV-6 antigens appear to be present in a majority of brains from adult persons [64,65].

HHV-6 antibody studies in MS
A possible association of HHV-6 and MS has been studied in the last few years. Higher levels of antibody to HHV-6 have been demonstrated in MS patients than in controls [66,67,68,70,72]. Studies have demonstrated HHV-6 IgG in serum from all persons studied [69,70,71], whereas HHV-6 IgM was observed more often in the MS groups [69–72]. HHV-6 IgG antibody was detected more often in CSF from MS patients than in neurologic controls [68,70]. A similar antibody pattern has been seen in patients with chronic fatigue syndrome [72,73].

HHV-6 in cerebrospinal fluid and serum
HHV-6 DNA has been detected in acellular CSF from a few MS patients but never from healthy controls [70,74], others have failed to detect HHV-6 DNA in CSF [31,75,76]. In contrast, HHV-6 DNA was demonstrated in serum significantly more frequent among MS patients [71], whereas two studies showed an insignificant increase in the number of MS patients with HHV-6 DNA [74,75], and one study failed to detect HHV-6 DNA in MS patients [31]. In a recent publication the results from one laboratory were summarized: 25 out of 108 MS patients were positive, in contrast to none of 70 patients without MS [71,77].

HHV-6 in peripheral blood mononuclear cells
Using PCR, peripheral blood mononuclear cells (PBMC) were examined for the presence of HHV-6 in MS patients and healthy controls. HHV-6 DNA was found in only 3 out of 56 MS patients [30]. In later studies, PBMC was found positive for HHV-6 DNA more frequent, although the percentage of positive MS patients and healthy controls did not differ [72,77,78]. Active replication of HHV-6 has been demonstrated by co-cultivating PBMC and PHA-stimulated human cord blood cells. Measured this way, PBMC from 80% of MS patients in active phases of the disease presented HHV-6 activity but nearly the same percentage of MS patients in remission or healthy controls presented the same activity [72].

HHV-6 in brain tissue
PCR amplification of DNA extracted from autopsy brain material demonstrated HHV-6 in more than 70% of both MS and control brain samples [32]. However, one study failed to amplify HHV-6 sequences from MS brain samples, including plaques, although HHV-6 was amplified in two samples from spinal and brain plaques derived from a patient with a very acute form of neuromyelitis optica and also from brain tissue in four out of eight normal adults [30]. Using immunocyto-

chemistry staining (ICC), HHV-6B specific antigens were found only in the nuclei of oligodendrocytes in MS patients, especially in those associated with the MS plaques [32]. This observation could not be confirmed in a separate ICC study, applying the same monoclonal antibody [80]. In a recent study using ICC, HHV-6 was demonstrated in brain tissue from 7 of 15 MS patients and not in brain tissue from controls [69].

HHV-6 and MS – conclusions
Despite the high variability in the results in the above mentioned studies, it can be concluded that HHV-6 is highly neurotropic and occurs to nearly the same degree in the brains and PBMC of MS patients and healthy controls. Apparently, reactivation and expression of HHV-6 are more frequent and more intensive in some MS patients than in healthy controls. An intensive reactivation and multiplication of HHV-6 may also occur in immunocompromised persons [81].

Indications for an involvement of HHV-6A in MS are scarce. HHV-6A DNA was detected more frequently in serum than HHV-6B DNA, but in peripheral blood mononuclear cells HHV-6A was barely detected and always in conjunction with HHV-6B DNA [77]. In a study where HHV-6 only were detected in PBMC from a few of the MS patients and not in healthy controls all genomic sequences were derived from HHV-6A [82]. In contrast to healthy controls, lymphoproliferative responses to crude HHV-6A and -6B preparations, showed a higher level of response to HHV-6A than to -6B in cells from MS patients [77]. Other studies suggest an involvement of HHV-6B in MS [32,76]. Antibody analysis to HHV-6A and -6B in CSF has only detected responses to HHV-6B in MS patients [83]. However, at the age of three years, approximately 100% are seropositive to HHV-6 in serum [84]. Thus, the accumulated information strongly suggests that infection with HHV-6B occurs early in life among widely distributed populations around the world [65]. If this is also true for HHV-6A, HHV-6 may not be able to explain the migration studies in MS.

Epstein-Barr virus

EBV is the most studied member of the herpes virus group because of its broad clinical importance. In contrast to the subclinical infections early in life, primary EBV infections around and after puberty cause infectious mononucleosis (IM) in approximately 50% of individuals. This is accompanied with a marked lymphocytosis, indicating an intensive cellular immune activation. Transmission of EBV infections occurs by the oral route and the primary site of viral replication is in the oro-pharynx, where the virus establish a persistent infection of epithelial cells [85,86]. Infection of B lymphocytes by EBV results in a persistent, latent infection with polyclonal B-cell activation and immunoglobulin secretion [85,87]. B cells can be immortalized *in vitro* through EBV infection and are then referred to as B-lymphoblastoid cell lines (LCLs). Once the cells are immortalized, they usually maintain the EBV genome in a latent state with no production of EBV particles.

LCLs can easily be established spontaneously from persons with IM, probably because of a high number of EBV infected B-cells. Spontaneous establishment of LCLs from healthy persons is on the other hand an uncommon phenomenon [87,88,89,90]. As described later, B cells from MS patients have a higher tendency to establish LCLs spontaneously than do B cells from healthy controls [91,92].

EBV and acute demyelination
Demyelinating disease after neurologically complicated primary EBV infection has been reported. Five patients were described, who developed progressive or relapsing neurologic deficits following a neurologically complicated primary EBV infection. Four patients followed up to 12 years eventually got the diagnosis of MS. The fifth patient presented with acute disseminated MS followed by persisting neurologic deficits [93]. Previously, in a case report on MS beginning in infancy, elevated EBV antibody titers during the clinical course have been reported [94].

EBV antibody studies and MS
Noticeably, following the implementation of sensitive and specific tests for detecting antibodies to EBV, virtually 100% of MS patients are seropositive for EBV. This is in contrast to healthy controls, where a significantly lower number has been found seropositive [41,95,96,97,98,99,100]. In several studies a higher antibody titer toward EBV has been documented [41,95,97,99]. Antibodies toward other viruses have also been analysed in some of these studies: the same frequency of seropositivity to HSV, VZV, measles and mumps was observed in MS patients and controls, whereas a significantly lower seropositivity for CMV was found in MS patients [41]. In another study the same frequency of seropositivity to HSV, VZV and CMV in MS patients and healthy controls was observed [99]. It is not clear why virtually all MS patients have antibodies to EBV. The treatment of MS patients with steroid cannot explain this observation: young MS patients and newly diagnosed MS patients, who have only in a few cases received steroid treatment, were all seropositive to EBV (Haahr et al., unpublished observations). The absence of recent EBV infections in MS patients was striking when compared with a significantly higher number of recently infected healthy controls [98,100]. Antibodies to EBV are infrequently present in CSF from MS patients [70].

EBV expression and MS
The absence of primary EBV infections in MS patients is consistent with an etiologic role of EBV in the development of MS [98]. Recent studies on the reactivation of EBV indicate that this occurs to the same degree in healthy controls and MS patients [98,100], but of particular interest is the potential association of EBV reactivation and disease activity in MS. In a longitudinal study of beta-interferon-treated MS patients, serologic association of EBV antibodies with acute attacks was observed, but the antibody responses were thought to be non-specific [101]. In a recent study 19 MS patients were followed monthly for one year [100] and active viral replication as measured by serology and the precence of serum EBV

DNA was seen in most patients with disease activity during the study period in contrast to patients with stable disease. In agreement with these observations, we have detected EBV DNA in plasma from a few patients with active disease (Kusk et al., unpublished data).

EBV-specific RNA has been looked for in the brains of 10 patients with MS by *in situ* hybridization, however, none of 21 plaques were positive [102]. In a later study using PCR, EBV was detected in about one quarter of brains from both MS patients and healthy controls. EBV was preferentially detected in inactive plaques as opposed to active plaques [29].

EBV and spontaneous B cell transformation
In a study from 1979 on spontaneous transformation of lymphocytes from MS patients and healthy controls, B lymphocytes from patients with clinically active MS were found to transform spontaneously into EBV containing LCLs more readily than lymphocytes from patients with inactive disease or those from normal subjects [91]. It was speculated that this phenomenon was caused by an increased load of EBV in B cells from MS patients. From a patient with an MS-like disease, a spontaneously developing LCL was established in 1992, which by electron microscopy displayed EBV particles and retrovirus-like particles, indistinguishable from C-type retrovirus [50]. Later in a consecutive study, LCLs were spontaneously established from 5 of 21 MS patients but only from 1 of 13 healthy controls. Again LCLs developed earlier from PBMC derived from MS patients, especially those patients with active disease, when compared to the healthy control. All LCLs were found to produce reverse transcriptase and express retrovirus particles as well as EBV particles [92]. Although it is a well known phenomenon that virus from the herpes group, especially EBV, and the retrovirus group have a mutually transactivating potential [103], the mechanism for the dual virus production in the LCLs is unknown. It was speculated that such a dual infection also takes place *in vivo* in MS, and a hypothesis was put forward that MS is caused, in predisposed persons, by a dual infection with late transmitted EBV and a retrovirus [104].

Infectious mononucleosis and MS
In a case-control study [39] significantly more MS patients had a history of IM, suggesting an older age at exposure to EBV in MS patients than in controls. In a historical prospective study nearly seven thousand previously diagnosed IM patients were searched in the nation-wide Danish MS Registry. The risk ratio for these persons to develop MS was 2.8. The time span from having IM to developing MS was 2–20 years, with a mean period around 10 years [105]. Previously a small number of IM patients were matched with patients in a local MS registry, and also here a higher risk of MS was found to occur after IM [106]. Another case-control study indicated that recall of IM was associated with a **high**er risk ratio for developing MS [107].

EBV and MS migration studies

The observations on infectious mononucleosis and MS may be important in explaining the migration studies in MS. Thus, individuals in high MS prevalence areas may be infected later in life with EBV (develop infectious mononucleosis) than individuals in low prevalence areas. Indeed, studies from low risk areas have shown an early age of infection with EBV [108].

EBV and MS geographical cluster studies

In recently reported clusters from Denmark [109] EBV subtyping was performed on members of one cluster comprising eight persons. The same subtype of EBV was observed in members of this cluster out of six different subtypes observed in the study. Although a preponderance of this subtype was found both in MS patients and in healthy controls, the presence of this subtype in all studied cluster members is significantly different from the finding in healthy controls (n=16), which include eight schoolmates to the cluster members and eight randomly selected healthy persons [110]. In addition, this finding was significantly different from all non-clustered individuals studied (n=44) [110]. It is tempting to speculate that specific virulent subtypes or strains of EBV have a higher tendency to induce MS and in this way may explain the development of clusters. If so, this might also explain the variability in prevalence in genetic stable populations [8,9]. As previously mentioned, cultural patterns, e.g. pre-chewed food for babies, used in some cultural settings, will also lead to early infection with EBV and perhaps to a lower incidence of MS.

Several "epidemics" of MS have been reported. Such "epidemics" may theoretically be explained by infection with EBV. It is known that primary EBV infections frequently evolve in military settings, and that this may be followed by months of high-level excretion of EBV from the oropharynx of these individuals [28]. It is likely that dissemination of EBV into isolated communities may lead to a higher number of infectious mononucleosis cases in a native adult population and subsequently to a higher number of MS as documented [105]. Importantly, the interval that elapsed from the arrival of the servicemen (exposure to EBV) to the development of MS in the "epidemics" are comparable with the interval between infectious mononucleosis and MS obtained from the historical prospective study (ten years) [105].

EBV and MS socio-economic status

The wider circulation of EBV in lower socio-economic areas with poorer sanitation and hygiene provides more children with immunising subclinical infections, at an earlier age, thereby reducing their risk for developing clinical disease. If EBV plays a role in the development of MS, a higher number of MS cases should then exist in higher socio-economic settings. Indeed, high socio-economic status is in some part of the world associated with a higher prevalence of MS [8]. This status is also associated with a higher prevalence of paralytic polio [26] and clinical symptoms following hepatitis A infection [27].

Table 1. The ability of candidate viruses to explain epidemiological findings in MS.

MS epidemiology	Virologic parameters	HHV–6A	HHV–6B	HCV	ERV9[a]	RGH–2[a]	EBV
Prevalence	Geographical distribution of virus and specific strains	n.i.[b]	n.i.	n.i.	–[c]	–	n.i.
Migration	Primary infections around and after puberty	(+)	–	–	–	–	++++
Clusters	Virulence; strain specificity among cluster members	(+)	(+)	(+)	–	–	++
Epidemics	Likely transmission from servicemen	(+)	(+)	(+)	–	–	+
Socio-economic conditions	Late transmission of virus	n.i.	–	–	–	–	++++

[a] Potential exogenous transmission is not considered.
[b] n.i.: no information.
[c] Ability to explain MS epidemiology: – to ++++.

EBV and MS – conclusions

Review of the epidemiology of EBV infections indicates that EBV is a candidate agent that may well explain the migration studies, the geographical cluster studies, and the socio-economic studies in MS and hypothetically the MS "epidemics". Antibody studies demonstrate virtually 100% seropositivity in MS patients and only 90–95% among healthy controls. However, antibodies against EBV is not a regular finding in CSF, and the EBV genome or its viral transcripts are rarely found in the CSF or CNS. Thus if EBV infections can trigger MS, the initiating events are most likely taking place in the periphery. This would be consistent with an autoimmune pathogenesis of the disease initiated by an exogenous agent that fulfil the criteria set by the epidemiological findings. The potential mechanisms for a virus-mediated activation of the peripheral immune system remain an open-ended question. The point to make here is, that the absence of a virus in the CNS does not disqualify the virus as a potential etiologic agent in MS.

Discussion

A summary of the epidemiological observations for the candidate viruses is shown in Table 1. Little epidemiological information is available on these viruses and only for EBV, studies exist about the epidemiological relationship to MS. Thus EBV seems most attractive as an explanation of the various epidemiological observations in MS. Hypothetically, part of the geographical gradients and the epidemics of MS may be explained by infection with EBV. The potential impact of EBV infections on the migration studies and the occurrence of clusters and especially

Table 2. Serological findings in MS patients and healthy controls in relation to candidate viruses.

	Coronavirus			ERV9/RGH2			HHV-6			EBV		
	MS	C	Ref.	MS	C	Ref.	MS	C	Ref.	MS	C	Ref.
Serum IgG: frequency	+++	+++	53	n.i.*)	n.i.	n.i.	+++	+++	69,70,71	+++	++	41,95-100
Serum IgG: titre	+++	+++	59	n.i.	n.i.	n.i.	+++	++	66,67,68,70,72	+++	++	41,95,97,99
Serum IgM	n.i.	–		n.i.	n.i.	n.i.	++	–	69,70,71,72	+	–	100
CSF IgG	+	–	59	n.i.	n.i.	n.i.	++	–	68,70	+	–	70

*n.i.: no information.

Table 3. Detection of candidate viruses in MS patients and healthy controls

Material	Method	Coronavirus MS	Coronavirus C	Coronavirus Ref.	ERV9/RGH-2 MS	ERV9/RGH-2 C	ERV9/RGH-2 Ref.	HHV-6 MS	HHV-6 C	HHV-6 Ref.	EBV MS	EBV C	EBV Ref.
PBMC	virusisolation or electron microsc.	n.i.[a]	n.i.		+++[b] +++	–[c] (+)	47[d] 49,50,92[d]	+++	+++	72	++	(+)	91,92[d]
PBMC	genomic PCR	n.i.	n.i.		n.i.	n.i.		+++ +	+++ –	72,77,78 30	n.i.	n.i.	
Serum	genomic PCR	n.i.	n.i.		++ ++	(+) –	48 52	++ (+) –	– – –	71,77 74,75 31	+	–	100
CNS	genomic PCR	+	–	60,61	n.i.	n.i.		++++ –	++++ +	32 30	++ –	++ –	29 102[e]
CNS	ICC[f]	n.i.	n.i.		n.i.	n.i.		++ –	– –	32,69 80	n.i.	n.i.	
CSF	genomic PCR	n.i.	n.i.		n.i.	n.i.		(+) –	– –	70,74 31,75,76	n.i.	n.i.	

[a] n.i.: no information.
[b] relative occurrence of positive results: (+) to ++++.
[c] – indicates negative result.
[d] detected by electron microscopy in spontaneously transformed B-cell lines.
[e] *in situ* hybridisation.
[f] immunocytochemistry staining.

the socio-economic observations in MS have been further supported by studies and observations.

A summary of antibody studies in MS for the candidate viruses is given in Table 2. The most consistent and important observation to be noted is the 100% seropositivity for EBV in MS patients in contrast to healthy controls. For all the other candidate viruses both healthy controls and MS patients most likely are seropositive. The occurrence of serum IgM in HHV-6 and EBV, although not observed in all studies performed, have to be further analyzed, preferable in longitudinal studies in MRI-evaluated patient groups. IgM antibodies in serum from a high number of MS patients have previously also been demonstrated against measles [11]. The higher titre of IgG observed for HHV-6 and EBV in MS patients in contrast to healthy controls has also been observed for other viruses. This is an unexplained phenomenon [11]. Demonstration of antibodies against coronavirus and HHV-6 in CSF is interesting, but antibodies against e.g. measles and vaccinia virus is also found in CSF from MS patients in contrast to controls [11].

The detection of candidate viruses in blood cells and serum besides CNS tissue is presented in Table 3. The demonstration of the same load of HHV-6 in PBMC from MS patients and healthy controls is essential, whereas the observation of HHV-6 DNA in serum is still rather controversial and needs further confirmation. The recent observation of EBV DNA in serum also needs further confirmation. For these two viruses, longitudinal quantitative PCR studies need to be applied on MRI-evaluated MS patients. The spontaneous transformation of B lymphocytes from patients with clinically active MS is provocative and the mechanism for this needs definitely to be explained. Using PCR on CNS material, no significant differences in virus load of the candidate virus have been demonstrated. The presence of HHV-6 antigens in brain samples has been demonstrated by ICC in two studies whereas one study were unable to confirm these observations.

The efficacy of specific anti-viral chemotherapy is the ultimate indication of an involvement of virus in MS pathogenesis. In all cases this means a reduction in the number of new MRI lesions. Specific chemotherapy does not exist for the viruses we have discussed here, hence it is not possible yet to perform these studies. HHV-6 and EBV belong to the herpes virus group and treatment with aciclovir has efficacy on certain members of this group. Whether or not aciclovir may ameliorate disease in a subgroup of MS patients is not clear yet [35]. However, aciclovir has little effect on EBV replication [111] and less, if any on HHV-6 replication [112].

Two treatment studies with valaciclovir, a prodrug to aciclovir, have been initiated. A randomized, dual-blinded treatment study has been performed at the University Hospital in Aarhus, Denmark, in collaboration with the Sahlgrenska Hospital in Göteborg, Sweden. Seventy MS patients have been radomized to receive either valaciclovir 1 g × 3 daily for 6 months or placebo. Patients have been followed clinically and by MRI scanning each month. The study is finished and is presently under evaluation. Another clinical trial is taking place in New York and is scheduled to end this year. 60 patients have been treated for 2 years in this study with the same dosage as in the Danish-Swedish treatment study. If encouraging results are obtained, virus serology and quantitative virus evaluation may potentially give information on the potential role of herpesviruses in MS.

Acknowledgement

The studies from our group referred to in this review were funded by the Danish MS Society, The John and Birthe Meyer Foundation, The Johnsen and Wife Foundation and Director E. Danielsen and Wife Foundation.

References

1. Marie P. Sclérose en plaques et maladies infectueuses. Prog Med 1884;12:287–289.
2. Charcot JM. Histologie de la sclerose en plaque. Gaz Hosp (Paris) 1868;41:551–555.
3. Dalgleish AG. Viruses and multiple sclerosis. Acta Neurol Scand 1997;95S:8–16.
4. Martin R, McFarland HF, McFarlin DE. Immunological aspects of demyelinating diseases. Annu Rev Immunol 1992;10:153–87.
5. Wucherpfennig KW, Strominger JL. Molecular mimicry in T-cell-mediated autoimmunity: viral peptides activate human T-cell clones specific for myelin basic protein. Cell 1995;80:695–705.
6. Hafler DA. The distinction blurs between an autoimmune versus microbial hypothesis in multiple sclerosis. J Clin Invest 1999;104:527–529.
7. Kurtzke JF. Epidemiological evidence for multiple sclerosis as an infection. Clin Microbiol Rev 1993;6:382–427.
8. Martyn CN. The epidemiology of multiple sclerosis. In: McAlpin´s Multiple Sclerosis. Matthews WB (ed) Churchill Livingston: Edinburg, London, Melbourne and New York, 1991;3–40.
9. Kinnunen E. Multiple sclerosis in Finland: Evidence of increasing frequency and uneven geographic distribution. Neurology 1984;34:457–61.
10. Gale CR, Martyn CN. Migrant studies in multiple sclerosis. Prog Neurobiol 1995;47:425–48.
11. Norrby E. Viral antibodies in multiple sclerosis. Prog med Virol 1978;24:1–39.
12. Hauser S. Risk of MS. Inside MS; Spring 1998:44–47.
13. Sawcer S, Robertson N, Compston A. Genetic epidemiology of multiple sclerosis. In: AJ Thompson, C Polman, R Hohlfeld (eds) Multiple Sclerosis. Clinical Challenges and Controversies. London: Martin Dunitz, 1997:13–34.
14. Ebers GC, Bulman DE, Sadovnick AD et al. A population-based study of multiple sclerosis in twins. N Engl J Med 1986;315:1638–42.
15. Eldridge R, Herndon CN. Multiple sclerosis in twins. New Engl J of Med 1987;317:50.
16. Herndon CN, Jennings RG. A twin-family study of susceptibility to poliomyelitis. Am J Hum Genet 1951;3:17–46.
17. Knox EG. Detection of clusters. In: P Elliott (ed) Methodology of Enquiries into Disease Clustering. Proceedings of a meeting held on 22 April 1988 at the London School of Hygiene and Tropical Medicine. London: Small Area Health Statistics Unit, 1989;17–21.
18. Riise T, Grønning M, Klauber MR, Barrett-Connor E, Nyland, Albrectsen G. Clustering of residence of multiple sclerosis patients at age 13–20 years in Hordeland, Norway. Am J Epidemiol 1991;133:932–939.
19. Riise T. Cluster studies in multiple sclerosis. Neurology 1997;49 (suppl 2) S27–S32
20. Kurtzke JF, Hyllested K Multiple sclerosis in the Faroe Islands. I: Clinical and epidemiological features. Ann Neurol 1979;79:6–21.
21. Kurtzke JF, Hyllested K. Multiple sclerosis in the Faroe Islands. II: Clinical update, transmission, and the nature of MS. Neurology 1986;36:307–328.
22. Kurtzke JF, Gudmundsson KR, Bergmann S. Multiple sclerosis in Iceland: 1. Evidence of a

postwar epidemic. Neurology ;1982;32:143–150.
23. Poskanzer DC, Prenney LB, Sheridan JL, yonKondy J. Multiple Sclerosis in the Orkney and Shetland Islands I: Epidemiology, clinical factors, and methodology. J Epidemiol Commun Health 1980;34:229–239.
24. Martin JR. Troop-related multiple sclerosis outbreak in the Orkneys. J Epidemiol Commun Health 1987;41:183–184.
25. Rosati G, Aiello I, Granieri E. Incidence of multiple sclerosis in Macomer, Sardinia 1912–1981: onset of the disease after 1950. Neurology 1986;36:14–19.
26. Melnick JL. Enterovirus. In: Evans AS (ed) Viral Infections of Humans. Epidemiology and Control. 3rd Edition. New York and London: Plenum Book Company, 1989:191–263.
27. Hadler SC and Margolis HS. Viral Hepatitis. In: Viral Infections of Humans. Epidemiology and Control. Editor: Evans AS. 3rd Edition. New York and London: Plenum Book Company, 1989;351–384.
28. Evans AS and Niederman JC. Epstein-Barr virus. In: Viral Infections of Humans. Epidemiology and Control. Editor: Evans AS. 3rd Edition. New York and London: Plenum Book Company, 1989;263–95.
29. Sanders VJ, Felisan S, Waddell A, Tourtellotte WW. Detection of herpesviridae in postmortem multiple sclerosis brain tissue and controls by polymerase chain reaction. J Neurovirol 1996;2:240–58.
30. Merelli E, Bedin R, Sola P, Barozzi P, Mancardi GL, Ficarra G, Franchina G. Human herpes virus 6 and human herpes virus 8 DNA sequences in brains of multiple sclerosis patients, normal adults and children. J Neurol 1997;244:450–4.
31. Martin C, Enbom M, Söderström M, Frederkson S, Dahl H, Lycke J, Bergström T, Linde A. Absence of seven human herpesviruses, including HHV-6, by polymerase chain reaction in CSF and blood from patients with multiple sclerosis and optic neuritis. Acta Neurol Scand 1997;95:280–283.
32. Challoner PB, Smith KT, Parker JD, MacLeod DL, Coulter SN, Rose TM, Schultz ER, Bennett JL, Garber RL, Chang M. Plaque associated expression of human herpesvirus-6 in multiple sclerosis. Proc Natl Acad Sci USA 1995;92:7440–4.
33. Abb J, Deinhardt F, Zander H, Tenser RB, Rapp F, Goust JM, Fuder HH, Vilcek J, Ho M, Merigan TC, Oldstone MB, Jackson GG. Trials of interferon therapy for multiple sclerosis. J Infect Dis 1982;146:109–15.
34. Lublin FD. How effective is interferon-beta in multiple sclerosis. In: AJ Thompson, C Polman, R Hohlfeld (eds) Multiple Sclerosis. Clinical Challenges and Controversies. London: Martin Dunitz, 1997;35–42.
35. Lycke J, Svennerholm B, Hjemquist E, Frisén, Badr G, Andersson M, Vahlne A, Andersen O. Acyclovir treatment of relapsing-remitting multiple sclerosis. A randomized, placebo controlled, double-blind study. J Neurol 1996;243:214–224.
36. Sibley WA, Bamford CR, Clark K. Clinical viral infections and multiple sclerosis. Lancet 1995;1:1313–1315.
37. Andersen O, Lygner P-E, Bergström T. Viral infections trigger multiple sclerosis relapses: a prospective seroepidemiological study. J Neurol 1993;240:417–422.
38. Alter M, Zhen-xin Z, Davanpour Z, Sobel E, Zibulewski J, Schwartz G, Friday G. Multiple sclerosis and childhood infections. Neurology1986;36:1386–98.
39. Operskalski EA, Visscher BR, Malmgren RM, Detels R. A case-control study of multiple sclerosis. Neurology 1989;39:825–39.
40. Haahr S, Møller-Larsen A, Pedersen E. Immunological parameters in multiple sclerosis patients with special reference to the herpes group. Clin exp Immunol 1983;51:197–206.
41. Bray PF, Bloomer LC, Salmon VC, Bagley MH, Larsen PD. Epstein-Barr virus infection and antibody synthesis in patients with multiple sclerosis. Arch Neurol 1983;40:406–8.
42. Black FL. Measles. In: Evans AS (ed) Viral Infections of Humans. Epidemiology and Control. 3rd Edition. New York and London: Plenum Book Company, 1989:351–384.

43. Hodge MJ, Wolfson C. Canine distemper virus and multiple sclerosis. Neurolopgy 1997;Suppl 2:S62–9.
44. Sigurdsson B, Pålsson PA, vanBogaert L. Pathology of Visna. Transmissible Demyelinating Disease in Sheep in Iceland. Acta Neuropathol 1962;1:343–362.
45. Höllsberg P, Hafler DA. Seminars in medicine of the Berth Israel Hospital,Boston. Pathogenesis of diseases induced by human lyphotropic virus type 1 infection. N Engl J Med 1993;328:1173–82.
46. Perron H, Geny C, Laurent A, Mouriquand C, Pellat I, Perret J, Seigneurin JM. Leptomeningeal cell lines from multiple sclerosis with reverse transcriptase activity and viral particles. Res Virol 1989;140:551–61.
47. Perron H, Garson JA, Bedin F, Beseme F, Paranhos-Baccala G, Komurian-Pradel F, Mallet F, Tuke PW, Voisset C, Blond JL, Lalande B, Seigneurin JM, Mandrand B and The Collaborative Research Group on Multiple Sclerosis. Molecular identification of a novel retrovirus repeatedly isolated from patients with multiple sclerosis. Proc Natl Acad Sci USA 1997;94:7583–88.
48. Garson JA, Tuke PW, Giraud P, Paranhoe-Buccala G, Perron H. Detection of virion-associated MSRV-RNA in serum of patients with multiple sclerosis. The Lancet 1998;351:33
49. Haahr S, Sommerlund M, Nielsen R, Møller-Larsen A, Hansen HJ. Just another dubious virus in cells from a patient with multiple sclerosis. Lancet 1991;337:863–864.
50. Sommerlund, M., Pallesen, G., Møller-Larsen, A., Hansen, H.J., Haahr, S.: Retrovirus-like particles in an Epstein-Barr virus-producing cell-line derived from a patient with chronic progressive myelopathy. Acta Neurol Scand 1993;87:71–76.
51. Haahr S, Sommerlund M, Christensen T, Jensen AW, Hansen HJ, Møller-Larsen A. A putative new retrovirus associated with multiple sclerosis and the possible involvement of Epstein-Barr virus in this disease. Ann New York Acad Sci 1994;724:148–156.
52. Christensen T, Dissing Sørensen P, Riemann H, Hansen HJ, Møller-Larsen A. Expression of sequence variants of endogenous retrovirus RGH in particle form in multiple sclerosis. Lancet 1998;352:1033
53. McIntosh K. Coronaviruses. In: Fields Virology 3rd edition. Editor Fields NB, Knipe M, Howley PM. Lippencott-Raven Publishers, Philadelphia 1996:1095–1103.
54. Nagashima K, Wege H, Meyermann R, ter Meulen V. Coronavirus induced subacute demyelinating encephalitis in rats. A morphological analysis. Acta Neuropathol 1978;44:63–70.
55. Erlich SS, Fleming JO, Stohlman SA, Weiner LP. Experimental neuropathology of chronic demyelination induced by a JHM virus variant. Arch Neurol 1987;44:839–42.
56. Burks JS, DeVald BL, Jankovsky LD, Gerdes JC. Two coronaviruses isolated from central nervous system tissue of two multiple sclerosis patients. Science 1980;209:933–934.
57. Gerdes JC, Klein I, DeVald BI, Burks JS. Coronavirus isolates SK and SD from multiple sclerosis patients are serologically related to murine coronaviruses A59 and JHM and human coronavirus OC43, but not to human coronavirus 229E. J Virology 1981;38:231–38.
58. Weiss SR Coronaviruses SD and SK share extensive nucleotide homology with murine coronavirus MHV-A59, more than that shared between human and murine coronavirus. Virology 1983;126:669–77.
59. Salmi A, Ziola B, Hovi T, Reunanen M. Antibodies to coronaviruses OC43 and 229E in multiple sclerosis patients. Neurology 1982;32:292–5.
60. Stewart JN, Mounir J, Talbot PJ. Human coronavirus gene expression in the brains of multiple sclerosis patients. Virology 1992;191:502–5.
61. Murray RS, Brown B, Brian D, Cabirac GF. Detection of coronavirus RNA and antigen in multiple sclerosis brain. Ann Neurol 1992;31:525–33.
62. Talbot PJ, Paquette JS, Ciurli C, Antel JP, Ouellet F. Myelin basic protein and human coronavirus 229E cross-reactive T cells in multiple sclerosis. Ann Neurol 1996;39:233–40.
63. Salahuddin SZ, Ablashi DV, Markham PD, Josephs SF, Sturzenegger S, Kaplan M, Halligan Biberfeld P, Wong-Staal F, Kramarsky B, Gallo RC. Isolation of a new virus, HBLV, in patients with lymphoproliferative disorders. Science 1986;234:596–601.

64. Salahuddin SZ, The discovery of human herpesvirus type 6. In DV Ablashi, GFR Krueger, and SZ Salahuddin (ed): Human herpesvirus 6: epidemiology, molecular biology, and clinical epidemiology, vol. 4. Amsterdam, The Netherland: Elsevier Biomedical Press, 1992;p.3–8.
65. Braun DK, Dominguez G, Pellett, PE. Human Herpesvirus 6. Clin Microbiol Rew 1997;10:521–67.
66. Wilborn F, Schmidt CA, Brinkmann V, Jendroska K, Oettle H, Siegert W. A potential role for human herpesvirus type 6 in nervous system disease. J Neuroimmunol 1994;49:213–214
67. Sola P, Merelli E, Marasca R, Poggi M, Luppi M, Montorsi M, Torelli G. Human herpesvirus 6 and multiple sclerosis: survey of anti-HHV-6 antibodies by immunoflourescense analysis and of viral sequences by polymerase chain reaction. J Neurol Neurosurg Psychiatry 1997;56:917–919.
68. Friedman JE, Lyons MJ, CU G, Ablasi DV, Whitman JE, Edgar M, Koskiniemi M, Vaheri A, Zabriskie JB. The association of the human herpesvirus 6 and MS. Mult Scler 1999;5:355–62.
69. Liedtke W, Malessa R, Faustmann PM, Eis-Hubinger AM. Human herpesvirus 6 polymerase chain reaction findings in human immunodeficincy virus associated neurological disease and multiple sclerosis. J Neurovirol 1995;1:253–8.
70. Ablashi DV, Lapps W, Kaplan M, Whitman JE, Richert JR, Pearson GR. Human herpes virus-6 infection in multiple sclerosis: a preliminary report. Mult Scler 1998;4:490–6.
71. Soldan SS, Berti R, Salem N, Secchiero P, Flamand L, Calabresi PA, Brennan MB, Maloni HW, McFarland HF, Lin HC, Patnaik M, Jacobson S. Association of human herpes 6 with multiple sclerosis: increased IgM response to HHV-6 early antigen and detection of serum HHV-6 DNA. Nat Med 1997;3:1394–7.
72. Ablashi DV, Eastman HB, Owen CB, Roman MM, Friedman J, Zabriskie JB, Peterson DL, Pearson GR, Whitman JE. Frequent (HHV-6) reactivation in multiple sclerosis (MS) and chronic fatigue syndrome (CFS) patients. J Clin Virol 2000;16:179–191.
73. Patnaik M, Komaroff L, Conley E, Ojo-Amaize EA, Peter JB. Prevalence of IgM antibodies to human herpesvirus 6 early antigen (p41/38) in patients with chronic fatigue syndrom. J Infect Disease 5;172:1364–67.
74. Fillet A-M, Lozeron P, Agut H, Lyon-Caen O, Libleu R. HHV-6 and multiple sclerosis. Nature Medicin 1997;4:537
75. Goldberg SH, Albright AV, Lisak RP, Gonzàlez-Scarano F. Polymerase chain reaction analysis of human herpesvirus-6 sequences in the sera and cerebrospinal fluid of patients with multiple sclerosis. J Neurovirol 1999;5:134–139.
76. Taus C, Pucci E, Cartechini E, Fie A, Giuliani G, Clementi M, Menzo A (2000) Absence of HHV-6 and HHV-7 in cerebrospinal fluid in relapsing-remitting multiple sclerosis. Acta Neurol Scand 2000;101:224–228.
77. Akhyani N, Berti R, Brennan MB, Soldan SS, Eaton JM, McFarland HF, Jacobsen S. Tissue Distribution and Variant Characterization of Human Herpesvirus (HHV)-6: Increased Prevalence of HHV-6A in Patients with Multiple Sclerosis. J Infect Dis 2000;182:1321–1325.
78. Mayne M, Krishnan J, Metz L, Nath A, Auty A, Sahai BM, Power C. Infrequent detection of human herpesvirus 6 in peripheral blood mononuclear cells from multiple sclerosis patients. Ann Neurol 1998;44:391–4.
79. Soldan SS, Leist TP, Juhng KN, McFarland HF, Jacobsen S. Increased lympho-proliferative respnse to human herpesvirus type 6A variant in multiple sclerosis patients. Ann Neurol 2000;47:306–313.
80. Coates ARM, Bell J. HHV-6 and multiple sclerosis. Nature Medicine 1997;4:537–8.
81. Blumberg BM, Mock DJ, Powers JM, Ito M, Assouline JG, Baker JV, Chen B, Goodman AD. The HHV-6 paradox: ubiquitus commensal or insidious pathogen? A two-step in situ PCR approach. J Clin Virol 2000;16:159–179.
82. Kim JS, Lee KS, Park JH, Kim MY, Shin WS. Detection of human herpesvirus 6 variant A in peripheral blood mononuclear cells from multiple sclerosis patients. Eur Neurol 2000;43:170–3.

83. Ongradi J, Rajda C, Marodi CL, Csiszar A, Vecsei L. A pilot study on the antibodies to HHV-6 variants and HHV-7 in CSF of MS patients. J Neurovirol 1999;5:529–532.
84. Nielsen L, Vestergaard BF. Competitive ELISA for detection of HHV-6 antibody: seroprevalence in a Danish population. J Virol Methods 1996;56:221–30.
85. Rickenson AB, Kieff E. Epstein-Barr virus. In: Fields NB, Knipe DM, Howley PM. (eds) Fields Virology 3rd edition. Lippencott-Raven Publishers, Philadelphia 1996:2397–2446.
86. Sixbey JW, Vesterinen EH, Nedrud JG, Raab-Traub N, Walton NLA, Pagano JS. Replication of Epstein-Barr virus in human epithelial cells infected *in vitro*. Nature 1983; 306:480–83.
87. Amen P, Lewin M, Nordström M, Klein G. EBV-activation of human lymphocytes. Curr Topic Microbiol Immunol 1986,132:266–71.
88. Pattengale PK, Gerber P, Smith RW. Selective transformation of B lymphocytes by EBV. Lancet 1973;2:1153–55.
89. Nilsson K. The nature of lymphoid cell lines and their relationship to the virus. In: Epstein MA, Achong BGT (eds The Epstein-Barr virus. Berlin: Springer-Verlag, 1979:225–62.
90. Yao , Rickinson AB, Epstein MA. A re-examination of the Epstein-Barr virus carrier state in healthy seropositive individuals. Int J Cancer 1985;35:35–42.
91. Fraser KB, Haire M, Millar JHD, McCrea S. Increased tendency to spontaneous in-vitro lymphocyte transformation in clinically active multiple sclerosis. The Lancet 1979;2:715–17.
92. Munch M, Møller-Larsen A, Christensen T, Morling N, Hansen HJ, Haahr S. B-lymphoblastoid cell-lines established from multiple sclerosis patients and a healthy control producing a putative new retrovirus and Epstein-Barr virus. Multiple Sclerosis 1995;1:78–81.
93. Bray PF, Culp KW, McFarlin DE, Panitch HS, Torkelsen RD, Schlight JP. Demyelinating disease after neurologically complicated primary Epstein-Barr virus infection. Neurology 1992;42:278–82.
94. Shaw CM, Alvord EC Jr. Multiple sclerosis beginning in infancy. J Child Neurol 1987;2:252–6.
95. Samaya CV, Myers LW, Ellison GW. Epstein-Barr virus antibodies in multiple sclerosis. Arch Neurol 1980;37:94–6.
96. Larsen PD, Bloomer LC, Bray PF. Epstein-Barr virus nuclear antigen and viral capsid antigen antibody titers in multiple sclerosis. Neurology 1985;35:435–8.
97. Sumaya CV, Myers LW, Ellison GW Ench Y. Increased prevalence and titer of Epstein-Barr virus antibodies in patients with multiple sclerosis. Ann Neurol 1985;17:371–7.
98. Munch M, Riisom K, Christensen, Møller-Larsen A, Haahr S. The significance of Epstein-Barr virus seropositivity in multiple sclerosis patients? Acta Neurol Scand 1998;97:171–4.
99. Myhr KM, Riise T, Barren-Connor E, Myrmel H, Vedeler C, Grønning M, Kalvenes MB, Nyland H. Altered antibody pattern to Epstein-Barr virus but not to other herpesviruses in multiple sclerosis: a population based case-control study from western Norway. J Neurol Neurosurg Psychiatry 1998;64:539–42.
100. Wandinger K-P, Jabs W, Siekhaus A, Bubel S, Trillenberg P, Wagner H-J, Wessel K, Kirchner H, Henning H Association between clinical disease activity and Epstein-Barr virus rectivations in MS. Neurology 2000;55:178–184.
101. Panitsch HS. Influence of infections on exacerbations of multiple sclerosis. Ann Neurol 1994;36:325—S28
102. Hilton DA, Love S, Fletcher A, Pringle JH. Absence of Epstein-Barr virus RNA in multiple sclerosis as assessed by in situ hybridization. J Neurol Neurosurg Psychiatry 1994;57:975–6.
103. Evermann JF, Derse D, Dern PL. Interactions between herpesviruses and retroviruses: implications in the initiation of disease. Microbial Pathogenesis 1991;10:1–9..
104. Haahr S, Sommerlund M, Møller-Larsen A, Mogensen S, Andersen HMK. Is multiple sclerosis caused by a dual infection with retrovirus and Epstein-Barr virus? Neuroepidemiology 1992;11:299–303.
105. Haahr S, Koch-Henriksen N, Møller-Larsen A, Eriksen LS, Andersen HMK. Increased risk of multiple sclerosis after late Epstein-Barr virus infection: a historical prospective study. Multiple

Sclerosis 1995;1:73–77.
106. Lindberg C, Andersen O, Vahlne A, Dalton M, Rumaker B. Epidemiological investigations of the association between infectious mononucleosis and multiple sclerosis. Neuroepidemiology 1991;10:62–65.
107. Martyn CN, Cruddas M, Compston DAS. Symptomatic Epstein-Barr virus infection and multiple sclerosis. J Neurol Neurosurg Psych 1993;56:167–68.
108. Leogrande G, Jirillo E. Studies on the epidemiology of child infections in the Bari area (South Italy) VII. Epidemiology of Epstein-Barr Virus infections. Eur J Epidemiology 1993;9:368–72.
109. Haahr S, Munch M, Christensen T, Møller-Larsen A, Hvas J. Cluster of multiple sclerosis patients from Danish community. Lancet 1998;349:923.
110. Munch M, Hvas J, Christensen T, Møller-Larsen A. Haahr S. A single subtype of Epstein-Barr virus in members of multiple sclerosis clusters. Acta Neurol Scand 1998;98:395–398.
111. Colby BM, Show JE, Elion GB, Pagano JS. Effect of acyclovir [9-(2-hydroxyethoxymethyl) guanine] on Epstein-Barr virus DNA replication. J Virol 1980;34:560–568.
112. Collins P. The spectrum of antiviral activities of acyclovir in vitro og in vivo. J Antimicrob Chemother 1983;12 Suppl B:19–27.

Active human herpesvirus six viremia in patients with multiple sclerosis*

Konstance K. Knox[†,1]; Joseph H. Brewer[2] and Donald R. Carrigan[1]

[1]*Institute for Viral Pathogenesis; 10437 Innovation Drive; Suite 417; Milwaukee, Wisconsin 53226*
and [2]*Division of Infectious Diseases; St. Luke's Hospital; Kansas City, Missouri*

Introduction

Human herpesvirus six (HHV-6) is perhaps the most neuroinvasive member of the human herpesvirus family [1]. Numerous studies have implicated CNS infection with HHV-6 as a major cause of seizures in children [2–6]. Other reports have suggested a role for HHV-6 in more severe neurological disease in children, including disseminated demyelination [7] and infarction of the basal ganglia [8]. Fatal HHV-6 encephalitis has been described in an HIV infected infant [9].

HHV-6 encephalitis has also been documented in bone marrow transplant patients [10–17], and the encephalitis can be fatal [10,11,14] although it is amenable in some cases to antiviral drug therapy [12,15]. Also, HHV-6 can cause encephalitis in liver transplant recipients [18,19] and chronic, demyelinating myelopathy in immunologically intact adults [20]. HHV-6 frequently infects the CNS tissues of patients with AIDS [21,22], and this infection has been proposed as the cause of some cases of AIDS dementia [21]. Recently, HHV-6 infections have been associated with focal encephalitis in immunologically intact adults and children with the clinical manifestations ranging from transient signs and symptoms of CNS dysfunction to death [23,24].

Interestingly, when the neuropathological changes associated with HHV-6 infections of the CNS have been analyzed, the most consistent findings have been demyelination ranging from diffuse and extensive loss of myelin [10,21] to sharply circumscribed foci of demyelination [20–24] combined with destruction of axons within areas of the most severe pathological changes [10]. These HHV-6 associated CNS disease patterns become especially interesting in the context of multiple sclerosis (MS) since MS is also associated with prominent demyelination combined with axonal destruction [25,26].

Previous work from our laboratory has documented the case of a young woman who died of clinically and histopathologically proven acute MS who had a dense and active HHV-6 infection of her brain and spinal cord [27]. The virus infected

---

*This work was supported by grant PP0543 from the National Multiple Sclerosis Society. Informed consent was obtained from the patients involved in these studies, and guidelines for human experimentation of the US Department of Health and Human Services were followed.

[†]*Correspondence address:* Dr. Konstance Knox, Institute for Viral Pathogenesis, 10437 Innovation Drive, Suite 417, Milwaukee, Wisconsin 53226, USA. Tel: (414) 774-8305. Fax: (414) 453-7295. E-mail: kknox@hhv6.com

cells were intimately associated with areas of active demyelination and were not seen in CNS areas free of demyelinative changes. In other work [28] active HHV-6 infections were detected in close association with active demyelination in the CNS tissues of 8 of 11 (73%) patients with definite MS. Further, active HHV-6 infections were present in the lymphoid tissues of 6 of 9 (67%) of MS patients, and blood samples from 22 of 41 (54%) definite MS patients were found to contain active HHV-6 infections compared with 0 of 61 normal control individuals.

These studies have now been extended by analyzing blood samples from additional patients with definite MS seen at three independent medical centers for the presence of active HHV-6 viremia. Additionally, one patient with MS has now been longitudinally followed with weekly or biweekly blood cultures for active HHV-6 for approximately one year.

Materials and methods

Patients and study design

This study was designed as a cross-sectional study of active HHV-6 viremia in three separate cohorts of patients with definite MS. Three medical centers provided the patient blood samples, and these were designated Center 1, Center 2 and Center 3. Blood samples were obtained from the patients during routine hospital visits and were shipped immediately to the laboratory for analysis. All specimens were analyzed within 48 hours of the time they were obtained. One patient with definite MS (30 year old male with relapsing/remitting disease) was longitudinally studied for approximately one year with weekly or biweekly blood samples. Control blood samples were obtained from 32 healthy laboratory and hospital workers and 48 normal blood donors.

Rapid HHV-6 culture assay

The rapid cell culture assay used in these studies has previously been used to diagnose active HHV-6 infections in bone marrow and liver transplant recipients and in patients with MS [28,29,30,31]. In this procedure, purified blood leukocytes from the patient are cocultivated with human diploid fibroblasts. Then, the fibroblasts are stained by indirect immunofluorescence with an affinity purified rabbit polyclonal antibody specific for the major immediate early protein of HHV-6. Positivity of a sample for an active HHV-6 infection is demonstrated by the presence of two or more fibroblasts with brilliant fluorescent staining restricted entirely to the cell nucleus. The infection in the patient's leukocytes must be active since the infection has to be transferred into the target fibroblasts. This technique has a sensitivity and specificity of 86% and 100%, respectively when compared to isolation of HHV-6 by cocultivation of patient samples with mitogen stimulated blood mononuclear cells [31].

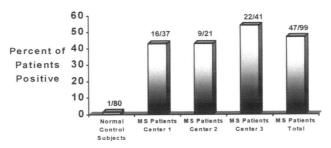

Fig. 1. Rate of positivity of patients with MS at three different medical centers for active HHV-6 viremia. Blood samples were analyzed for active HHV-6 infection by a rapid culture assay. Differences in positivity rates were statistically identical for the three centers, but all were significantly higher than the rate seen in normal controls ($p < 0.0001$ by two sided Fisher's Exact Test).

Critical review of literature concerning role of HHV-6 in MS

In order to evaluate the results of these studies and place them in the context of other published reports on the role of HHV-6 in the pathogenesis of MS, a MEDLINE search of the National Library of Medicine for two MESH headings [(1) Multiple Sclerosis and (2) Herpesvirus Six, Human] was performed. A total of 39 articles were identified by this search, and these were classified as follows: (1) 15 review articles/commentaries, (2) 2 case reports, (3) 5 basic science reports and (4) 17 diagnostic technique applications. The articles describing diagnostic technique applications were retrieved and analyzed with respect to the methodologies used and the conclusions drawn.

Results

Cross sectional study of active HHV-6 viremia in patients with MS

The results obtained in the cross sectional study of active HHV-6 viremia in patients with MS are summarized in Fig. 1. The patients with MS seen at all three medical centers had similar rates of active HHV-6 viremia with the percent positive ranging from 43% to 54%, and all three percentages were statistically indistinguishable from one another (p values of 1.00, 0.38 and 0.59). In contrast, the active infection rates at all three centers, as well as that of the total MS patients, were significantly higher ($p < 0.0001$ by two sided Fisher's Exact Test) than that observed with the normal control subjects.

Longitudinal study of active HHV-6 viremia in a patient with MS

The pattern of active HHV-6 viremia observed in a patient with MS over one year is shown in Fig. 2. The viremia was intermittently present during the observation period. Overall, 40% (16/40) blood samples were positive for active HHV-6 viremia, with little discernible pattern. However, there was an extended period

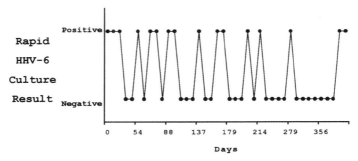

Fig. 2. Longitudinal course of active HHV-6 viremia in a 30 year old male with relapsing/remitting MS. Blood samples were obtained weekly or biweekly for the period of time shown and were analyzed for active HHV-6 infection by a rapid culture assay. Positivity rate during interval Day 0 through Day 98 (67%; 8/12) was significantly ($p < 0.01$ by two sided Fisher's Exact Test) higher than that seen for interval Day 221 through 377 (8%; 1/12).

(156 days, day 221 through day 377)) in which only 1 of 12 (8%) of the blood samples tested was positive which contrasts with an earlier period of increased viral activity (98 days; day 0 through 98) in which 8 of 12 (67%) of the samples tested were positive. The blood sample positivity rates between these two time periods were significantly ($p < 0.01$ by two sided Fisher's Exact Test) different from one another.

Critical review of literature concerning role of HHV-6 in MS

Numerous previous studies have investigated the possible role of HHV-6 in MS with contradictory results. The results of these investigations can be best interpreted in the context of the diagnostic technology used. A summary of these studies is shown in Table 1.

In studies using diagnostic technologies that could not distinguish between active and latent HHV-6 infections, i.e. PCR analysis of blood leukocytes, CSF containing cells or CNS tissue, essentially no differences were found between samples from patients with MS and control individuals. Interpretation of these negative data with respect to the pathogenesis of MS is unclear and is made even more uncertain by the wide range of positive results seen with normal cells and tissues, e.g. normal leukocyte positivity rates of 5% to 95% and normal CNS tissue positivity rates of 15% to 85%.

Studies in which classical serological methods were used, i.e. viral specific serum IgG titers and viral specific serum IgM detection, suggest a special role for HHV-6 in MS. Six of 8 studies showed either increased HHV-6 IgG titers or a higher rate of positive HHV-6 IgM in MS patients compared to controls. An additional study detected HHV-6 IgG in samples of CSF from patients with MS but not in controls, supporting the idea of a special role for the virus. It should be noted that, while the serological methods used in these studies suggest a special role for HHV-6 in MS, results obtained for any one individual must be interpreted

Table 1. Summary of Current Literature Regarding the Role Of HHV-6 in MS.

Interpretation of Positive Result	Method	Patient Sample	Supportive of Role for HHV-6 in MS	Not Supportive of Role for HHV-6 in MS	Comment
Technique cannot distinguish between latent and active infection [60]	PCR[a]	Blood Leukocytes	None	28[b],57,58, 59	Controls: 14% (13/92) + MS: 15% (21/143) + Range with normal controls is 5% [61] to 95% [62] +
	PCR	Cellular CSF[c] and CNS Tissues	63	27,57,64, 65,66,67	Controls: 43% (67/157) + MS:44% (60/136) + Range with normal controls is 15% [68] to 85% [69] +
May or may not indicate active infection	HHV-6 IgG Titers	Serum	28,29,64	71	Studies used both ELISA and immunofluorescence procedures
	HHV-6 IgM Detection	Serum	30,33,63	64	Controls: 16% (25/156) + MS: 32% (24/74) +
	HHV-6 IgG Detection	CSF	72	None	OND Controls: 0% (0/20) + MS: 27% (15/55) +
Only positive when active infection is present	PCR	Serum	33,67	29,31	Controls: 0% (0/77) + MS: 16% (16/101) + Non-supportive studies used no + serum control Non-supportive studies used no control for inhibition of PCR by substances in serum
	PCR	Acellular CSF[d]	29,30,64	70	Controls: 0% (0/59) + MS: 9% (9/101) + Non-supportive study controlled for inhibition with sample spike at 500 times lower limit of sensitivity
	IHC Staining	CNS Tissue	27,35,63	None	Controls: 0% (0/81) + MS: 48% (26/54) +

[a]Polymerase chain reaction.
[b]Number of reference.
[c]Untreated cerebrospinal fluid (CSF).
[d]CSF centrifuged prior to processing to remove cells present.

with caution since healthy individuals can have high titers of HHV-6 IgG and can occasionally be positive for HHV-6 IgM. Conversely, patients with MS who have an active HHV-6 viremia may have low titers of HHV-6 IgG and may be negative for HHV-6 IgM [Knox and Carrigan, unpublished observations].

PCR analysis of serum samples yielded mixed results with some studies suggesting a special role for HHV-6 in MS and others failing to show such an association. Importantly, the two negative serum PCR studies failed to include positive serum controls, raising the question of whether their procedures could detect HHV-6 DNA in serum. Also, neither of these two studies controlled for inhibition of the PCR reaction by substances in the patient specimens.

Three of four studies using PCR analysis of acellular CSF samples demonstrated a clear increased positivity of the MS patients compared to other neurologic disease (OND) controls. While the negative acellular CSF study used a control for PCR inhibition, the control DNA used was 500 times higher than the lower limit of sensitivity of the PCR technique. Significant inhibition of the PCR reaction could have occurred that would not have been detected.

The three positive investigations using PCR analysis of acellular CSF samples deserve special comment. In the three studies, 14% (3/21) [47], 11% (4/36) [41] and 17% (2/12) [49] of CSF samples from patients with MS were positive for HHV-6 DNA demonstrating active infection within their CNS tissues [54,55,56,57]. This reduced rate of positivity compared to that observed with immunohistochemical staining of CNS tissues [28,39,40] probably reflects the relatively low level of active infection present in the CNS tissues of MS patients [28] compared to that seen in the brains of immunocompromised patients with HHV-6 encephalitis [10,21]. It is well known that analysis of CSF samples by PCR in cases of herpes simplex encephalitis can give false negative results if the sample is obtained too early in the disease course when the infection is focal and limited in size [54,55,56,57].

All three independent studies using immunohistochemical staining of CNS tissues from patients with MS and controls, demonstrated active HHV-6 infections only in the MS patients. Also, striking is the fact that the percentage of MS patients who were positive in the three studies were quite similar, i.e. 50% (16/32) [40], 73% (8/11) [28] and 47% (7/15) [39].

Thus, in summary, when appropriate diagnostic technologies are used that are restricted to the detection of active HHV-6 infections, a strong relationship between HHV-6 and the pathogenesis of MS is reproducibly observed.

Discussion

HHV-6 has become well established as an important neuropathogen and has been strongly associated with acute, and frequently fatal, encephalitis in both immunocompromised and immunologically intact individuals [1–24]. The major pathologic change seen in CNS tissues actively infected with HHV-6 is various degrees of demyelination [7,10,21,23,24]. HHV-6 can cause a persistent demyelinating myelitis similar to tropical spastic paraparesis [20], and at least some cases of acute MS can more accurately be described as subacute HHV-6 leukoencephalitis [27].

Numerous studies have sought to explore the possibility of HHV-6 playing a special role in the pathogenesis of MS (reviewed in Table 1). Investigations employing inappropriate diagnostic technologies, such as cellular DNA based

PCR analyses, which cannot distinguish between active and latent HHV-6 infections have failed to demonstrate such a role. In contrast, studies using diagnostic techniques that can only detect active HHV-6 infections and that use specimens derived from the CNS of patients with MS, such as PCR of acellular CSF or immunohistochemical staining of CNS tissues, have almost uniformly demonstrated a special role of the virus in MS.

The studies described here have extended previous investigations [28] into the occurrence of active HHV-6 infections in the periphery of patients with MS. Analysis of blood sample from patients with MS seen at three separate medical centers demonstrated active HHV-6 viremia in 47% (47/99) of the patients. Normal control individuals had only a very low incidence of active HHV-6 viremia, i.e. approximately 1% (1/80).

In patients with MS, the relationship between active HHV-6 viremia and the active HHV-6 infections observed in close association with demyelination in CNS tissues remains to be defined. However, the episodic nature of the HHV-6 viremia in an MS patient with relapsing/remitting disease (Fig. 2) suggests that the active infection in the peripheral blood of the patients may be responsible for dissemination of the infection into the CNS. If this proves to be the case, effective intervention with antiviral drug therapy may be able to suppress the peripheral infection and thereby, decrease or prevent new disease activity within the CNS of the patients.

References

1. Kimberlin DW and Whitley RJ. Human herpesvirus 6: neurologic implications of a newly described viral pathogen. J Neurovirol 1998;4:474–485.
2. Hall CB, Long CE, Schnabel KC, Caserta MT et al. Human herpesvirus 6 infection in children: A prospective study of complications and reactivations. N Engl J Med 1994;331:432–438.
3. Caserta MT, Hall CB, Schnabel K et al. Neuroinvasion and persistence of human herpesvirus 6 in children. J Infect Dis 1994;170:1586–1589.
4. Portolani M, Cermelli C, Moroni A, Bertolani MF, Di Luca D, Cassai E, Sabbatini AM. Human herpesvirus 6 infections in infants admitted to hospital. J Med Virol 1993;39:146–151.
5. Kondo K, Nagafuji H, Hata A, Tomomori C, Yamanishi K. Association of human herpesvirus 6 infection of the central nervous system with recurrence of febrile convulsions. J Infect Dis 1993;167:1197–1200.
6. Ward KN and Gray JJ. Primary human herpesvirus 6 infection is frequently overlooked as a cause of febrile fits in young children. J Med Virol 1994;42:119–123.
7. Kamei A, Fujiwara T, Hiraga S, Onuma R and Ichinohe S. Acute disseminated demyelination due to primary human herpesvirus-6 infection. Eur J Pediatr 1997;156(9):709–712.
8. Webb DW, Thomas EE, Hukin J, Sargent MA and Bjornson BH. Basal ganglia infarction associated with HHV-6 infection. Arch Dis Child 1997;76(4):362–364.
9. Knox KK, Harrington D and Carrigan DR. Fulminant human herpesvirus six (HHV-6) encephalitis in an HIV infected infant. J Med Virol 1995;45:288–292.
10. Drobyski WR, Knox KK, Majewski D and Carrigan DR. Fatal encephalitis due to variant B human herpesvirus 6 infection in a bone marrow transplant recipient. N Engl J Med 1994;330:1356–1360.
11. Yanagihara K, Tanaka-Taya K, Itagaki Y, Toribe Y, Arita K, Yamanishi K and Okada S. Human herpesvirus 6 meningoencephalitis with sequelae. Pediatr Infect Dis J 1995;14:240–242.

12. Mookerjee BP and Vogelsang G. Human herpes virus 6 encephalitis after bone marrow transplantation: successful treatment with ganciclovir. Bone Marrow Transpl 1997;20:905–906.
13. Rieux C, Gautheret-Dejean A, Challine-Lehmann D, Kirch C, Agut H and Vernant JP. Human herpesvirus 6 meningoencephalitis in a recipient of an unrelated allogeneic bone marrow transplantation. Transplantation 1998;65:1408–1411.
14. Bosi A, Zazzi M, Amantini A, Cellerini M, Vannucchi AM, De Milito A, Guidi S, Saccardi R, Lombardini L, Laszlo D and Rossi Ferrini P. Fatal herpesvirus 6 encephalitis after unrelated bone marrow transplant. Bone Marrow Transpl 1998;22:285–288.
15. Cole PD, Stiles J, Boulad F, Small TN, O'reilly RJ, George D, Szabolcs P, Kiehn TE and Kernan N. Successful treatment of human herpesvirus 6 encephalitis in a bone marrow transplant recipient. Clin Infect Dis 1998;27:653–654.
16. Tsujimura H, Iseki T, Date Y, Watanabe J and Kumagai K. Human herpesvirus 6 encephalitis after bone marrow transplantation: magnetic resonance imaging could identify the involved sites of encephalitis. Eur J Haematol 1998;61:284–285.
17. Wang FZ, Linde A, Hagglund H, Testa M, Locasciulli A and Ljungman P. Human herpesvirus 6 DNA in cerebrospinal fluid specimens from allogeneic bone marrow transplant patients: does it have clinical significance? Clin Infect Dis 1999;28:562–568.
18. Singh N, Gayowski T, Carrigan DR and Singh J. Human herpesvirus 6 associated febrile dermatosis with thrombocytopenia and encephalopathy in a liver transplant recipient. Transplantation 1995;60:1355–1357.
19. Paterson D, Singh N and Carrigan DR. Human herpesvirus six encephalitis and liver transplantation. Liver Transplant Surgery 1999;in press.
20. MacKenzie I, Carrigan DR, Wiley CA. Chronic myelopathy associated with human herpesvirus six. Neurology 1995;45:2015–2017.
21. Knox KK and Carrigan DR. Active human herpesvirus six (HHV-6) infection of the central nervous system in patients with AIDS. J Immune Defic Syndr and Hum Retrovir 1995;9:69–73.
22. Saito Y, Sharer LR, Dewhurst S, Blumberg BM, Hall CB and Epstein LG. Cellular localization of human herpesvirus 6 in the brains of children with AIDS encephalopathy. J Neurovirol 1995;1:30–39
23. McCullers JA, Lakeman FD and Whitley RJ. Human herpesvirus 6 is associated with focal encephalitis. Clin Infect Dis 1995;21:571–576.
24. Novoa LJ, Nagra RM, Nakawatase T, Edwards-Lee T, Tourtellotte WW and Cornford ME. Fulminant demyelinating encephalomyelitis associated with productive HHV-6 infection in an immunocompetent adult. J Med Virol 1997;52:301–308.
25. Raine CS. The Norton Lecture: a review of the oligodendrocyte in the multiple sclerosis lesion. J Neuroimmunol 1997;77(2):135–152.
26. Trapp BD, Peterson J, Ransohoff RM, Rudick R, Mork S and Bo L. Axonal transection in the lesions of multiple sclerosis. N Engl J Med 1998;338:278–285.
27. Carrigan DR, Harrington D and Knox KK. Subacute leukoencephalitis caused by CNS infection with human herpesvirus six manifesting as acute multiple sclerosis. Neurology 1996;47:145–148.
28. Knox KK, Brewer JH, Henry JM, Harrigton DJ and Carrigan DR. Human herpesvirus six and multiple sclerosis: Systemic active infections in patients with early disease. Clin Infect Dis, submitted for publication.
29. Singh N, Gayowski T and Carrigan DR. Human herpesvirus six (HHV-6) infection in liver transplant recipients: Documentation of pathogenicity. Transplantation 1997;64:674–678.
30. Singh N and Carrigan DR. Human herpesvirus 6 in transplantation: An emerging pathogen. Ann Intern Med 1996;124:1065–1071.
31. Carrigan DR, Milburn G, Knox K, Kernan N, Papadopoulos E, and Singh N. Diagnosis of active human herpesvirus six (HHV-6) infections in immunocompromised patients with a rapid shell-vial assay. Amer Soc Microbiol, 1996.
32. Carrigan DR. Human herpesvirus six (HHV-6) and bone marrow transplantation. Blood

1995;85:294–295.
33. Sola P, Merelli E, Marasca R, Poggi M et al. Human herpesvirus 6 and multiple sclerosis: survey of anti-HHV-6 antibodies by immunofluorescence analysis and of viral sequences by polymerase chain reaction. J Neurol Neurosurg Psychiatry 1993;56:917–919.
34. Merelli E, Bedin R, Sola P, Barozzi P, Mancardi GL, Ficarra G and Franchini G. Human herpesvirus 6 and human herpesvirus 8 DNA sequences in brains of multiple sclerosis patients, normal adults and children. J Neurol 1997;244:450–454.
35. Mayne M, Krishnan J, Metz L, Nath A, Aury A, Sahai BM and Power C. Infrequent detection of human herpesvirus 6 DNA in peripheral blood mononuclear cells from multiple sclerosis patients. Ann Neurol 1998;44:391–394.
36. Rotola A, Cassai E, Tola MR, Granieri E and Di Luca D. Human herpesvirus 6 is latent in peripheral blood of patients with relapsing remitting multiple sclerosis. J Neurol Neurosurg Psych 1999;67:529–531
37. Wilborn F, Schmidt CA, Zimmerman R, Brinkmann V, Neipel F and Siegert W. Detection of herpesvirus type 6 by polymerase chain reaction in blood donors: random tests and prospective longitudinal studies. Brit J Haematol 1994;88:187–192.
38. Cone RW, Huang MLW, Ashley R and Corey L. Human herpesvirus 6 DNA in peripheral blood cells and saliva from immunocompetent individuals. J Clin Microbiol 1993;31:1262–1267.
39. Friedman JE, Lyons MJ, Cu G, Ablashi DV, Whitman JE, Edgar M, Koskiniemi M, Vaheri A and Zabriskie JB. The association of the human herpesvirus 6 and MS. Mult Scler 1999;5:134–139.
40. Challoner PB, Smith KT, Parker JD et al. Plaque associated expression of human herpesvirus 6 in multiple sclerosis. Proc Natl Acad Sci 1995;92:7440–7444.
41. Liedtke W, Malessa R, Faustmann PM and Eis-Hubinger AM. Human herpesvirus 6 polymerase chain reaction findings in human immunodeficiency virus associated neurological disease and multiple sclerosis. J Neurovirol 1995;1:253–258.
42. Sanders VJ, Felisan S, Waddell A and Tourtelotte W. Detection of herpesviridae in postmortem multiple sclerosis brain tissue and controls by polymerase chain reaction. J Neurovirol 1996;2:249–258.
43. Martin C, Enbom M, Soderstrom M, Fredrikson S, Dahl H, Lycke J, Bergstrom T and Linde A. Absence of seven human herpesviruses, including HHV-6, by polymerase chain reaction in CSF and blood from patients with multiple sclerosis and optic neuritis. Acta Neurol Scand 1997;95:280–283.
44. Goldberg SH, Albright AV, Lisak RP and Gonzales-Scarano F. Polymerase chain reaction of human herpesvirus 6 sequences in the sera and cerebrospinal fluid of patients with multiple sclerosis. J Neurovirol 1999;5:134–139.
45. Liedtke J, Trubner K and Schwechheimer. On the role of human herpesvirus 6 in viral latency in nervous tissue and cerebral lymphoma. J Neurol Sci 1995;134:184–188.
46. Luppi M, Barozzi P, Maiorana A, Marasca R and Torelli G. Human herpesvirus 6 infection in normal brain tissue. J Infect Dis 1994;169:943–944.
47. Wilborn F, Schmidt CA, Brinkmann V, Jendroska K et al. A potential role for human herpesvirus type 6 in nervous system disease. J Neuroimmunol 1994;49:213–214.
48. Nielson L, Moller-Larsen A, Munk M and Vestergaard BF. Human herpesvirus 6 immunoglobulin G antibodies in patients with multiple sclerosis. Acta Neurol Scand 1997;Supplement 169:76–78.
49. Ablashi DV, Lapps W, Kaplan M, Whitman JE, Richert JR and Pearson GR. Human herpesvirus 6 (HHV-6) infection in multiple sclerosis: a preliminary report. Multiple Sclerosis 1998;490–496.
50. Soldan SS, Berti R, Salem N, Secchiero P, Flammand L, Calabresi P, Brennan MB, Maloni HW, McFarland HF, Lin HC, Patnaik M and Jacobson S. Association of human herpes virus 6 (HHV-6) with multiple sclerosis: Increased IgM response to HHV-6 early antigen and detection of serum HHV-6 DNA. Nature Med 1997;3:1394–1397.
51. Enbom M, Wang FZ, Fredrikson S, Martin C, Dahl H and Linde A. Similar humoral and cel-

lular immunological reactivities to human herpesvirus 6 in patients with multiple sclerosis and controls. Clin Diag Labor Immunol 1999;6:545–549.
52. Martin C, Linde A, Bergstrom T, Lycke J, Dahl H, Fredrikson S, Soderstrom M, and Enbom M. Absence of seven human herpesviruses, including HHV-6, by polymerase chain reaction in CSF and blood from patients with multiple sclerosis and optic neuritis. Acta Neurol Scand 1997;95(5):280–283.
53. Mirandola P, Stefan A, Brambilla E, Campadelli-Fiume G and Grimaldi LME. Absence of human herpesvirus 6 and 7 from spinal fluid and serum of multiple sclerosis patients. Neurology 1999;53:1367–1368.
54. Kimberlin DW, lakeman FD, Arvin AM, Prober CG, Corey L, Powell DA, Burchett SK, Jacobs RF, Starr SE, Whitley RJ and the National Institute of Allergy and Infectious Disease Collaborative Antiviral Study Group. Application of the polymerase chain reaction to the diagnosis and management of neonatal herpes simplex virus disease. J Infect Dis 1996;174:1162–1167.
55. Atkins JT. HSV PCR for CNS infections: pearls and pitfalls. Pediatr Infect Dis 1999;18:823–824
56. Tang YW, Mitchell PS, Espy MJ, Smith TF and Pershing DH. Molecular diagnosis of herpes simplex virus infections in the central nervous system. J Clin Microbiol 1999;37:2127–2136.
57. Zunt JR and Marra CM. Cerebrospinal fluid testing for the diagnosis of central nervous system infection. Neurol Clin 1999;17:675–689.

Retroviruses in multiple sclerosis

T. Christensen[1,*], P. Dissing Sørensen[2], H.J. Hansen[3] and A. Møller-Larsen[1]

[1]*Department of Medical Microbiology and Immunology, University of Aarhus, DK-8000 Aarhus C, Denmark;* [2]*Biotechnological Institute, Kogle Alle 2, DK-2970 Hørsholm, Denmark;* [3]*Department of Neurology, Aarhus University Hospital, DK-8000 Aarhus C, Denmark*

Introduction

The aetiology of the neurodegenerative disease multiple sclerosis (MS) remains uncertain. Both with respect to genetic susceptibility and to pathology, MS appears to be heterogenic. Among the circumstantial evidence pointing towards an infectious implication in MS is the intrathecal synthesis of IgG/the presence of oligoclonal bands in the cerebrospinal fluid; the increased titers to virus/bacteria; the recruitment of activated T cells, B cells and macrophages to the central nervous system, and epidemiological studies demonstrating only 30% concordance in homozygotic twins as well as the influence of pre-pubertal migration on the risk of later development of MS [1]. The environmental agents contributing to the development of MS in genetically predisposed individuals could presumably be one or more (retro)viruses [2].

Following the observation of retrovirus-like particles in short-term cultured lymphocytes from MS patients, spontaneously formed long-term cell cultures from peripheral blood mononuclear cells were established from several MS patients. These cells produce type C-like retrovirus particles in addition to expression of Epstein-Barr virus (EBV) proteins and – rarely – mature EBV particles[3,4]. The retrovirus particles are reverse transcriptase (RT) positive [5,6] and share a few antigenic determinants with HTLV-1, but are distinct from the known exogenous retroviruses at the antigenic level [5].

We have optimized a method for ultracentrifugation-based purification of retroviral particles in self-forming Optiprep gradients. RT activity, presence of retroviral particles in EM, and RNA content co-locate in the gradients [7]. This method enables purification of intact virions without loss of surface glycoproteins or hyperosmotic damage and has facilitated both the sequence analyses and immunologic characterization presented here.

Sequence variants of the human endogenous retrovirus HERV-H/RGH-2 were demonstrated by RT-PCR on virion genomic RNA and thus represent the possibly replication-competent retrovirus particles produced by MS cell lines. Variant HERV-H sequences with high homology to HERV-H/RGH-2 were also found specifically at particle level in cell-free plasma from MS patients [8].

Correspondence address: Tove Christensen, Department of Medical Microbiology and Immunology, University of Aarhus, DK-8000 Aarhus C, Denmark. Tel: +45 89 42 17 68. Fax: +45 86 19 12 77.

The recent chromosomal meta-analyses for MS susceptibility indicate that a number of chromosomal regions may be involved. We performed a database search through the currently known human genomic sequences and found several representative copies of both HERV-H and the endogenous retrovirus HERV-W [9] in chromosomal regions of interest.

The HERV-H/RGH-2 virion proteins could theoretically be encoded by other retroviral sequences in the genome; we therefore immunized Wistar rats with purified virions and could demonstrate, that a specific serologic response developed towards HERV-H/RGH-2 encoded synthetic peptides, indicating that at least some virion proteins are encoded by HERV-H.

We have previously demonstrated increased seroreactivity in MS sera towards a HTLV-1-derived polypeptide (HTLV-1 is related to HERV-H) [5].

We suggest that MS is associated with replication of otherwise quiescent endogenous retroviruses. This could be a phenomenon parallel to the association of HTDV/HERV-K with male germ cell tumours[10,11] and in accordance with the recent report of expression of another putatively endogenous retrovirus in MS[12,13], and of the putative expression of a superantigen, encoded by a HERV-K-variant (IDDMK1,222) in type 1 diabetes [14].

Materials and methods

Plasma

All blood samples were obtained with informed consent. The samples were from healthy volunteers or patients from the Neurology Dept., Aarhus University Hospital; the Dermatology Dept., Marselisborg Hospital, Aarhus; the MS Hospital in Ry; or the Dept. of Internal Medicine, Middelfart Hospital (all in Denmark). Samples were drawn in the respective clinics and delivered by hand to us immediately after drawing. All samples were processed in our laboratory. Cell- and debris-free plasma samples were obtained from 40 ml citrate blood samples (apr. 10 ml plasma). After an initial low speed centrifugation, the plasma was aspirated and subjected to another centrifugation at 4°C, 30 min., 2500 x g to pellet remaining cellular debris. Following aspiration, plasma samples were treated as described below.

MS cell lines

All blood samples were obtained with informed consent. The blood samples were from MS patients and long-term cell cultures from relapsing-remitting, primary progressive, or secondary progressive types of MS were established as described previously [4].

Particle RNA purification

Retroviral RNA was purified from cell culture supernatants and from plasma samples using the Optiprep method as described elsewhere; Optiprep is a iodinated, nonionic density gradient medium (Nycomed Pharma, Norway) which, in contrast to sucrose, maintains intact retroviral particles. Briefly, the method involves low speed centrifugation followed by ultracentrifugation of the supernatant onto an Optiprep cushion, filtration through a 0.45µm filter to remove debris, and finally Optiprep gradient centrifugation [7].

RT activity was measured for each gradient fraction, and poly-A RNA was purified by lysing the 5–6 fractions containing major RT activity directly in 5 vol lysis-binding buffer (Dynal, Norway), adding 100 µl beads/sample (mRNA Direct kit, Dynal, Norway) and following the manufacturers instruction.

Cell-free *plasma samples* were layered on 50% Optiprep, ultracentrifuged and filtered as described above. The retrovirus/Optiprep mixture was directly lysed in 5 vol. lysis-binding buffer (Dynal, Norway) and poly-A RNA was purified on 50 µl beads/sample using the mRNA-Direct kit (Dynal, Norway) as above.

If not used directly, the RNA-coupled beads were stored in 80% EtOH at –80°C. Before use, each RNA sample was treated with amplification grade DNAse (Life Technologies) according to the manufacturers instructions. After cooling on ice the DNAse was removed by exchanging the buffer containing the RNA-complexed paramagnetic beads with DEPC-treated ddH$_2$O.

Product enhanced reverse transcriptase (PERT) assay

Reverse transcriptase activity was monitored via ultrasensitive RT assays (PERT (Product Enhanced Reverse Transcriptase)). PERT assays were performed directly on 5 µl gradient fractions, essentially as described [6,15].

RT-PCR, cloning and sequencing

RT-PCR was performed on virion RNA using the GeneAmp RNA PCR kit (Perkin Elmer) according to the manufacturers suggestions, with 1 û of Taq polymerase per reaction. First strand synthesis was primed with random hexamers, in conjunction with the bead-oligo-dT complexed RNA-template. In some experiments, we ensured the integrity of the RNA template by cDNA synthesis, specifically primed with the downstream primer; in this case, control cDNA synthesis primed with the corresponding upstream primer was negative.

Please refer to Christensen et al.[8, 16] for the RT-PCR conditions and primers.

In every assay, water controls corresponding to each step were included together with duplicates of each reaction without RT, to ensure that no DNA-templates were amplified. PCR products were analyzed by agarose gel electrophoresis, and each product was cloned in pUC using the SureClone Ligation kit (Pharmacia) and sequenced using the ABI Prism kit (Applied Biosystems) on an automatic

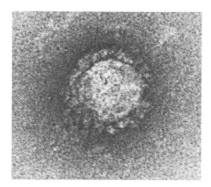

Fig. 1. An Optiprep purified virion from the MS cell line MS1851. The size is about 100 nm. Note the intact envelope glycoproteins.

sequenator (ABI 377).

Wistar rat immunization

Virus suspension for immunization was prepared from 800 ml supernatant from the cell line MS1859. The virus were purified and concentrated in Optiprep gradients as previously described [7]. The purified virions were suspended in 1.75 ml TNE buffer and mixed with equal amounts of Freunds adjuvant, using the complete adjuvant for the initial injections and incomplete adjuvant for the succeeding boosting. Preimmune sera were collected before immunization. Eight weeks old female Wistar rats were injected subcutaneously with 0.5 ml of the virus-adjuvant mixture. The rats were reinjected after 4 weeks and again after further 2 weeks. 10 days after the final boost 1.5 ml of peripheral blood was collected from each rat for preparation of serum samples and subsequent measuring of antiviral antibodies. Peptides were synthesized by Genosys Sigma, UK. Peptide-based capture ELISAs were performed essentially as described by Lombardi et al. [17].

Results

Virions

Particles were purified from the long-term cell culture supernatants by ultracentrifugation in Optiprep gradients. Their retroviral origin was confirmed by negative staining electron microscopy (EM) on each gradient fraction, and by the sensitive PERT (Product Enhanced Reverse Transcriptase) assays[6,15] also performed on each gradient fraction. The presence of RNA in the particles was demonstrated by chasing the culture with [5-^{3}H]-uridine(Amersham) for 24 hrs before harvest, followed by TCA precipitation of each gradient fraction, and counting in a β-counter. PERT activity, presence of retroviral particles in EM, and RNA content co-located in the gradients [7]. Fig. 1 shows an Optiprep purified retrovirus particle from the

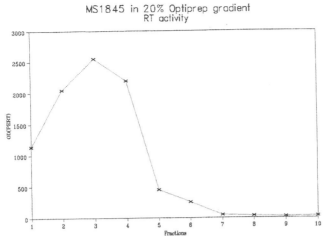

Fig. 2. Reverse transcriptase activity (PERT assay) on gradient fractions from the MS cell line MS1845.

cell line MS1851, consistent with a type C morphology. Most of the surface glycoproteins are still attached to the virions/viral particles purified by this method.

The presence of the characteristic retroviral enzyme reverse transcriptase (RT), an RNA-dependent DNA polymerase, is a major criterium for the verification of the retroviral origin of the particles. RT activity can be monitored by conventional assays but due to the low virus production more easily by highly sensitive polymerase chain reaction (PCR) based assays. These assays can detect less than 100 molecules of RT. Fig. 2 demonstrates the distribution of RT activity in the 10 first gradient fractions of a standard Optiprep purification (MS1845 in this case). An 11.2 ml gradient is harvested as 21 fractions. The fractions from 10–21 are RT negative (not shown).

We recently published detailed descriptions of RT analyses and virion purification[7,16].

HERV-H/RGH-2 sequences in retroviral particles produced by MS-derived cell lines

Using RT-PCR on retroviral RNA templates isolated from cell culture supernatants, with subsequent sequencing of the cloned amplicons, we initially identified several *gag* and *env* fragments with a high degree of homology to HERV-H/RGH-2 in the retroviral particles produced by the MS cell lines. The range of homology to RGH-2 was between 80–95%[8,16].

Furthermore, we tested MS cell line particle RNA's, purified by our standard procedure, using the reported ERV-9/HERV-W related nested primerset ST1-1/-2 and RT-PCR conditions [12]. The results were negative.

Table 1. RT-PCR for HERV-H/RGH-2 in particles in plasma samples.

Plasma sample type	n	MS Diagnosis	% positive by RT-PCR
Multiple sclerosis	9	RR[a]	78
Multiple sclerosis	7	PP[b]	57
Multiple sclerosis	15	SP[c]	73
Total MS	31	na[d]	71
Other diseases	29	na	0
Healthy control	18	na	0

All MS patients had clinically definite MS. [a]RR: relapsing-remitting, [b]PP: primary progressive, [c]SP: secondary progressive, [d]na: not applicable. The other disease group contain 27 autoimmune disease patients with diabetes mellitus, prurigo nodularis Hide, colitis ulcerosa, SLE, arteritis temporalis, rheumatoid arthritis, thyreotoxicosis, sarcoidosis, collagenosis, polymyositis, myelomatosis, and polymyalgia rheumatoides. Two patients had leukaemia.

RGH sequences specifically present in plasma from MS patients

To determine if HERV-H/RGH-2 sequences at particle level could also be associated with MS *in vivo*, we subsequently performed RT-PCR analyses on clinical specimens, i.e. cell-free, filtered, ultracentrifuged plasma samples from MS patients, from patients with autoimmune diseases, and from healthy controls. RNA extracted from particles isolated from plasma was assayed by RT-PCR with subsequent sequencing of the cloned amplicons. The results of the plasma RT-PCR and sequencing analyses are presented in Table 1. Variants of HERV-H/RGH-2 was demonstrated in 22 of 31 MS patients but could not be demonstrated in 29 plasma samples from patients with autoimmune or other diseases nor in 18 plasma samples from healthy controls [8,16]. 9 of the MS patients had active MS defined as chronic progressive MS with no stationary phase, or exacerbations in relapsing/remitting MS and all of these were positive. Expression of the sequences at particle level was thus found both in the virus particles from cell culture supernatants and in the particulate fraction of plasma from different MS patients and was specific for MS.

Genomic HERV-H/RGH-2 sequences

Chromosomal meta-analyses for MS susceptibility genes have shown that no single gene or gene family is responsible for MS development, but that several chromosomal regions may be contributing to MS susceptibility. As endogenous retroviruses represent putative aethiological factors both as genes and as viruses in MS, an updated search in the human genome databases was performed to localize genomic copies of HERV-H. We found several homologous copies on for example chromosome 2,5,7,10,11,14,16,19, and X and Y. The published HERV-H/HERV-H19 clone with an intact Env ORF [18] has a very high degree of homology to HERV-H/RGH-2 and was included in the database search. HERV-H/RTVL-H copies contain pol and pol-env deletions and are not included here. Figs. 3a and b

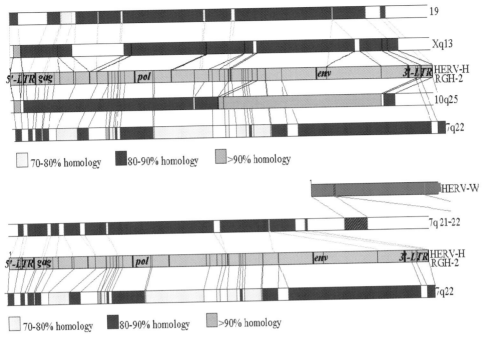

Fig. 3. Genomic alignments of HERV-H/RGH-2. a) shows alignment with regions on chromosome 10, 19, and X. b) shows alignment with two regions on chromosome 7. The position of an adjacent potential copy of HERV-W is also indicated. Hatched areas indicate homology.

show genomic alignments of some of the HERV-H/RGH-2 copies with the highest homology scores. Fig. 3b also includes alignment with a region on chromosome 7q21-22 where a HERV-H copy is found in close proximity (less than 2 kb) to a putative HERV-W copy. However, the HERV-H copy contains a large pol-env deletion and the HERV-W copy contains insertions; and it should be noted, that none of the available chromosomal sequences corresponds to versions of either HERV-H or HERV-W with intact open reading frames for Gag, Pol and Env together. It is presently unknown how many of the genomic copies are transcribed or translated or if complementation occurs between the molecules.

Virion proteins: immunologic characterization

The packaging of HERV-H/RGH-2 RNA at particle level is clearly demonstrated in the material from cell cultures by the co-migration in Optiprep gradients of retrovirus particles, morphologically identifiable by EM, and RT activity. It remains, however, to be assessed to whether all the retroviral proteins in the virions are encoded by HERV-H.

Indications that some of the virion proteins are encoded by HERV-H-like sequences were obtained through the following experiments: Wistar rats, immu-

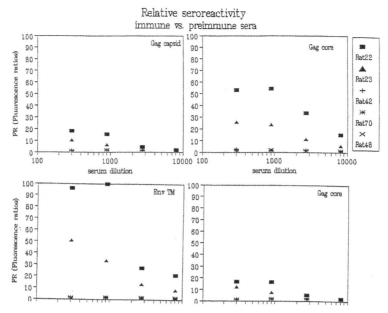

Fig. 4. Shows the relative seroreactivity of sera from two Wistar rats (rat 22 and rat 23) immunized with virions from the MS cell line MS1859 as compared to the seroreactivity of sera from 3 Wistar rats (rat 42, rat 48, rat 70) immunized with irrelevant antigens (components of the complement activation system in this case, rat 42 and 70 were immunized with recombinant proteins, rat 48 was immunized with a peptide). The 4 different peptides used for capture in the ELISAs are derived from the HERV-H/RGH-2 regions indicated.

nized with Optiprep purified MS cell culture virions in Freunds adjuvant, developed a specific response towards synthetic peptides, derived from HERV-H/RGH-2 sequences. The serologic response of the virion-immunized rats was analyzed in conjunction with the serologic response of Wistar rats immunized by the same procedure but with irrelevant antigens. Examples are shown in Fig. 4.

Discussion

Endogenous retroviruses are primarily transmitted vertically, and are found ubiquitously in mammalian genomes of which they constitute several percent. It has been suggested, that endogenous retroviruses may play a role in autoimmune diseases [19,20]. Apart from MS, in addition to the previously mentioned implication of a HERV-K variant in insulin dependent diabetes mellitus (IDDM) [14], reports have been published on a putatively exogenous retrovirus in Sjögren's syndrome [21,22], and on HERVs in SLE (systemic lupus erythematosus) [23], and in RA (rheumatoid arthritis) [24]. Only once before has a human endogenous retrovirus, HTDV/HERV-K, been clearly demonstrated as particles with possible pathological connotation – in teratocarcinoma cell lines [10]. It has not yet been determined, whether HTDV/HERV-K occurs as particles in patient plasma. Table 2 gives an

Table 2. Human endogenous retroviruses in autoimmune disease.

HERV	Copies/haploid genome	Size, kb	PBS	Structure	Autoimmune disease association
Class I					
HRES-1	1	?	His	LTR-gag-pol	systemic lupus erythematosus
ERV-9	40	9.6	Arg	LTR-gag-pol-env-LTR	multiple sclerosis, rheum. arthritis
HERV-W	100?	7.6	Trp	LTR-gag-pol-env-LTR	multiple sclerosis, rheum. arthritis
HERV-H: RTVL-H	900	5.8	His	LTR-gag- pol-LTR	
RGH	**50–100**	**8.7**	**His**	**LTR-gag-pol-env-LTR**	**multiple sclerosis**
Class II					
HERV-K	50	9.2	Lys	LTR-gag-pol-env/rev-LTR	type 1 diabetes, rheumatoid arthritis
HERV-L	200	6.5	Leu	LTR-?gag-pol-?env-LTR	rheumatoid arthritis

It should be noted that the class I HERVs implied in autoimmune diseases are related to HTLV-1/-2.

overview of the HERVs implied.

The HERV-H (priming of transcription with tRNA^{His}) family of type-C like endogenous retrovirus is related to the onco-retroviruses human T-cell leukaemia virus (HTLV-1/-2), to bovine leucosis virus (BLV), and to ERV-9/HERV-W. A HERV-H clone with an Env ORF was reported recently [18] and HERV-H mRNA is found in normal lymphocytes, placental tissue and some neoplasias[25–28], but HERV-H/RGH virions as such have not been reported previously. RGH constitute a subfamily as there are differences in genome size and LTR organization as compared with other members of the HERV-H group. Originally, two RGH clones were published: a truncated form (RGH-1) lacking *gag* and part of *pol*, and a full-length form (RGH-2) with complete coding potential: 5'LTR *gag-pol-env* 3'LTR. RGH-like sequences were reported present in about 100 copies/haploid genome [29].

Here, we document a specific association between MS and HERV-H. HERV-H sequence variants, highly homologous to RGH-2, were demonstrated at particle level in supernatants from MS-derived cell cultures and in particulate form specifically in the plasma from about 70% of MS patients. A database search on the available human genome sequences demonstrated several HERV-H/RGH-2 copies in chromosomal regions implied in MS susceptibility in the United Kingdom genome screen [30].

Previously, we have also demonstrated that in about 70% of MS sera, there is serologic cross-reactivity towards an HTLV-1-derived Env polypeptide (HERV-H is related to HTLV-1). This serological study included 48 MS patients, 19 patients with other neurological diseases and 57 healthy controls [5].

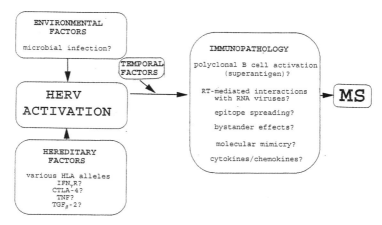

Fig. 5. Flow chart illustrating hypotheses for HERV activation as the cause of MS.

Both the PCR-based study and the serological study indicate a correlation between MS disease activity and HERV-H expression.

The specific serological response towards HERV-H/RGH-2 encoded peptides which developed in Wistar rats immunized with purified virions indicate that some of the virion proteins are encoded by HERV-H, but part of the virion-proteins may also be encoded by other retroviral sequences in the genome, as heterologous RNA co-packaging in virions [31] is a well-known phenomenon. Whether the HERV-H/RGH-2 variants are transmissible – behave as exogenous retroviruses as is known for some murine endogenous retroviruses – is presently unclear.

It was reported that ERV-9/MSRV/HERV-W sequences are present in a high proportion of plasma samples from MS patients and rarely found in controls [13]. Even though we could not demonstrate such sequences in particles from the MS cell cultures, the fact that HERV-H/RGH-2 and ERV-9/MSRV/HERV-W belong to the same subsection of the type-C oncovirinae may indicate that we are looking at two aspects of the same phenomenon.

We suggest, that MS is associated with production of otherwise replicatively quiescent endogenous retroviruses. Whether this represent a causal factor by eliciting an autoimmune response is unknown. In the study presented here, the presence of HERV-H/RGH-2 sequences at particle level is clearly MS specific. This could indicate a direct and specific implication in the disease process, also because all plasma samples from MS patients with active disease were positive. The known human retroviruses can elicit neurological signs and symptoms similar to MS. Certainly, activated HERVs fits most of the conventional epidemiological causality criteria: by association (presence in most cases), by generalizability (found in both PP, SP and RR MS), by temporality, by specificity (the association of HERV-H/RGH-2 in virion form with MS is specific), and by coherence.

A causal effect could be mediated through several mechanisms, as outlined in Fig. 5.

These mechanisms have all been proposed in conjunction with various model systems for MS: microbial(retroviral) encoded superantigens as reviewed by Torres and Johnson [32], epitope spreading as demonstrated by the Theiler's murine encephalomyelitis virus-induced demyelinating disease in SLJ/J mice [33], autoimmune responses through molecular mimicry as shown by analyses of peptide activation of myelin basic protein-specific T cell clones [34], mediation by proinflammatory cytokines/chemokines: we have demonstrated, that the MS long-term cell cultures actively produce a scewed repertoire of proinflammatory cytokines, probably due the virus [35] – and the levels of these cytokines are altered in MS – or by interaction through RT with non-retroviral RNA viruses as shown by the persistence of an RNA virus (lymphocytic choriomeningitis virus) in DNA form in previously infected adult mice [36].

Retroviruses are known to be immuno-suppresive, they can elicit polyclonal B-cell activation, and induce CNS-damage and autoimmune reactions. Thus all aspects in the development of MS could be due to infection with/(re)activation of retrovirus.

Acknowledgement

M.Sc. S.V. Pedersen is gratefully acknowledged for making the rat control sera available to us. We are indebted to Dr. S. Thiel and to Dr. P. Andersson for many interesting discussions. We also thank M. Schjerven, M. Aagaard, R. Nielsen, B.T. Sørensen and B. Smith for their excellent technical assistance.

This work was supported by a grant from the Danish Multiple Sclerosis Society, by Dir. J Madsen's Foundation, by The Beckett-foundation, by Lykfeldts Foundation, by AP Møller and Chastine Mc-Kinney Møllers Foundation for Medical Science, and by Vækstfonden. The automatic sequenator in Aarhus was kindly donated by the Velux Foundation.

References

1. Martyn C. The epidemiology of multiple sclerosis. In: Matthews WB (ed) McAlpines's Multiple Sclerosis. 2nd ed. Churchill Livingstone: Edinburgh, 1991:3–40.
2. Christensen T, Møller-Larsen A, Haahr SA retroviral implication in multiple sclerosis?. Trends Microbiol 1994;9:332–336.
3. Sommerlund M et al. Retrovirus-like particles in an Epstein-Barr virus-producing cell line derived from a patient with chronic progressive myelopathy. Acta Neurol Scand 1993;87:71–76.
4. Munch M et al. B-lymphoblastoid cell lines established from multiple sclerosis patients and a healthy control producing a putative new human retrovirus and Epstein-Barr virus. Multiple Sclerosis 1995;1:78–81.
5. Christensen T et al., Characterization of retrovirus from multiple sclerosis patients. Acta Neurol Scand 1997;Suppl 169:49–58.
6. Christensen T et al. RT Activity and Particle Production in B-lymphoblastoid Cell Lines established from lymphocytes from MS patients. AIDS Res Hum Retroviruses 1999;15:285–291.
7. Møller-Larsen A, Christensen T. Isolation of a retrovirus from multiple sclerosis patients in self-generated Iodixanol Gradients. J Virol Methods 1998;73:151–161.

8. Christensen T et al. Expression of the Endogenous Retrovirus RGH in Particle Form in Multiple Sclerosis. The Lancet 1998;352:1033.
9. Blond JL et al. Molecular characterization and placental expression of HERV-W, a new human endogenous retrovirus family. J Virol 1999;73:1175–1185.
10. Boller K et al. Characterization of the antibody response specific for the human endogenous retrovirus HTDV/HERV-K. J Virol 1997;71:4581–4588.
11. Tönjes RR et al. HERV-K: The biologically most active human endogenous retrovirus family. J Acq Imm Def Syndr Hum Retrovirol 1996; Suppl 13:261–267.
12. Perron H et al. Molecular identification of a novel retrovirus repeatedly isolated from patients with multiple sclerosis. Proc Natl Acad Sci 1997;94:7583–7588.
13. Garson JA et al. Detection of virion-associated MSRV-RNA in serum of patients with multiple sclerosis. The Lancet 1998;351:33.
14. Conrad B et al. A human endogenous retroviral superantigen as candidate autoimmune gene in type 1 diabetes. Cell 1997;90:303–313.
15. Silver J, Maudru T, Fujita K, Repaske R. An RT-PCR assay for the enzyme activity of reverse transcriptase capable of detecting single virions. Nucl Acids Res 1993;21:3593–3594.
16. Christensen T et al. Molecular characterization of HERV-H variants associated with multiple sclerosis. Acta Neurol Scand Scand 2000;101:229–238.
17. Lombardi VRM et al. Detection of antibodies to human immunodeficiency virus type by using synthetic peptides and time-resolved fluoroimmuno assay. Eur J Epidemiol 1992; 8: 298–304.
18. Lindeskog M, Mager DL, Blomberg J. Isolation of a human endogenous retroviral HERV-H element with an open *env* reading frame. Virology 1999;258:441–450.
19. Hohenadl C, Leib-Mösch C, Hehlmann R, Erfle V. Biological significance of human endogenous retroviral sequences. J Acquired Immune Defic Syndr Hum Retrovirol 1996;Suppl 1:5268–5273.
20. Dalgliesh AG. Viruses and multiple sclerosis. Acta Neurol Scand 1997;Suppl 169:8–15.
21. Yamano S et al. Retrovirus in salivary glands from patients with Sjögren's syndrome. Clin Pathol 1997;50:223–229.
22. Griffiths DJ, Venables PJW, Weiss RA, Boyd MT. A novel exogenous retrovirus sequence identified in humans. J Virol 1997;71:2866–2872.
23. Bengtsson A et al. Selective antibody reactivity with peptides from human endogenous retroviruses and nonviral poly(amino acids) in patients with systemic lupus erythematosus. Arthritis Rheum 1996;39:1654–1663.
24. Nakagawa K, Brusic V, McColl G, Harrison LC. Direct evidence for the expression of multiple endogenous retroviruses in the synovial compartment in rheumatoid arthritis. Arthritis Rheum 1997;40:627–638.
25. Medstrand P, Lindeskog M, Blomberg J. Expression of human endogenous retroviral sequences in peripheral blood mononuclear cells of healthy individuals. J Gen Virol 1992;73;2463–2466.
26. Mager DL, Freeman JD. Human endogenous retrovirus-like genome with type C *pol* sequences and *gag* sequences related to human T-cell lymphotropic virus. J Virol 1987;61:4060–4067.
27. Johansen J, Holm T, Bjorklid E. Members of the RTVL-H family are expressed in human placenta. Gene 1989;79:259–267.
28. Goodchild NL, Wilkinson DA, Mager DL. Recent evolutionary expansion of a subfamily of RTVL-H human endogenous retrovirus-like elements. Virology 1993;196:778–788.
29. Hirose Y, Takamatsu M, Harada F. Presence of env genes in members of the RTVL-H family of human endogenous retrovirus-like elements. Virology 1993;192:52–61.
30. Chataway J et al. The genetics of multiple sclerosis: principles, background and updated results of the United kingdom systematic genome screen. Brain 1998;121:1869–1887
31. Linial ML, Miller AD. Retroviral RNA packaging: sequence requirements and implications. Curr Top Microbiol Immunol 1990;157:125–152.
32. Torres BS, Johnson HM Modulation of disease by superantigens. Curr Opin Immunol 1998;10:465–470.

33. Vanderlugt CL et al. The functional significance of epitope spreading and its regulation by co-stimulatory molecules. Immunol Rev 1998;164:63–72.
34. Wucherpfennig KW, Strominger JL Molecular mimicry in T-cell mediated autoimmunity: viral peptides activate human T cell clones specific for myelin basic protein. Cell 1995;80:695–705.
35. Christensen T et al. Cytokine production in lymphoblastoid cell lines producing both Epstein-Barr virus and retrovirus-like particles: possible implications for multiple sclerosis. Immunol Infect Dis 1995;5:224–229.
36. Klenerman P, Hengartner H, Zinkernagel RM A non-retroviral RNA virus persists in DNA form. Nature 1997;390:298–301.

The role of neuroinvasive human coronaviruses in autoimmune processes associated with multiple sclerosis

Annie Boucher[1], Mélanie Tremblay[1], Nathalie Arbour*, Julie Edwards[1], Robert Day[2], Jia Newcombe[3], Pierre Duquette[4], François Denis[1] and Pierre J. Talbot[†]

[1]Laboratory of Neuroimmunovirology, Human Health Research Center, INRS-Institut Armand-Frappier, Laval, Québec, Canada H7V 1B7; [2]Department of Pharmacology, Sherbrooke University, Sherbrooke, Québec, Canada J1H 5N4; [3]NeuroResource, Institute of Neurology, University College London, London WC1N 1PJ, England; [4]MS Clinic, Hôpital Notre-Dame, Montréal, Québec, Canada H2L 4K8

Introduction

Multiple sclerosis (MS) is a central nervous system (CNS) chronic autoimmune disease characterized by inflammation and T cell-mediated myelin destruction. While precise disease etiology is still unknown, it appears to be multifactorial, involving both genetic (such as HLA haplotype) and environmental factors [1]. Epidemiological studies have suggested that exposure to infectious pathogens prior to puberty might increase the risk of developing MS [2]. While several infectious agents have been associated with MS [3–5], it has been difficult to draw conclusive evidence on etiological association. However, factors that are difficult to assess in epidemiological studies are pathogen adaptation to the host such as changes in cellular tropism, and pathogen entry and persistence within the CNS. Moreover, viruses proposed to be involved in MS initiation or progression, might act through direct cytopathology, indirect virus-induced autoimmunity, or a combination of both, all resulting in the CNS immunopathology associated with MS. Obviously, several pathogens could be involved in MS but act through common immunopathogenic pathways.

Coronaviruses are attractive candidate infectious actors in MS since murine coronaviral infection can induce in rodents a pathology similar to MS [6]. Coronaviruses are enveloped positive-stranded RNA viruses that infect cells of the respiratory tract in humans [7]. Interestingly, murine coronaviruses can be neurotropic and neuroinvasive: they can enter the CNS from the periphery via the olfactive nerve and infect neural cells [8,9]. In genetically susceptible rodents, persistent coronaviral CNS infection leads to a neurological disease characterized by immune system-dependent chronic and recurrent demyelination, highly reminis-

---

Present address: Neuropharmacology Department, The Scripps Research Institute, 10550 North Torrey Pines Road, La Jolla, CA 92037, USA.

[†]*Correspondence address:* Dr Pierre J. Talbot, Centre de recherche en santé humaine, INRS-Institut Armand-Frappier 531, boulevard des Prairies, Laval (Québec), Canada H7V 1B7. Tel: (450) 686-5515. Fax: (450) 686-5566. E-mail: Pierre.Talbot@inrs-iaf.uquebec.ca

cent of MS [10]. Therefore, coronaviruses provide a unique model for the study of viral involvement in autoimmune-mediated neurological diseases.

Our laboratory studies human coronaviruses (HCoV) as possible etiological agents of MS. Human viral isolates are divided into two serotypes, represented by reference strains 229E and OC43. These pathogens are the causative agents of up to one third of common colds and nearly 100% of the population is seropositive for HCoV by age five [11]. While HCoV mostly infect the respiratory tract, several lines of evidence suggest a possible neurological involvement. For instance, coronavirus-like particles were detected in brains of MS patients [12], increased antibody titers against both HCoV serotypes were found in the cerebrospinal fluid of MS patients when compared to controls [13], coronaviruses were isolated from MS brains [14,15], and coronaviral gene products were detected in MS brains [4,16].

Thus, these viruses appear to be neuroinvasive and we have shown that they are neurotropic, in that they have the capacity to replicate in human neuronal and glial cells, including in primary cultures [17]. Moreover, we have demonstrated viral persistence in human neural cell lines infected with either HCoV serotypes [18-20]. Thus, the ability of HCoV to enter and replicate in the CNS could possibly represent a contributing factor in MS pathology.

However, the mechanisms involved in neurovirulence after neuroinvasion and infection of neural cells remain to be elucidated, but may involve local virus-induced inflammatory pathways. This could involve the induction of pro-inflammatory molecules that have been associated with CNS pathologies and have been detected within MS lesions [21-23]. For instance, glial cells can secrete matrix metalloproteinases (MMPs) and several MMPs can cleave myelin basic protein (MBP) within the encephalitogenic portion of the protein [24]. Also, astrocytes and microglial cells can produce nitric oxide (NO), which mediates oligodendrocyte cell death [25,26]. Interestingly, active NO synthase was detected in astrocytes located at the edges of MS plaques, where demyelination occurs [26]. Finally, cytokines and chemokines released by infected CNS cells might contribute to localized T-cell activation and serve as T-cell attractants.

Given that T-cell chemoattractants can be produced by infected CNS cells, that MS appears to be a T cell-mediated autoimmune disease and that one of the major genetic factor predisposing to MS is the HLA haplotype of patients, molecular mimicry between viral and self antigens represents an attractive hypothesis to explain autoimmune attacks [27]. Our laboratory has shown that peripheral T-cells cross-reacting with coronaviral and myelin antigens are frequently found in MS patients but rarely in controls [28].

Herein, we summarize current evidence from our laboratory that HCoVs are neuroinvasive in humans, that coronavirus infection of astrocytic and microglial cell lines leads to increased expression of pro-inflammatory molecules and that T-cell clones derived from MS patients can be activated by both viral antigens and myelin components. The possible contribution of coronaviral infections to MS pathology is discussed.

Materials and methods

Brain samples

Frozen coded brain samples from MS patients and control donors were received from the following brain banks: The Multiple Sclerosis Society Tissue Bank, Institute of Neurology, University of London, London (Dr. Jia Newcombe); Laboratoire de neuropathologie Raymond Escourolle, Groupe hospitalier Pitié-Salpêtrière, Paris, France (Drs. Jean-Jacques Hauw and Danielle Seilhean); Rocky Mountain Multiple Sclerosis Center Tissue Bank, Englewood, CO, U.S.A. (Drs. Catalin Butunoi and Ronald S. Murray); Multiple Sclerosis Human Neurospecimen Bank, West Los Angeles VA Medical Center, Los Angeles, CA, U.S.A. (Drs. Virginia Sanders and Wallace W. Tourtellotte); Douglas Hospital Research Center Brain Bank, McGill University, Verdun, Québec, Canada (Ms. Danielle Cécyre, Mr. Yvan Dumont and Dr. Rémi Quirion). Samples were stored at –70°C.

RNA extraction, gene amplification and hybridization

Total cellular RNA from human brain tissue was extracted and amplified as described [18,19]. Amplification of HCoV-OC43, HCoV-229E, and MBP (internal gene control) RNA and subsequent Southern hybridization of PCR amplicons was carried out as reported [18,19]. *In situ* hybridization of autopsy brain sections were performed with a slight modification of a published protocol [29,30].

Virus and cell line

HCoV-OC43 and HCoV-229E were originally obtained from the American Type Culture Collection (ATCC, Manassas, VA). The U-373MG cell line was obtained from ATCC and was originally derived from a grade III human astrocytoma.

Production of pro-inflammatory molecules by infected glial cells

Monolayers of U-373MG cells underwent two distinct treatments. First, 100 U/mL of IFN-γ was added to the culture to determine the influence of exogenous IFN-γ on release of inflammatory mediators. Another treatment was a HCoV-OC43 infection at a multiplicity of infection (MOI) of 0.25. Treatments were carried-out for up to 72 hours at 33°C. Total cellular RNA was obtained using TRIzol™ (Gibco), reverse-transcribed and amplified as described elsewhere [31]. Amplicon signals on agarose gels were normalized against the GAPDH housekeeping gene signal of uninfected cells to determine a stimulation index (semi-quantitative RT-PCR). Secretion and gelatinase activity of MMP-2 and MMP-9 was measured by zymography [32]. The specificity of both metalloproteinases was confirmed by the use of appropriate inhibitors.

Donors

Patients diagnosed with MS and healthy controls were selected at random with prior informed consent and approval of the experimental protocol from institutional ethics committees. Confidentiality was preserved at all times by a coding system.

Antigens

Human MBP was prepared from a normal adult brain (Montreal Brain Bank, Douglas Hospital, Verdun, Québec, Canada [28]. Proteolipid protein (PLP), which was purified by chromatography, was graciously provided by Dr. Mario Moscarello (Department of Biochemistry Research, The Hospital for Sick Children, Toronto, Ontario, Canada). HCoV-229E and HCoV-OC43 were propagated and viral and control antigens were prepared as already described [28,33,34].

Generation and maintenance of T-cell clones

Human blood samples were collected from human donors and buffy coat prepared. Peripheral blood mononuclear cells (PBMC) were separated by Ficoll/Hypaque (Pharmacia) and resuspended at 1×10^6 cells/mL and seeded as described elsewhere [28]. After 7 days of incubation, 100 units/mL of human recombinant interleukin-2 (Hoffman La Roche) was added. At day 12, primary T-cell lines were tested for antigen specificity by an antigen-specific [^{3}H]thymidine incorporation assay [28]. Antigen-specific primary T-cell lines were cloned on day 16 by limiting dilution in Terasaki wells, in the presence of autologous irradiated PBMC as antigen-presenting cells (APC) and selecting antigen. After incubation, each Terasaki well was checked for clonal growth until day 24–26. Wells showing positive growth at the cell dilution level yielding less than 33% positive wells were expanded in 96-well flat-bottom plates (the Poisson distribution suggests that a clone is considered as such when the cloning efficiency obtained is less then 33%). At day 30, clones were tested as described above. Clones selected for proliferation assays were also expanded by restimulation with antigen. Antigen specificity was concluded when stimulation indices were more than 3.0 and [^{3}H] incorporation was at least 800 cpm.

Results

Human coronavirus RNA in human brains

To provide definitive evidence for invasion and infection of human CNS by human coronaviruses and for a possible association of viral persistence with MS, a large panel of coded brain samples from donors with MS, other neurological diseases (OND) and normal controls, was analyzed for the presence of HCoV RNA. The

Table 1. Detection of HCoV-OC43 and HCoV-229E RNA in autopsy brain samples from donors with multiple sclerosis (MS), other neurological diseases (OND) and normal controls, using RT-PCR combined with Southern hybridization.

Donor group	HCoV detection			
	n	229E	OC43	Both
MS	39	51.3 %	35.9 %	28.2 %
Normal	25	44.0 %	20.0 %	6.0 %
OND	26	34.6 %	7.7 %	7.7 %
All donors	90	44.4 %	23.3 %	18.8 %

rationale for detecting RNA rather than proteins was that during establishment of persistent infection in the murine model, coronaviral RNA was still detectable long after viral proteins were below detection levels [35]. We amplified a portion of the nucleocapsid (N) protein gene since this RNA is expressed in large amounts during infection and its sequence is present on every genomic and subgenomic viral RNAs [7]. Amplification was carried out by very stringent RT-PCR on two different regions of the N gene and from two different RNA preparations. Specificity and sensitivity were enhanced by Southern hybridization with a radiolabeled internal oligonucleotide. Brain samples were considered positive only when a positive signal was detected from both virus-specific RT-PCR-Southern hybridization reactions or when two distinct RNA preparations from the same brain yielded a positive signal. As shown in Table 1, a surprisingly high proportion of samples were positive for HCoV RNA. Overall, 44.4% of the samples were positive for HCoV-229E and 23.3% for HCoV-OC43. Statistical analysis showed that HCoV-OC43 was preferentially found in MS patients versus other neurological disease controls ($p=0.0169$) or both OND and normal controls ($p=0.0137$). Such association was not observed for HCoV-229E, which appeared more prevalent overall than HCoV-OC43,

To confirm that RT-PCR signals represented viral replication in neural cells, frozen sections were analyzed by in situ hybridization. The detection threshold and RNA quantity allowed observation of HCoV-OC43 RNA in the brain parenchyma, outside blood vessels (data not shown), which confirmed neuroinvasion [30].

Activation of glial cells by coronavirus infection

Neuroinvasion of HCoV and infection of glial cells in the CNS could lead to their activation and release or inflammatory mediators. Moreover, the OC43 strain of HCoV is closely related at the molecular and antigenic levels to murine coronaviruses that are able to induce an MS-like disease in rodent [6,10]. Thus, we investigated the effect of a HCoV-OC43 infection of the human astrocytic cell line U373-MG [19] on the production of MS-associated inflammatory mediators (cytokines, chemokines, MMPs, nitric oxide). Treatment with INF-γ that could be released by infiltrating T cells was also evaluated. The expression of IL-6,

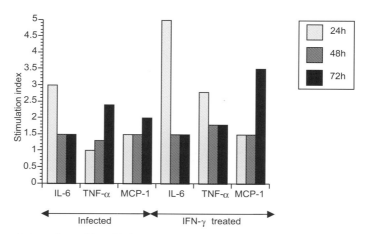

Fig. 1. Upregulation of cytokine (Il-6 and TNF-α) and chemokine (MCP-1) mRNA expression, detected by semi-quantitative RT-PCR, in the human astrocytic U-373MG cell line after either infection with HCoV-OC43 or treatment with IFN-γ.

TNF-α and MCP-1 mRNA was measured by semi-quantitative RT-PCR. As shown in Fig. 1, a 3- or 5-fold upregulation of IL-6 mRNA levels was observed 24 hours after HCoV-OC43 infection or addition of exogenous IFN-γ, respectively. HCoV-OC43 infection also led to an increase in TNF-α mRNA expression, reaching a maximum stimulation index of 2.4 at 72 hours post-infection and 2.8 at 24 hours after treatment with IFN-γ. The expression of the MCP-1 chemokine mRNA was also upregulated, with stimulation indices reaching 2.0 and 3.5 at 72 hours after virus infection or INF-γ treatment, respectively.

The secretion of MMP-2 and MMP-9 (type IV collagenases) was evaluated by zymography after infection of U-373MG cells with HCoV-OC43. Expression of MMP-2 was increased 24 hours post-infection when compared to untreated cells [31]. Pro-MMP-9 was detected at 24 hours post-infection, and active MMP-9 at 48 hours post-infection [31].

Given that astrocytes appear to represent the main producers of nitric oxide (NO) in the CNS [36], we also verified whether this pro-inflammatory molecule could be induced after a coronavirus infection. Untreated, unstimulated cells produced about 2.4 μM of NO [31]. As shown in Table 2, this level was increased 2-fold by HCoV-OC43 infection for 48 and 72 hours.

Myelin-coronavirus cross-reactive T-cell clones

Combined with viral persistence in the CNS and activation of glial cells, the specific activation of T cells by virus and/or myelin may contribute to MS immunopathology in genetically-susceptible individuals. We have previously reported, mainly in MS patients, T-cell lines that were cross-reactive for HCoV and myelin antigens. Overall, 29% of T-cell lines isolated from MS patients but only 1.3%

Table 2. Nitric oxide (NO) production in U-373MG astrocytic cell line, as compared to non-treated cells. + : NO level similar to non-treated cells, ++ : two-fold increase in NO level.

Stimulus	24h	48h	72h
None	+	+	+
PMA	++	+	+
HCoV-OC43 infection	+	++	++

Table 3. Typical antigen-specific proliferation profile of a myelin-coronavirus cross-reactive T-cell clone from an MS patient. This clone was selected with HCoV-OC43 and then tested for reactivity against either HCoV-OC43 or MBP, using a [^{3}H]thymidine incorporation assay.

Antigen	HCoV-OC43	MBP
CPM	**6284**	2628
S.I.[1]	18.3	3.8

[1] Stimulation index.

of T-cell lines from healthy control subjects showed HLA-DR-restricted reactivities to both HCoV-229E and MBP [28]. To confirm that this phenomenon represented recognition of both antigens by the same T-cell and could also be observed with other coronaviral and myelin antigens, we generated T-cell clones specific to HCoV-229E and HCoV-OC43 as well as MBP and PLP antigens and verified their activation by viral and myelin antigens. Table 3 shows a typical clonal coronavirus-myelin T-cell cross-reactivity pattern, which was observed for both HCoV strains and myelin antigens tested, MBP and PLP (data not shown).

Discussion

Our data clearly demonstrate that human coronaviruses are neuroinvasive for human brains and appear to replicate and most likely persist in neural cells, as *in situ* hybridization in brain parenchyma outside blood vessels [30] combined with our cell culture data [18,19] suggests. Indeed, our *in situ* data are consistent with a previously published study [16]. Further analysis will be required to characterize the CNS persistence of HCoV-229E. Our *in vitro* viral replication studies have suggested an apparent molecular adaptation of HCoV after persistence in neural cells, as shown by the appearance of point mutations in the viral S protein gene [18,19]. This was also observed after sequencing the viral N amplicons from human brains [30] and needs to be investigated for other viral genes expressed in human brains. Presumably, such adaptation could favor persistence in the CNS. Whether these mutations are related to a pathologic state remains to be studied. Despite the known high prevalence of HCoV in the human population, it had so far been mainly associated with infections of the upper respiratory tract. Therefore, the observed high proportion of infected human brains was surprising. This suggests that HCoV are part of the brain viral flora and that their presence may only lead to neurovirulence in some individuals that are susceptible through unknown factors, such as genetic susceptibility genes that may lead to aberrant responses to viral persistence or to replication in specific brain areas.

Having shown neuroinvasion of HCoV in human brains, we next demonstrated that infection of astrocytes by HCoV-OC43 led to their activation and production of inflammatory molecules that may be toxic to myelin sheaths or be involved in leukocyte recruitment and amplification of the local immunopathological inflammatory response [31]. It has been reported that bystander virus-mediated inflammation could trigger autoimmunity [37]. Moreover, infiltrating activated lymphocytes [38] could release INF-γ that we showed could also activate astrocytes [31]. Also, MMP secretion resulting from infection could have an effect on myelin protein integrity since it has been showed that MBP is cleaved by MMP-9, releasing immunogenic peptides [23].

The production of IL-6 in the CNS can be induced by IFN-γ and its upregulation is also shown following infection of astrocytes [39]. It has been reported to modulate MMP expression [32]. An immunosuppressive role has also been ascribed to IL-6, but this apparently did not suppress *in vitro* TNF-α production in our study. Indeed, we provide evidence for upregulation of TNF-α mRNA expression after coronavirus infection. Inflammation in the CNS, myelin damage, MMP modulation and induction of inducible nitric oxide synthase (iNOS), have been associated with TNF-α [32,40,41]. Moreover, astrocytes are also the main CNS producers of the MCP-1 chemokine [42]. This chemoattractant plays a key role in inflammation, promoting leukocyte infiltration. The expression of MCP-1 could give rise to an inflammatory state attracting autoreactive or virus-reactive lymphocytes to the CNS, thus promoting pathologies such as MS. Additive or synergistic effects of inflammatory cytokines and activated T-cells may exacerbate virus-induced immunopathology.

Enhancement of MMP-2 and MMP-9 secretion was observed in infected cells. Upregulation of MMPs after a coronavirus infection has potential pathological significance since MMP-2 cleaves pro-MMP-9 [43], and MMP-9 can promote the degradation of MBP, thus potentially harming the oligodendrocyte-myelin unit [44]. Furthermore, viral infection also enhanced nitric oxide (NO) production. This is consistent with the observation that astrocytes may be the main source of NO in the CNS [45]. These observations are compatible with the notion that a coronavirus infection of astrocytes might mediate inflammation, which could be maintained by NO secretion and TNF-α secreted from activated astrocytes [46]. These pro-inflammatory molecules may be involved in a self-amplification loop.

Finally, cross-reactive T-cell clones were generated from the peripheral blood of MS patients. The same cross-reactivity patterns we had previously reported with T-cell lines [28], were obtained at the clonal level and involved the two major myelin antigens and the two HCoV strains. This strengthens our hypothesis of a possible coronavirus-induced activation of T-cell autoimmunity by molecular mimicry. Presumably, lymphocytes attracted to the CNS to clear a virus infection could also react against myelin epitopes that could now be exposed in significant concentration after damage resulting from inflammation brought about by virus-mediated activation of glial cells. Alternatively, autoreactive T-cells activated in the periphery by coronavirus infection may penetrate the CNS and recognize myelin and/or viral antigens. It will be important to demonstrate the presence in MS brains of such T-cells and their potential to be directly involved in CNS pathology.

Even though coronaviruses remain attractive MS candidates, it should be borne in mind that several different viruses could be involved in such virus-induced autoimmunity. For example, measles virus has also been considered a potential candidate for mimicry-induced T-cell activation based on amino acid homologies between the virus and MBP [47]. Moreover, it must be kept in mind that molecular mimicry in the context of T lymphocytes will involve conformational homologies or motifs within the trimolecular TCR-MHC-peptide complex that may be impossible to predict trough classical sequence alignment [48].

The understanding of the triggering events in MS and/or its relapses would obviously favor the design of rational therapeutic strategies. For instance, peripheral tolerance or immunoregulation of cross-reactive T-cell clones could be induced by altered peptide ligands for HLA-DR2+ MS patients [49]. However, this potential therapeutic avenue remains complex and the induction of antigen-specific suppression with synthetic peptides could result in immunosuppression through the generation of T-cells secreting suppressive cytokines [50]. Obviously, for such a therapeutic strategy to be helpful, the relevant mimicry peptides must first be identified. Hence, molecular analysis will be necessary to determine the fine specificity of the cross-reactive T-cell clones and to search for their presence in the CNS and participation in local pathology.

In summary, our study suggests that human coronaviruses are neuroinvasive and have an autoimmunogenic potential through the triggering of a variety of specific and non-specific immune responses that may lead to CNS immunopathology in genetically-susceptible individuals.

Acknowledgements

We thank Francine Lambert for expert technical help and Tina Miletti for critically reading the manuscript. We are deeply grateful to J.-J. Hauw, D. Seilhean, C. Butunoi, R.S. Murray, V. Sanders, W.W. Tourtellotte, D. Cécyre, Y. Dumont and R. Quirion for providing brain samples used in these studies.

This work was funded by the Medical Research Council of Canada (neuroinvasion and glial cell activation) and the Multiple Sclerosis Society of Canada. A. Boucher and N. Arbour acknowledge studentship support from the Multiple Sclerosis Society of Canada and M. Tremblay from the Natural Sciences and Engineering Research Council of Canada.

References

1. Ebers GC, Sadovnick AD The role of genetic factors in multiple sclerosis susceptibility. J Neuroimmunol 1993;54:1–17.
2. Kurtzke JF. Epidemiologic evidence for multiple sclerosis as an infection. Clin Microbiol Rev 1993;6:382–427.
3. Challoner PB, Smith KT, Parker JD, Macleod DL, Coulter SN, Rose TM, Schultz ER, Bennett JL, Garber RL, Chang M, Schad, PA, Sewart PM, Nowinski RC, Brown JP, Burmer GC. Plaque-associated expression of human herpesvirus 6 in multiple sclerosis. Proc Natl Acad Sci USA 1995;92:7440–7444.
4. Stewart JN, Mounir S, Talbot P. Human coronavirus gene expression in the brains of multiple sclerosis patients. Virology 1992;191:502–505.
5. Perron H, Garson JA, Bedin F, Beseme F, Paranhos-Baccala G, Komurian-Pradel F, Mallet F, Tuke PW, Voisset C, Blond JL, Lalande B, Seigneurin JM, Mandran B. Molecular identification of a novel retrovirus repeatedly isolated from patients with multiple sclerosis. Proc Natl Acad Sci USA 1997;94:7583–7588.
6. Lane TE, Buchmeier MJ. Murine coronavirus infection: a paradigm for virus-induced demyelinating disease. Trends Microbiol 1997;5:9–14.
7. Lai MMC, Cavanagh D. The molecular biology of coronaviruses. Adv Virus Res 1997;48:1–100.
8. Lavi E, Fishman PS, Highkin MK, Weiss SR. Limbic encephalitis after inhalation of a murine coronavirus. Lab Invest 1988;58:31–36.
9. Perlman S, Evans G, Afifi A. Effect of olfactory bulb ablation on spread of a neurotropic coronavirus into the mouse brain. J Exp Med 1990;172:1127–1132.
10. Fazakerley JK, Buchmeier MJ. Pathogenesis of virus-induced demyelination. Adv Virus Res 1993;42:249–324.
11. Myint SH. Human coronaviruses - A brief review. Rev Med Virol 1994;4:35–46.
12. Tanaka R, Iwasaki Y, Koprowski H. Intracisternal virus-like particles in brain of a multiple sclerosis patient. J Neurol Sci 1976;28:121–126.
13. Salmi A, Ziola B, Hovi T, Reunanen M. Antibodies to coronaviruses OC43 and 229E in multiple sclerosis patients. Neurology 1982;32:292–295.
14. Burks JS, DeVald BL, Jankovsky LD, Gerdes JC. Two coronaviruses isolated from central nervous system tissue of two multiple sclerosis patients. Science 1980;209:933–934.
15. Fleming JO, El Zaatari AK, Gilmore W, Berne JD, Burks JS, Stohlman SA, Tourtelotte WW, Weiner LP. Antigenic assessment of coronaviruses isolated from patients with multiple sclerosis. Arch Neurol 1988;45:629–633.
16. Murray RS, Brown B, Brian D, Cabirac GF. Detection of coronavirus RNA and antigen in multiple sclerosis brain. Ann Neurol 1992;31:525–533.
17. Bonavia A, Arbour N, Yong VW, Talbot PJ. Infection of primary cultures of human neural cells by human coronaviruses 229E and OC43. J Virol 1997;71:800–806.
18. Arbour N, Ekandé S, Côté, G, Lachance C, Chagnon F, Tardieu M, Cashman NR, Talbot PJ. Persistent infection of human oligodendrocytic and neuroglial cell lines by human coronavirus 229E. J Virol 1999;73:3326–3337.
19. Arbour N, Côté G, Lachance C, Tardieu M, Cashman NR, Talbot PJ Acute and persistent infection of human neural cell lines by human coronavirus OC43. J Virol 1999;73:3338–3350.
20. Lachance C, Arbour N, Cashman NR, Talbot PJ. Involvement of aminopeptidase N (CD13) in infection of neural cells by human coronavirus 229E. J Virol 1998;72:6511–6519.
21. Benveniste EN. Role of macrophages/microglia in multiple sclerosis and experimental allergic encephalomyelitis. J Mol Med 1997;75:165–173.
22. Munoz-Fernandez MA, Fresno M. The role of tumor necrosis factor, interleukin 6, interferon-γ and inducible nitric oxide synthase in the development and pathology of the nervous system. Prog Neurobiol 1998;56:307–340.

23. Chandler S., Miller KM, Clements JM, Lury J, Corkill D, Anthony DCC, Adams SE, Gearing AJH. Matrix metalloproteinases, tumor necrosis factor and multiple sclerosis: An overview. J Neuroimmunol 1997;72:155–161.
24. Cossins JA, Clements JM, Ford J, Miller KM, Pigott R, Vos W, Van Der Valk P, De Groot, CJA. Enhanced expression of MMP-7 and MMP-9 in demyelinating multiple sclerosis lesions. Acta Neuropathol 1997;94:590–598.
25. Mitrovic B, Parkinson J, Merrill JE. An *in vitro* model of oligodendrocyte destruction by nitric oxide and its relevance to multiple sclerosis. Methods 1996;10:501–513.
26. Lee SC, Brosnan CF. Cytokine regulation of iNOS expression in human glial cells. Methods 1996;10:31–37.
27. Oldstone MBA. Molecular mimicry and autoimmune disease. Cell 1987;50:819–820.
28. Talbot PJ, Paquette JS, Ciurli C, Antel JP, Ouellet F. Myelin basic protein and human coronavirus 229E cross-reactive T cells in multiple sclerosis. Ann Neurol 1996;39:233–240.
29. Schäfer MKH, Day R. In situ hybridization techniques to study processing enzyme expression at the cellular level. Methods Neurosci 1994;23:16–44.
30. Arbour NA, Day R, Newcombe JA, Talbot PJ. Neuroinvasion by human respiratory coronaviruses. J Virol 2000;74:8913–8921.
31. Edwards JE, Denis F, Talbot PJ. Activation of glial cells by human coronavirus OC43 infection. J Neuroimmunol 2000;108:73–81.
32. Giraudon P, Buart S, Bernard A, Bélin MF. Cytokines secreted by glial cells infected with HTLV-1 modulate the expression of matrix metalloproteinase (MMPs) and their natural inhibitor (TIMPs): possible involvement in neurodegenerative process. Mol Psych 1997;2:107–110.
33. Jouvenne P, Mounir S, Stewart JN, Richardson CD, Talbot PJ. Sequence analysis of human coronavirus 229E RNAs 4 and 5 – evidence for polymorphism and homology with myelin basic protein. Virus Res 1992;22:125–141.
34. Mounir S, Talbot PJ. Sequence analysis of the membrane protein gene of the human coronavirus OC43 and evidence for O-glycosylation. J Gen Virol 1992;73:2731–2736.
35. Fleming JO, Houtman JJ, Alaca H, Hinze HC, McKenzie D, Aiken J, Bleasdale T, Baker S. Persistence of viral RNA in the central nervous system of mice inoculated with MHV-4. Adv Exp Med Biol 1993;342:327–332.
36. Robbins DS, Shirazi Y, Drysdale BE, Lieberman A, Shin HS, Shin ML. Production of cytotoxic factor for oligodendrocytes by stimulated astrocytes. J Immunol 1987;139:2593–2597.
37. Horwitz MS, Bradley LM, Harbertson J, Krahl T, Lee J, Sarvetnick N. Diabetes induced by coxsackie virus : initiation by bystander damage and not molecular mimicry. Nature Med 1998;4:781–785.
38. Glabinsky AR, Ransohoff RM. Chemokines and chemokine receptors in CNS pathology. J Neurovirol 1999;5:3–12.
39. Beneviste EN. Cytokine actions in the central nervous system. Cyt Growth Fact Rev 1998;9:259–275.
40. Probert L, Akassoglou K, Pasparakis M, Kontogeorgos G, Kollias G. Spontaneous inflammation demyelinating disease in transgenic mice showing central nervous system-specific expression of tumor necrosis factor alpha. Proc Natl Acad Sci USA 1995;92:11294–11298.
41. Gottschall PE, Yu X. Cytokines regulate gelatinase A and B (matrix metalloproteinase 2 and 9) activity in cultured rat astrocytes. J Neurochem 1995;64:1520–1531.
42. Sun N, Grzybick D, Castro RF, Murphy S, Perlman S. Activation of astrocytes in the spinal cord of mice chronically infected with a neurotropic coronavirus. Virology 1995;213:482–493.
43. Fridman R, Toth M, Pena D, Mobashery S. Activation of progelatinase B (MMP-9) by gelatinase A (MMP-2). Cancer Res 1995;55:2548–2555.
44. Chandler S, Coates R, Gearing A, Lury J, Wells G, Bone E. Matrix metalloproteinases degrade myelin basic protein. Neurosci Lett 1995;201:223–226.
45. OguraT, Esumi H. Nitric oxide synthase expression in human neuroblastoma cell line induced by cytokines. Neuroreport 1996;7:853–856.

46. Lee SC, Brosnan SC. Cytokine regulation of iNOS expression in human glial cells. Methods 1996;10:31–37.
47. Liebert UG, Hashim GA, ter Meulen V. Characterisation of measles virus-induced cellular autoimmune reactions against myelin basic protein in Lewis rats. J Neuroimmunol 1990;29:139–147.
48. Wucherpfennig KW, Strominger JL. Molecular mimicry in T cell-mediated autoimmunity: viral peptides activate human T cell clones specific for myelin basic protein. Cell 1995;80:695–705.
49. Vergelli M, Hemmer B, Utz U, Vogt A, Kalbus M, Tranquill L, Conlon P, Ling N, Steinman L, McFarland HF, Martin R. Differential activation of human autoreactive T cell clones by altered peptide ligands derived from myelin basic protein peptide (87–99). Eur J Immunol 1996;11:2624–2634.
50. Anderson S, Burkhart C, Metzler B, Wraith D. Mechanisms of central and peripheral T-cell tolerance: lessons from experimental models of multiple sclerosis. Immunol Rev 1999;169:123–137.

Association of multiple sclerosis with Chlamydia pneumoniae: demonstration of the 16S rRNA gene and immunoreactivity of CSF cationic antibodies against C. pneumoniae antigens

S. Sriram[1], C.W. Stratton[2], S. Yao[1], J.D. Bannan[2], W.M. Mitchell[2]

[1]*Departments of Neurology and* [2]*Pathology, Vanderbilt School of Medicine, Nashville, Tennessee, USA*

Introduction

We previously have demonstrated a strong correlation between the presence of *Chlamydia pneumoniae* in the CSF of patients with MS by culture, PCR, and immunological methods [1,2]. In these first reports, *C. pneumoniae* was isolated from CSF cultures in MS patients. In addition, we have used PCR methodology that detected the major outer membrane protein (MOMP) gene (*ompA*) of *C. pneumoniae*. This gene was selected because DNA sequence analysis of *ompA* indicates that the amino acid sequence of MOMP in all *C. pneumoniae* strains varies <6%, while other species are >30% different from *C. pneumoniae* [3–5]. Finally, we have demonstrated elevation of CSF IgM and IgG titers against *C. pneumoniae* antigens by ELISA methodology; the specificity of this antibody response to *C. pneumoniae* has been confirmed by Western blot assays [2].

A reclassification of chlamydiae based on 16S rRNA sequence recently has been reported and provides a consistent method for identifying present taxa using criteria that are comparable to those established for other major bacterial lineages [2,3,6]. In particular, the coherence of the *C. pneumoniae* taxon has been demonstrated by DNA sequence analysis of rRNA with partial analysis of this gene being used to identify this species. Accordingly, for this study a nested PCR procedure was developed in which the outside 16S rRNA primers amplify members of the *Chlamydia* genus while the inner 16S rRNA primers are specific for *C. pneumoniae* and amplify a 446 base pair sequence. Southern hybridization using a labeled probe was used to provide additional specificity for the PCR products.

MS patients are known to have an increase in CSF immunoglobulins in which a portion of this increase is seen as oligoclonal bands on isoelectric focusing gels [7]. These oligoclonal bands represent antibodies that have isoelectric points in the cathodal region of the gel. In this study, the immunoreactivity of these cationic antibodies in the CSF of MS patients was evaluated using affinity-driven immunoblot assays [8,9].

Correspondence address: Subramaniam Sriram, MD, Multiple Sclerosis Research Laboratory, 1222H Vanderbilt Stallworth Rehabilitation Hospital, 2201 Capers Avenue, Nashville, TN 37212, USA. Tel: 615-963-4042. Fax: 615-321-5247. E-mail: Srirams@ctrvax.vanderbilt.edu

Materials and methods

Patients and patient selection

Seventeen patients with relapsing-remitting MS, six patients with progressive disease (5 secondary progressive, one primary progressive) who satisfied the Poser criteria for definite MS, were selected for the study. Age and gender matched controls were recruited from 13 patients with other neurologic diseases (OND) in whom CSF was being obtained for diagnostic studies. The patients tested were among those previously reported [2]. In addition CSF from two patients with subacute sclerosing pan encephalitis were examined in the immunoblot assays.

PCR amplification of 16S rDNA gene of C. pneumoniae

PCR methodology for DNA amplification of the 16S rRNA gene of *C. pneumoniae* from CSF specimens was done as follows: To at least 300 µl of CSF sample in its original collection tube, 200 µl of trypsin (0.25%) – ethylenediaminetetraacetic acid (1mM) (EDTA; GIBCO BRL, Gaithersburg, MD) in Hank's balanced salt solution (HBSS; GIBCO) at pH 7.2 was added to achieve a final concentration of 0.1% trypsin. (All reagents used were confirmed to be *C. pneumoniae*-free by quality control testing with the same PCR procedure.) The sample was then vortex mixed and incubated at 37°C for 30 minutes. Following incubation, the sample was again vortex mixed and centrifuged for 45 minutes at 12,000 X g in a microcentrifuge at ambient temperature. The pellet was resuspended in 20 µl of lysis buffer (0.5% sodium dodecylsulfate [SDS], 1% NP40, 0.2 M NaCl, 10 µM dithiothreitol [DTT], 10 mM EDTA, and 20 mM Tris-HCl at pH 7.5). Following perturbation of the chlamydial surface by reduction of its extensive disulfide bonding, 8 µl of Proteinase K (20 µg/ml; Boehringer-Mannheim, Indianapolis, IN) was added. The specimen was mixed and incubated at 37°C overnight. Purified DNA was then extracted from the aqueous fraction of the specimen with Na acetate (1:10 dilution by volume of a 3 M solution; Fisher Scientific, Pittsburgh, PA) and mixing/precipitation with 2:2.5 dilution by volume of absolute ethanol at 4°C after performing initial extraction with a mixture of phenol: chloroform: isoamyl alcohol (25:24:1; Sigma Chemical, St. Louis, MO) followed by 2 extractions with chloroform. The DNA precipitate is washed with 70% ethanol in water, centrifuged at 600 X g for 5 minutes at ambient temperature, and resuspended in 20 µl of water.

Nested PCR was carried out to detect the 16S rRNA gene of *C. pneumoniae* as follows: The 16S rRNA gene primers used were outer forward (TTT AGT GGC GGA AGG GTT AGT A), outer reverse (CAC ATA TCT ACG CAT TTC ACC G), inner forward (CTT TCG GTT GAG GAA GAG TTT ATG C), and inner reverse (TCC TCT AGA AAG ATA GTT TTA AAT GCT G). With this nested PCR procedure, the outside 16S rRNA primers amplify members of the *Chlamydia* genus while the inner 16S rRNA primers are specific for *C. pneumoniae* and amplify a

446 base pair sequence. Reaction mixtures for the outer primer reaction contain 20 µl of target DNA, 200 pM of each outer primer, 200 µM of each dNTP, and 1 unit of Deep Vent polymerase in the manufacturers buffer (New England Biolabs, Boston, MA) with no additional $MgCl_2$. (total volume). The PCR reaction was performed for 35 cycles at 94°C for 1 minute, 55°C for 2 minutes, and 74°C for 3 minutes. Five µl of the reaction mix is removed from the reaction tube and placed in a second tube containing the same components with the exception that inner primers are used instead of the outer primers. Reaction conditions for the second nested phase were 35 cycles at 94°C for 1 minute, 50°C for 2 minutes, and 74°C for 3 minutes and subjected to electrophoresis in 1% agarose gels for 45 minutes at 95 V. Amplified DNA was transferred from the agarose gel to positively charged nylon membranes (Boehringer-Mannheim) by capillary blotting. The 16S rRNA gene homologous to the inner primers first was obtained by PCR from the TWAR strain of *C. pneumoniae* (VR-1310, American Type Culture Collection [ATCC], Manassas, VA). This inner primer product was then labeled with digoxigenin (DIG, Boehringer-Mannheim) following the manufacturers directions. DIG-labeled inner product was used as a probe for membranes prehybridized at 65°C in hybridization buffer (10% dextran sulfate, 1 M NaCl, 1% SDS) by adding 100 ng of DIG-labeled probe in fresh hybridization buffer and incubated overnight at 65°C overnight. Blots were washed three times in 2X saline-sodium citrate containing 1% SDS at ambient temperature and an additional three times at 65°C followed by high stringency washes in 0.1X saline-sodium citrate containing 0.1% SDS at ambient temperature. Membranes were blocked in 5% w/v dehydrated nonfat milk in phosphate buffered saline (PBS), 0.2% Tween 20 for 1 hour at ambient temperature, then incubated in anti-DIG-alkaline phosphatase-conjugated Fab fragments (Boehringer-Mannheim) diluted 1:5000 in PBS, 0.2% Tween 20, and developed with nitro blue tetrazolium (NBT)/5-bromo-4-chloro-3-indoyl phosphate (BCIP) substrate.

Affinity-driven immunoblot assays for CSF cationic antibodies

Affinity-driven immunoblot assays for CSF cationic antibodies were performed using modifications from published methods as follows: IEF was performed with 0.3 µg of immunoglobulin from CSF obtained from MS patients and OND controls [8]. Focusing was carried out under constant voltage conditions in a stepped fashion (100V for 15 minutes, 200V for 15 minutes, and 450V for 20 minutes). The IEF gel was transferred to nitrocellulose paper (Bio-Rad, Trans-blot, 0.45 µm) that was precoated with *C. pneumoniae* EB antigens by overnight incubation with gentle rocking at 4° C. EB antigens are prepared from concentrated *C. pneumoniae* EBs by treating them with 25 mM DTT, 2% 2-mercaptoethanol, and 2% SDS for 5 minutes at 100°C as previously described [2]. Treated EBs are sonicated and then resuspended (2 mg/l protein) in PBS pH 7.4. Unoccupied sites are blocked with 3% BSA. Antibody bound to antigen is probed with peroxidase-conjugated goat anti-human IgG (Sigma) using a chemiluminiscent detection assay (Pierce,

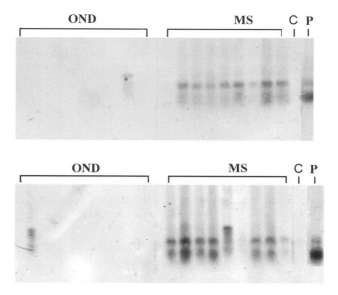

Fig. 1. Visualization of a 446 base pair region of the 16S rRNA gene of *C. pneumoniae* amplified by a nested PCR procedure followed by Southern hybridization with a digoxigenin-labeled specific probe. The gels represent CSF from 17 patients with relapsing remitting MS (Table 2) versus 13 OND controls (patients 4–17, Table 3). The gels include quality control markers. Lane P represents a positive control of *C. pneumoniae* (VR1310, ATCC) while lane C represents a distilled water negative control that has been subjected to the entire PCR procedure.

Rockford, IL).

Results

Fifteen of 17 (88%) of the CSF samples from MS patients contained DNA specific for the 16S rRNA gene of *C. pneumoniae* (Fig. 1). Of these 15 CSF samples that were positive by PCR, 8 yielded strongly positive signals for the 446 base pair *C. pneumoniae* product. Because the 643 base pair outer primer product also was present in the sample used in the nested PCR, labeling of it was seen as would be expected. One of the CSF samples from MS patients read as negative (number 5, lower gel) contained hybridization-specific products with a significantly higher size of the major band. This may represent a chlamydial species with a mutation within the inner primers yielding a product of approximately 520 base pairs. In contrast, only two of the 13 CSF samples from the OND control group (number 3, upper gel and number 1, lower gel) yielded some weak hybridization product of the incorrect base pair size and were scored as PCR negative.

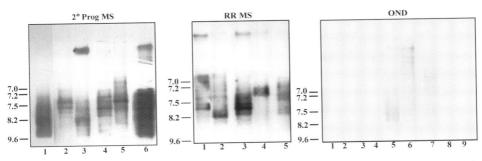

Fig. 2. (Panels a–c) Affinity-driven immunoblot studies on 4 MS patients. Lanes 1–4 represent the banding pattern of oligoclonal antibodies following affinity-driven transfer onto *C. pneumoniae* antigen-coated nitrocellulose membranes and probed with anti-human Ig antibody. Panel a, secondary progressive MS, (Table 1), Panel b, relapsing remitting MS (patient numbers 15, 14, 2, 6, and 7; Table 2) and C, OND controls (patient numbers 1–5, 10, 11, 13, and 6; Table 3).

Table 1. Clinical and CSF profiles on 17 patients with relapsing remitting MS.

Patient number	Age/sex	Time of onset of 1st symptom to time of dx	EDSS	CSF protein concentration; CSF cell count	CSF Ig index*	Oligoclonal bands	CSF PCR/S
1	26/F	1 year	1.5	32 mg/dl; 4 cells/µl	1.28	present	positive
2	47/F	2 months	3.0	47 mg/dl; 7 cells/µl	1.09	present	positive
3	22/F	2 years	1.5	40 mg/dl; 11 cells/µl	0.60	present	positive
4	51/F	3 years	1.5	32 mg/dl; 0 cells/µl	0.61	present	positive
5[b]	22/F	6 months	6.0	34 mg/dl; 10 cells/µl	0.90	present	positive
6	20/F	3 months	1.0	24 mg/dl; 9 cells/µl	1.42	present	positive
7	49/M	1 year	3.5	57 mg/dl; 4 cells/µl	1.44	present	positive
8	50/M	6 months	3.5	86 mg/dl; 2 cells/µl	1.22	present	positive
9	39/F	6 months	2.0	53 mg/dl; 0 cells/µl	0.55	absent	positive
10	29/F	12 years	3.5	22 mg/dl; 0 cells/µl	0.48	absent	positive
11	28/F	4 months	1.5	68 mg/dl; 1 cell/µl	0.5	absent	positive
12	27/F	4 years	2.5	45 mg/dl; 2 cells/µl	1.97	present	positive
13	49/F	1.5 years	2.5	55 mg/dl; 2 cells/µl	1.60	present	positive
14[b]	47/M	2 years	5.5	82 mg/dl; 14 cells/µl	0.67	present	positive
15	26/F	6 months	3.0	20 mg/dl; 6 cells/µl	1.19	present	positive
16	40/F	22 years	2.0	33 mg/dl; 1 cell/µl	0.88	present	positive
17	44/M	6 months	3.0	24 mg/dl; 3 cells/µl	0.46	absent	positive

*The Ig Index is the ratio of CSF/serum immunoglobulin to CSF/serum albumin and is an indication of intrathecal immunoglobulin production.
[b]Patients were taking interferon-ß at the time the spinal tap were performed.
CSF = cerebrospinal fluid; MS = multiple sclerosis; EDSS = Expanded Disability Status Score; PCR/S = polymerase chain reaction/Southern hybridization.

Table 2. Clinical and CSF profiles on 6 patients with progressive MS.

Patient number	Age/ sex	Age of onset	EDSS	CSF protein concentration; CSF cell count	CSF Ig index*	Oligo- clonal bands	Affinity-driven immunoblot
1	40/M	38	6.5	85 mg/dl; 4 cells/µl	0.64	present	positive
2	42/F	38	6.0	18 mg/dl; 3 cells/µl	1.73	present	weakly positive
3	35/F	29	7.0	48 mg/dl; 2 cells/µl	3.69	present	positive
4	27/M	24	6.0	155 mg/dl; 57 cells/µl	0.75	positive	positive
5	56/M	32	8.5	57 mg/dl; 5 cells/µl	not done	present	positive
6	54/M	44	8.5	82 mg/dl; 15 cells/µl	1.39	present	positive

Affinity-driven immunoblotting for the presence of anti-chlamydia cationic antibodies in CSF of MS patients and OND controls

MS patients have an increase in immunoglobulins in the CSF, and part of the increase is represented as oligoclonal bands on isoelectric focusing gels. Oligoclonal bands represent cationic antibodies that focus at their isoelectric points to the anodic region of the gel. The relationship between anti-EB antibodies in CSF and their isoelectric points was examined in 11 MS patients and 8 OND controls. These experiments were carried out by initially performing IEF gels on equal amounts of immunoglobulin from spinal fluid, followed by an affinity transfer onto nitrocellulose membranes incubated with sonicated EB antigens with unbound sites blocked with dehydrated milk. The presence of binding of the *C. pneumoniae* EB antigens to cationic antibodies from the CSF was determined by probing with labeled anti human antibody. Fig. 2 (Panels a–c), shows a representative visualization of the immunoglobulins in 11 MS patients and 8 controls. In all eleven MS patients, antibodies that focussed to the cathodal regions and recognized chlamydial antigens were detected. Anti-*C. pneumoniae* antibodies that migrated to the cathodal region of the gel were not seen in any of the 8 OND controls. When CSF was transferred to membranes coated with myelin basic protein, no bands were seen in either the MS patients or controls (data not shown). These results argue that cationic anti-EB antibodies are present in patients with MS and may represent in part the specificity of the oligoclonal bands.

Discussion

The development of DNA-based classification methods during the 1980s has provided new techniques for differentiating chlamydial groups. Progress in molecular biology using PCR and sequencing of DNA encoding 16S rRNA has enabled determination of the precise phylogenetic position of each bacterial species, particularly those obligate intracellular bacteria that express few phenotypic characters. The primers are chosen from sequences that are highly conserved among the phylogenetic group referred to as the eubacteria, but which are not found in eukaryotes, archaea or mitochondria [10]. It is therefore possible to amplify only

Table 3. Clinical and CSF profiles on patients with other neurologic diseases.

Patient number	Age/sex	Neurologic diagnosis	CSF protein concentration; CSF cell count	CSF Ig index*	Oligoclonal bands	CSF PCR/S	CSF Ig vs EBs by Western Blot
1	NA	SSPE	Not Available	not done	present	not done	negative
2	32/M	CNS Vasculitis	121 mg/dl; 0 cells/μl	2.6	absent	negative	negative
3	25/M	Chronic Meningitis	155 mg/dl; 227 cells/μl	0.5	present	negative	negative
4	80/F	CNS Wegener's	63 mg/dl; 11 cells/μl	not done	not done	negative	weak positive
5	38/F	HSV-2 Myelitis	74 mg/dl; 4 cells/μl	0.2	not done	negative	negative
6	34/M	Thoracic Myelitis	59 mg/dl; 21 cells/μl	0.48	absent	negative	negative
7	36/F	Brain Tumor	137 mg/dl; 6 cells/μl	0.48	absent	negative	negative
8	52/F	Myelitis	51 mg/dl; 2 cells/μl	0.50	absent	positive	negative
9	36/F	Aseptic Meningitis	39mg/dl; 13 cells/μl	0.38	absent	negative	negative
10	41/F	HSV-2 Myelitis	58 mg/dl; 0 cells/μl	0.54	absent	positive	weak positive
11	36/F	CNS Lupus	57 mg/dl; 0 cells/μl	0.46	absent	negative	negative
12	28/F	Vasculitis	58 mg/dl; 1 cell/μl	0.49	absent	negative	negative
13	38/M	Lumbrosacral Plexopathy	62 mg/dl; 0 cells/μl	0.47	absent	negative	negative
14	62/M	W-K Encephalopathy	50 mg/dl; 0 cells/μl	not done	absent	negative	not done

*The CSF Ig Index is the ratio of CSF/serum immunoglobulin (Ig) to CSF/serum albumin and is an indication of intrathecal immunoglobulin production
W-K = Wernicke-Korsakoff.

bacterial 16S rRNA gene sequences even in the presence of nucleic acids from other types of organisms. Because the nucleotide sequences found in 16S rRNA genes vary in an orderly fashion throughout the phylogenetic tree, they have been useful for the study of molecular evolution [10,11]. Thus 16S rRNA gene sequencing is one of the most powerful and precise methods for determining the distant as well as close (intrageneric) genealogical relationships of bacteria.

These results of 16S rRNA PCR testing therefore confirm our earlier identification of *C. pneumoniae* in the CSF of MS patients. Moreover, these results corroborate those from our previous reports in which PCR data with MOMP sequencing (GenBank AF131889) clearly demonstrated an association of *C. pneumoniae* with MS, but not with OND controls. This 16S rRNA method provides an independent PCR procedure with a smaller end product [1,2].

Our results demonstrate that the development of an intrathecal immune response to *C. pneumoniae* is a common occurrence in patients with MS. Although epidemiologic studies have shown the presence of antibody titers to *C. pneumoniae* in over 50% of individuals over 40 years of age, the elevated levels in the CSF of MS patients over that of OND controls strongly argues for a compartmentalized increase within the CNS [12].

In most diseases other than MS in which intrathecal synthesis of antibodies are seen, they represent an immune response to infectious agents. Chronic bacterial and fungal CNS infections such as TB, syphilis, neuroborreliosis, coccidiodo-domycosis, and cryptococcal meningitis are all characterized by the presence of increased intrathecal synthesis of antibody [12]. Similarly, oligoclonal bands constitute the major immune response to the respective viral antigens in chronic viral infections such as SSPE, rubella panencephalitis, and HTLV-1 myelitis [13–15]. The specificity of the intrathecal antibodies to infectious agents has been proven with adsorption studies in CNS syphilis, HTLV-1 myelopathy and neuroborreliosis [13,16,17].

Prior studies have attempted to define the nature of the putative infectious agent in MS by determining the antigenic specificity of the intrathecal immune response [13,16–19]. Using IEF and immunoblot techniques, investigators have noted that CSF antibodies react to a number of infectious agents and self-antigens (e.g., heat shock protein) that focus at the cathodal region of the gel. However, these bands did not constitute the major oligoclonal bands, since incubation of CSF immunoglobulins with these antigens did not adsorb out the oligoclonal bands. The results of IEF/affinity-driven assays demonstrate that a CNS immune response specifically to *C. pneumoniae* EB antigens is present in patients with MS. More specific adsorption studies with infectious antigens of *C.pneumoniae* are likely to show the exact specificity of the oligoclonal bands and further confirm the specificity of the intrathecal immune response.

Acknowledgement

This study was supported in part by grants from the National MS Society and the E. Proctor and W. Weaver Fund, and a contract with Merlin Technologies, Inc.

References

1. Sriram S, Stratton C, Mitchell W. Multiple sclerosis associated with *Chlamydia pneumoniae* infection of the central nervous system. Neurol 1998;50;571–572.
2. Sriram S, Stratton CW, Yao S, et al. *Chlamydia pneumoniae* infection of the central nervous system in patients with multiple sclerosis. Ann Neurol 1999;46;6–14.
3. Everett KDE, Anderson AA. The ribosomal intergenic spacer and domain 1 of the 23S rRNA gene are phylogenetic markers for the *Chlamydia* species. Int J Syst Bacteriol 1997;47;461–473.
4. Melgoza-Perez PM, Kuo CC, Campbell LA. Sequence analysis of the MOMP membrane proteins of *Chlamydia pneumoniae*. Infect Immun 1991;59;2195–2199.
5. Gaydos CA, Eiden JJ, Oldbach D. Diagnosis of *Chlamydia pneumoniae* infection in patients with community acquired pneumonia by polymerase chain reaction. Clin Infect Dis 1994;19;157–160.
6. Everett KDE, Bush RM, Anderson AA. Emended description of the order *Chlamydiales*. Int J Syst Bacteriol 1999;49;415–440.
7. Tourtelotte WW, Tumani H. Multiple sclerosis cerebrospinal fluid. In: Raine CS, McFarland HF, editors. Multiple Sclerosis: Clinical and Pathogenic Basis, 1997;57–79.
8. Dorries R, Muelen VT. Detection and identification of virus specific oligoclonal IgG in unconcentrated cerebrospinal fluid by immunoblot techniques. J Neuroimmunol 1984;7;77–89.
9. Sindic CJ, Monteyne P, Laterre EC. The intrathecal synthesis of virus-specific oligoclonal IgG in multiple sclerosis. J Neuroimmunol 1994;54;75–80.
10. Wilson KH, Blitchington RB, Greene RC. Amplification of bacterial 16S ribosomal DNA with polymerase chain reaction. J Clin Microbiol 1990;28;942–946.
11. Woese CR. Bacterial evolution. Mcrobiol Rev 1987;51;221–271.
12. Kuo CC, Jackson LA, Campbell LA, Grayston JT. *Chlamydia pneumoniae* (TWAR). Clin Microbiol Rev 1995;8;451–461.
13. Link H, Cruz M, Gessain A, Gout O, de The G, Kam-Hansen S. Chronic progressive myelopathy associated with HTLV-1: oligoclonal IgG and anti-HTLV-1 IgG antibodies in cerebrospinal fluid and serum. Neurol 1989;39;1566–1572.
14. Vandvik B, Nilsen RE, Vartdal F, Norrby E. Mumps meningitis: specific and non-specific antibody responses in the central nervous system. Acta Neurol Scand 1982;65;468–487.
15. Vartdal F, Vandvik B. Multiple sclerosis. Electrofocused "bands" of oligoclonal cerebrospinal fluid IgG do not carry antibody activity against measles, varicella-zoster or rotaviruses. J Neurol Sci 1982;54;99–107.
16. Vartdal F, Vandvik B, Michaelsen TE, Loe K, Norrby E. Neurosyphilis: intrathecal synthesis of oligoclonal antibodies to *Treponema pallidum*. Ann Neurol 1982;11;35–40.
17. Wang ZY, Hansen K, Siden A, Cruz M. Intrathecal synthesis of anti-*Borrelia burgdorferi* antibodies in neuroborreliosis: a study with special emphasis on oligoclonal IgM antibody bands. Scand J Immunol 1993;37;369–376.
18. Prabhakar S, Kurien E, Gupta RS, Zielinski S, Freedman MS. Heat shock protein immunoreactivity in CSF: correlation with oligoclonal banding and demyelinating disease. Neurol 1994;44;1644–1648.
19. Rand KH, Houck H, Denslow ND, Heilman KM. Molecular targets for oligoclonal bands in multiple sclerosis. J Neurol Neurosurg Psych 1998;65;48–55.

© 2001 Elsevier Science B.V. All rights reserved.
Genes and Viruses in Multiple Sclerosis.
O.R. Hommes, H. Wekerle, M. Clanet, editors.

A rationale for antiviral therapy in multiple sclerosis

Oluf Andersen*

Institute of Clinical Neuroscience, Sahlgrenska Department of Neurology, Sahlgrenska University Hospital, SE 413 45 Gothenburg, Sweden

Introduction

While the mainstay of MS therapy is suppression of the central nervous system (CNS) inflammation by immunomodulation, an attractive complementary method would be to reduce the activity of the systemic immunity by infection prophylaxis, thereby reducing the number of activated T cells crossing the blood-brain barrier. We examine a possible rationale for such intervention under four headings.

Rationale (1): Candidate self-replicating agents

A direct causal role of self-replicating viral infection of the CNS infection in MS has largely been discounted. Certainly, a number of virus isolations from CNS were reported in MS patients. However, subsequently negative results were reported from polymerase chain reaction (PCR) studies for candidate childhood diseases in MS brain autopsy material (Godec et al., 1992) and (using RT-PCR) for entero- and cardioviruses (used in experimental MS models) in specimens from 25 MS patients (Dessau et al., 1997). PCR studies in MS autopsy material were also essentially negative for coronavirus, as PCR studies were for coronavirus in the CSF of patients with optic neuritis (Dessau et al., 1999). Human herpes 6 (HHV-6) DNA was reported to be present in active MS plaques but also in controls and was therefore considered to be commensal in the human brain. However, it was reported to have a specific nuclear localisation in MS oligodendrocytes (Challoner et al., 1995), but this finding and the later report of HHV-6 DNA in serum samples from MS patients (Soldan et al., 1997) seem to be unconfirmed. The rationale for antiherpes therapy in MS was based on the neuronotropism of herpes simplex virus (HSV), particularly the capacity of HSV-1 to propagate axonally not only in the PNS but also in the CNS, and on its intermittent reactivation in vitro and in vivo. Furthermore, we isolated an atypical HSV from the CSF of a MS patient during her first bout (Bergström et al., 1989). Subsequently, a 2-year randomized study with the anti-herpetic drug Acyclovir in 60 patients with relapsing-remitting MS showed a reduction in the number of relapses by 34% in the treatment group as compared to the placebo group. This reduction was not statistically significant (p = 0.083). Meanwhile, the reduction in relapse rate was significant

---

Correspondence address: Sahlgrenska University Hospital, Institute of Clinical Neuroscience, Department of Neurology, SE 413 45, Göteborg, Sweden. E-mail: oluf.andersen@neuro.gu.se

in a subgroup of patients who had a 2-year baseline observation before the trial allowing for individually paired tests (Lycke *et al.*, 1989). This promising result prompted the Scandinavian Study Group of Antiviral Treatment in MS to perform a trial with Valacyclovir, a prodrug to Acyclovir known to have better tissue penetration than Acyclovir, and a similar low toxicity. In a two-center study (Universities of Århus, Denmark and Göteborg, Sweden) 75 patients were treated with Valacyclovir 1000 mg thrice a day or placebo for 24 weeks. The primary outcome parameter, the total number of active lesions on monthly MRI, did not show any useful therapeutic effect (Bech E et al, publication in progress).

Rationale (2): Candidate childhood/adolescence initiating infections

A higher proportion of children showed positive titers to many viral diseases early in life in areas where MS is rare compared with areas where MS is common (Alter *et al.*, 1986). A number of studies within high-risk areas showed that the age at which several childhood infections are contracted tends to be higher in persons who subsequently develop MS (Alvord *et al.*, 1987) (Gronning *et al.*, 1993). A recent Swiss study confirmed a significant shift to higher ages for several childhood diseases retrospectively studied in MS patients (Bachmann and Kesselring, 1998).This included an individual relationship between infectious mononucleosis (IM) and subsequent MS, the EBV infection more often manifesting as IM when the infection occurs in adolescence. The first indication of this relationship came from a case-control study (Operskalski *et al.*, 1989). We cross-matched our MS-register with the diagnosis register of the Department of Infectious Diseases in Göteborg. There was a 3 times increased risk of MS onset subsequent to infectious mononucleosis (IM). The latency was on average 12 years (Lindberg *et al.*, 1991). The 3 times surplus risk was confirmed in a larger study, combining the central Danish MS and EBV registers (Haahr *et al.*, 1995). A further study confirmed this general risk, with a teenage subgroup having an eight times surplus risk (Martyn *et al.*, 1993). Furthermore, 100% of MS patients (as compared with 162/170 controls) had IgG antibodies to EB virus (Myhr *et al.*, 1998). However, the relatively late occurrence of childhood infections, which seems to determine a higher risk for subsequent MS, may result from a genetic protective or hyperimmune state in the future MS patient that delays the common infectious events through adolescence. Acyclovir was used in IM and reported to inhibit oropharyngeal shedding, but three trials failed to show any benefit of acyclovir in IM (Andersson *et al.*, 1986). Measles cannot be a prerequisite for MS, as MS onset sometimes precedes measles (Ryberg, 1979). However, the national vaccination programs implemented to eradicate measles and two other childhood diseases may alter the disease spectrum sufficiently to reduce MS incidence. By scrutinizing school doctors' records in Göteborg we found a sharp division between the measles unvaccinated birth cohorts, until 1968, and the partially measles vaccinated cohorts from 1969 on. The Swedish birth cohorts from 1981 were vaccinated at both 18 months and 12 years of age, thus entering one of the first two measles combined (two age) eradication programs implemented in the world. The age-specific MS incidence at 15–19 years is so far essentially unchanged, and the measles eradicated 20+ years age cohort is starting to appear now. So far, there is no indication of any abrupt change. A dimensioning computation shows that a Swedish nationwide (10 million population base) survey would be needed to ascertain a theoretical 20% reduction in the completely vaccinated cohort during the next five years (Ahlgren et al., unpublished data).

Rationale (3): Candidate systemic trigger infections

There is experimental proof that non-replicating antigen can persist locally in the CNS. However, an appropriate systemic infection may later elicit a specific local CNS inflammation.
- Heat-killed Calmette bacillus were sequestered behind the blood-brain barrier. Two to three months after subcutaneous challenge with the bacillus there was local CNS demyelination (Matyszak et al., 1997).
- Defective (E1-deleted) adenovirus used as a vector in the CNS did no harm, but after subcutaneous injection of the adenovirus cellular infiltration was seen in brain areas containing neurons capable of retrograde transport of the adenovirus vector, supporting the notion that viral antigens that are retrogradely transported by neurons can also be the target of a T cell attack (Byrnes et al., 1996).
- In rodent experiments, lymphocytic choriomeningitis virus (LCMV) was inactive in oligodendrocytes. When replicating LCMV was given parenterally to these animals, LCMV-specific T lymphocytes were activated systemically, crossed the blood-brain barrier and caused disease (Selin et al., 1994; von Herrath and Oldstone, 1996).

In addition there is evidence that memory cells may be long-lived and persist in the absence of antigen (Murali-Krishna et al., 1999). The epidemiology of MS is compatible with a persisting antigen or immunological memory in the CNS. Five clinical studies indicate that relapses in established MS cases are triggered by systemic viral infections. An "at risk" (AR) period was defined in relation to the onset day of a common viral infection, usually an upper respiratory tract infection (URTI), but sometimes gastrointestinal. The AR period is usually computed as one week before and 3 weeks after the onset of an URTI, sometimes as 2 weeks before and 5 weeks after the onset. The remaining observation time is "not at risk" (NAR). The studies (Table 1) show an increased risk of a relapse during the AR period. The maximum observed risk occurred approximately 2 weeks after the URTI. We confirmed the relationship between URTI and MS relapses using a 4 weeks AR period. There was a significant periannual variation with a winter maximum in AR MS relapses, but no significant periannual variation of the NAR MS relapses. There was a bimodal periannual distribution for the AR relapses, compatible with e.g. rhinovirus infections. However, no rhinovirus diagnostics were available at that time, and the serological data suggested that there was a significantly increased risk for an adenovirus infection to be followed by an AR relapse. There was no

Table 1. Five studies showing increased risk of a relapse of MS associated with an upper respiratory tract infection (URTI). For explanation of the AR/NAR periods, see text.

(Sibley and Foley, 1965)	33/69 relapses associated with infection
(Sibley et al., 1985)	AR/NAR relapse rate 0.64/0.23 p<0.001
(Andersen et al., 1993)	AR/NAR relapse rate 1.7/1.29 p=0.048
(Panitch, 1994)	AR/NAR relapse rate 2.92/1.16 p<001
(Edwards et al., 1998)	AR/NAR relapse rate 3.3/1.6 p=0.004
	AR/NAR serology 3.4 p=0.006

significant relationship between serologically confirmed influenza virus infections and MS relapses. For the other viral serologies tested, parainfluenza, respiratory syncytial virus (RSV) and mycoplasma, there was not enough AR relapses for conclusive statistics. The inference from this data set is that adenovirus has, or at least belongs to a group of viruses that has, a relationship to MS relapses, while influenza virus does not (Andersen et al., 1993). Adenovirus establishes long or latent infections in the human host. Thirty percent of healthy adults excrete adenovirus by feces, the virus probably originating from lymphoid tissues in Peyer's plaques (Allard et al., 1992).

The five studies in Table 1 were fairly consistent in showing an increased MS relapse risk immediately after common viral infections, although the relative risk was lower in the Göteborg study. One possible explanation could be differences in the neurological tradition of evaluation of MS exacerbations as bouts as opposed to (febrile) pseudobouts, but the findings may be explained by heterogeneity in genetics and infection spectra. The serological parts of the studies provide divergent, mostly vague results. Complement fixation methods were generally used, which are insensitive. One further reason for the divergence may be that the specimens for serology were obtained at pre-set regular intervals in four of the studies. In the Göteborg study the time of obtaining blood specimens for serology was determined by the time of infections (as acute and convalescence specimens). Patients were instructed to come to the clinic for venipuncture immediately at the appearance of the signs of a common cold. The healthy control group in the second study (Table 1) importantly demonstrated a higher infection frequency than in MS patients, compatible with the notion of particularly efficient immunity in MS patients (Sibley et al., 1985). The periannual distribution with a summer trough differs from most other studies (Jin et al., 1999). Again, this may be due to heterogeneity in immunogenetics and infection spectra. It was shown that beta-interferon does not exert its effect on MS relapses by reducing the number of infections. Rather, it dissociates the relapses from the infections (Panitch, 1994).

A new development is the capability to diagnose almost all URTI, with a diagnostic yield of 60–85%, using nasal washings and PCR (Arruda et al., 1997; Mäkelä et al., 1998). The use of these methods for surveillance should enable us to diagnose almost all viral infections to define the AR periods. There are many emerging therapies for URTI, using principles like nucleoside analogues, local interferons, vaccinations, and other principles (Johnston, 1997). A phase III clinical trial is ongoing for the treatment of rhinovirus with an agent which stiffens the capside, thus inhibiting the release of RNA from the rhinovirus particle (Vaidehi and Goddard, 1997) (Hadfield et al., 1999). Also, development of inhibitors of the protease that cleave the polyprotein precursor to rhinoviruses is in progress (Wang, 1999). Antiviral agents are available for short periods of prophylaxis and treatment of respiratory syncytial virus and influenza virus. The epidemic pattern of adenovirus type-4 infections in army trainees indicated that vaccination was effective as prophylaxis (Hendrix et al., 1999). Monoclonal antibodies may be used to prevent mucosal transmission of UTRI (Zeitlin et al., 1999).

Rationale (4): Candidate balanced heteromorphisms

The large genetic component in MS, as demonstrated in twin and adoption studies and genome scans, seems to be polygenic, probably related to immune defence genes. A body of evidence indicates that evolution of immune defence genes occurs under the selection pressure of changing infections (Hill, 1998), and they mutate much faster than most other genes. Infections belong to the most heritable of human diseases, as shown in a study on several causes of death in adoptees (Sörensen et al., 1988). Specific HLA, TNF or chemokine genes were associated with disease severity in widely different human infections caused by hepatits B virus (HBV), human immunodeficiency virus (HIV), and malaria, notably the protection against HIV by a variant of the chemokine CCR5 (Martin et al., 1998; Thursz and Thomas, 1997). In a human twin study on HBV carrier state, monozygotic twins were concordant for HBsAg (HBV soluble antigen) carriage in 50% of cases, whereas only 20% of dizygotic twins and untwinned siblings were concordant (Lin et al., 1989). The ABO blood group system is associated with infection susceptibility in several ways. Secretor status of blood group antigens is associated with increased risk of several URTI (Raza et al., 1991). In some instances a relationship between specific immune protection and the risk for autoimmunity was revealed. The natural resistance-associated macrophage protein (Nramp1) gene determines the ability of mice to resist infection with salmonella and mycobacteria. In the Nramp1 gene there are two common alleles, one of which is associated with activation of macrophages and linked to protection against these disorders, but also to autoimmune disease, while the other allele has the opposite effects (Searle and Blackwell, 1999). Several genes, each with minor effects, may augment or suppress autoimmunity, as was shown experimentally (Morel et al., 1999). However, the occurrence of certain human autoantibodies was also related to previous exposure to measles and other infections (Lindberg et al., 1999). It was suggested that the infrequent occurrence of autoimmunity in tropical Africa is related to multiple infections (Greenwood, 1968). Mortality rates from a variety of infectious diseases correlated negatively with the MS mortality (Alter et al., 1986). We found that MS is associated with significantly decreased fertility already in the preclinical stage (Runmarker and Andersen, 1995). As the population incidence of MS is not decreasing, this finding of low fertility is compatible with balanced heteromorphism, conceivably driven by an infection spectrum. A protective component in this disposition may explain the low infection frequency in MS (Sibley et al., 1985), and local variations in this driving spectrum may explain heterogeneity of predisposing HLA genetics. While HLA-DR2 is associated with the highest relative risk factor for MS (3–4 times increase) in northern Europe and America, HLA-DR4 was reported to prevail or to be important in some other areas such as Sardinia (Marrosu et al., 1998), Turkey (Saruhan-Direskeneli et al., 1997), and in Japanese patients with "Western"-type MS without oligoclonal bands (Fukazawa et al., 1998). Probably, immune defence genes are important but must be specifically activated in order to contribute to MS.

Are these selection pressures still working, and if so, could they be modified? Immunological specificity has both genetic and somatic components: Immunoglobulin and T-cell receptor genes, rearrangement of genes, and for B-cells: Hypermutation, affinity maturation, that in principle does not occur until the cell has met its specific antigen. A central question is whether the antibody response against measles and other early virus disorders is genetically determined (Ochsenbein and Zinkernagel, 2000), or whether it is the result of an infection (notwithstanding complicated immunopathology in either case). Results from studies of the measles antibody avidity using increasing salt concentrations showed that the CSF antibodies in MS had low affinity, different from encephalitis where specific antibodies were high avidity (Luxton et al., 1995). This may indicate that the CSF measles antibodies are natural (genetically determined). However, recent studies of somatic mutations showed a pattern of hypermutation indicating antigen-driven response (Smith-Jensen et al., 2000), tending to weight the interpretation of rationale (2) above from genetics to (possibly abnormal) early infectious events. Also, a divergence in the vigorous antibody response to two different glycoproteins in rubella virus in MS suggests that the antibody response is not simply a non-specific polyvalent B-cell response (Nath and Wolinsky, 1990). Subclinical CNS involvement occurs regularly in common childhood or adolescence diseases: A CSF pleocytosis was described in 30% of uncomplicated measles cases (Hänninen et al., 1980), and in 25% of uncomplicated mononucleosis cases. In an experimental MS model, CNS persistence of the mouse hepatitis virus was dependent on anti-viral antibody.

A crucial event in MS is the migration of activated T cells across the blood-brain-barrier. T cells specific for foreign and those for non-tolerant peripheral self do not differ with respect to induction requirements. The therapeutic principle in MS has been immunosuppression or at least immunomodulation intending to inhibit the spread of inflammation in the CNS. An alternative therapeutic strategy MS would be to reduce the number of circulating activated T cells by reducing the peripheral viral activation of the immune system by diminishing the intensity and frequency of systemic infections. However, to accomplish this we need detailed knowledge on the rationale 1 through 4, and perform sensible clinical trials using the essential information in these four elements of rationale.

References

Allard A, Albinsson B, Wadell G. Detection of adenoviruses in stools from healthy persons and patients with diarrhoea by two-step polymerase chain reaction. J Med. Virol. 1992;37:149–157.

Alter M, Zhang Z, Davanipour Z, Sobel E, Zibulewski J, Schwartz G, et al. Multiple sclerosis and childhood infections. Neurology 1986;36:1386–1389.

Alvord E, Jahnke U, Fischer E, Kies M, Driscoll B, Compston D. The multiple causes of multiple sclerosis: The importance of age of infections in childhood. J Child Neurol 1987;2:313–321.

Andersen O, Lygner P-E, Bergström T, Andersson M, Vahlne A. Viral infections trigger multiple sclerosis relapses: a prospective seroepidemiological study. J Neurol 1993;240:417–422.

Andersson J, Britton S, Ernberg I, Andersson U, Henle W, Sköldenberg B, et al. Effect of acyclovir on infectious mononucleosis: a double-blind, placebo-controlled study. J Infect Dis 1986;153:283–90.

Arruda E, Pitkaranta A, Witek TJJ, Doyle CA. Frequency and natural history of rhinovirus infections in adults during autumn. J Clin Microbiol 1997;35:2864–2868.

Bachmann S, Kesselring J. Multiple sclerosis and infectious childhood diseases. Neuroepidemiology 1998;17:154–160.

Bergström T, Andersen O, Vahlne. Isolation of herpes virus type 1 during first attack of multiple sclerosis. Ann Neurol 1989;26:283–285.

Byrnes AP, MacLaren RE, Charlton HM. Immunological instability of persistent adenovirus vectors in the brain: Peripheral exposure to vector leads to renewed inflammation, reduced gene expression, and demyelination. Journal of Neuroscience 1996;16:3045–3055.

Challoner PB, Smith KT, Parker JD, MacLeod DL, Coulter SN, Rose TM, et al. Plaque-associated expression of human herpesvirus 6 in multiple sclerosis. Proc Natl Acad Sci U S A 1995;92:7440–7444.

Dessau R, Lisby G, Frederiksen J. Coronaviruses in spinal fluid of patients with acute monosymptomatic optic neuritis. Acta Neurol Scand 1999;100:88–91.

Dessau RB, Nielsen LP, Frederiksen JL. Absence of entero- and cardioviral RNA in multiple sclerosis brain tissue. Acta neurol Scand 1997;95:284–286.

Edwards S, Zvartau M, Clarke H, Irving W, Blumhardt LD. Clinical relapses and disease activity on magnetic resonance imaging associated with viral upper respiratory tract infections in multiple sclerosis. J Neurol Neurosurg Psychiatry 1998;64:736–741.

Fukazawa T, Kikuchi S, Sasaki H, Hamada K, Hamada T, Miyasaka K, et al. The significance of oligoclonal bands in multiple sclerosis in Japan: relevance of immunogenetic backgrounds. J Neurol Sci 1998;158:209–214.

Godec MS, Asher DM, Murray RS, Shin ML, Greenham LW, Gibbs CJ, et al. Absence of measles, mumps, and rubella viral genomic sequences from multiple sclerosis brain tissue by polymerase chain reaction. Ann Neurol 1992;32:401–404.

Greenwood B. Autoimmune disease and parasitic infections in Nigerians. Lancet 1968;II: 380–382.

Gronning M, Riise T, Kvale G, Albrektsen G, Midgard R, Nyland H. Infections in childhood and adolescence in multiple sclerosis. Neuroepidemiology 1993;12:61–69.

Haahr S, Koch-Henriksen N, Möller-Larsen A, Eriksen LS, Andersen HMK. Increased risk of multiple sclerosis after late Epstein-Barr virus infection: a historical prospective study. Multiple Sclerosis 1995;1:73–77.

Hadfield A, Diana G, Rossman M. Analysis of three structurally related antiviral compounds in complex with human rhinovirus 16. Proc Natl Acad Sci USA 1999;96:14730–14735.

Hendrix R, Lindner J, Benton F, Monteith S, Tuchscherer M, Gray G, et al. Large, persistent epidemic of adenovirus type 4-associated acute respiratory disease in US army trainees. Emerg Infect Dis 1999;5:798–801.

Hill A. The immunogenetics of human infectious diseases. Annu Rev Immunol 1998;16:593–617.

Hänninen P, Arstila P, Lang H, Salmi A, Panelius M. Involvement of the central nervous system in acute, uncomplicated measles virus infection. Journal of Clinical Microbiology 1980;11:610–613.

Jin Y-P, de Pedro-Cuesta J, Söderström M, Link H. Incidence of optic neuritis in Stockholm, Sweden, 1990-1995. Arch Neurol 1999;56:975–980.

Johnston SL. Problems and prospects of developing effective therapy for common cold viruses. Trends in Microbiology 1997;5:58–63.

Lin TM, Chen CJ, Wu MM, Furthermore AA. Hepatitis B virus markers in Chinese twins. Anticancer research 1989;9:737–741.

Lindberg B, Ahlfors K, Carlsson A, Ericsson U, Landin-Olsson M, Lernmark Å, et al. Previous exposure to measles, mumps, and rubella – but not vaccination during adolescence – correlates to the prevalence of pancreatic and thyroid autoantibodies. Pediatrics 1999;104:e12.

Lindberg C, Andersen O, Vahlne A, Dalton M, Runmarker B. Epidemiological investigation of the association between infectious mononucleosis and multiple sclerosis. Neuroepidemiology 1991;10:62–65.

Luxton R, Zeman A, Holzel H, Harvey P, Wilson J, Kocen R, et al. Affinity of antigen-specific IgG distinguishes multiple sclerosis from encephalitis. J Neurol Sci 1995;132:11–19.

Lycke J, Andersen O, Svennerholm B, Appelgren L, Dahlöf C. Acyclovir concentrations in serum and cerebrospinal fluid at steady state. J Antimicrob Chemother 1989;24:947–954.

Marrosu MG, Murru MR, Costa G, Murru R, Muntoni F, Cucca F. DRB1-DQA1-DQB1 loci and multiple sclerosis predisposition in the Sardinian population. Hum Mol Genet 1998;7:1235–1237.

Martin MP, Dean M, Smith MW, Winkler C, Gerrard B, Michael NL, et al. Genetic acceleration of AIDS progression by a promoter variant of *CCR5*. Science 1998;282:1907–1911.

Martyn CN, Cruddas M, Compston DAS. Symptomatic Epstein-Bar virus infection and multiple sclerosis. J Neurol Neurosurg Psychiatry 1993;56:167–168.

Matyszak MK, Townsend MJ, Perry VH. Ultrastructural studies of an immune-mediated inflammatory response in the CNS parenchyma directed against a non-CNS antigen. Neuroscience 1997;78:549–560.

Morel L, Tian X, Croker B, Wakeland E. Epistatic modifiers of autoimmunity in a murine model of lupus nephritis. Immunity 1999;11:131–139.

Murali-Krishna K, Lau L, Sambhara S, Lemonnier F, Altman J, Ahmed R. Persistence of memory CD8 cells in MHC Class1-deficient mice. Science 1999;286:1377–1381.

Myhr K-M, Riise T, Barrett-Connor E, Myrmel H, Vedeler C, Grönning M, et al. Altered antibody pattern to Epstein-Barr virus but not to other herpesviruses in multiple sclerosis: a population based case-control study from western Norway. J Neurol Neurosurg Psychiatry 1998;64:539–542.

Mäkelä MJ, Puhakka T, Ruuskanen O, Leinonen M, Saikku P, Kimpimäki M, et al. Viruses and bacteria in the etiology of the common cold. Journal of Clinical Microbiology 1998;36:539–542.

Nath A, Wolinsky JS. Antibody response to rubella virus structural proteins in multiple sclerosis. Ann Neurol 1990;27:533–536.

Ochsenbein A, Zinkernagel R. Natural antibodies and complement link innate and acquired immunity. Immunol Today 2000;21:624–630.

Operskalski EA, Visscher BR, Malmgren RM, Detels R. A case-control study of multiple sclerosis. Neurology 1989;39:825–829.

Panitch HS. Influence of infection on exacerbations of multiple sclerosis. Ann Neurol 1994;36:S25–S28.

Raza M, Blackwell C, Molyneaux P, James V, Ogilvie M, Inglis J, et al. Association between secretor status and respiratory viral illness. BMJ 1991;303:815–818.

Runmarker B, Andersen O. Pregnancy is associated with a lower risk of onset and a better prognosis in multiple sclerosis. Brain 1995;118:253–261.

Ryberg B. Acute measles infection in a case of multiple sclerosis. Acta Neurol Scand 1979;59:221–224.

Saruhan-Direskeneli G, Esin S, Baykan-Kurt B, Ornek I, Vaughan R, Eraksoy M. HLA-DR and -DQ associations with multiple sclerosis in Turkey. Hum Immunol 1997;55:59–67.

Searle S, Blackwell J. Evidence for a functional repeat polymorphism in the promotor of the human NRAMP1 gene that correlates with autoimmune versus infectious disease susceptibility. J Med Genet 1999;36:295–9.

Selin L, Nahill S, Welsh R. Cross-reactivities in memory CTL recognition of heterologous viruses. J Exp Med 1994;179:1933–1943.

Sibley WA, Bamford CR, Clark K. Clinical viral infections and multiple sclerosis. Lancet 1985;1:1313–1315.

Sibley WA, Foley JM. Infection and immunization in multiple sclerosis. Ann N Y Acad Sci 1965;122:457–468.

Smith-Jensen T, Burgoon M, Anthony J, Kraus H, Gilden D, Owens G. Comparison of immunoglobulin G heavy-chain sequences in MS and SSPE brains reveals an antigen-driven response. Neurology 2000;54:1227–1232.

Soldan SS, Berti R, Salem N, Secchiero P, Flamand L, Calabresi PA, et al. Association of human herpes 6 (HHV-6) with multiple sclerosis: Increased IgM response to HHV-6 early antigen and detection of serum HHV-6 DNA. Nature Medicine 1997;3:1394–1397.

Sörensen T, Nielsen G, Andersen P, Teasdale T. Genetic and environmetal influences on premature death in adult adoptees. NEJM 1988;318:727–732.

Thursz MR, Thomas HC. Host factors in chronic viral hepatitis. Seminars in liver disease 1997;17:345–350.

Vaidehi N, Goddard WA. The pentamer channel stiffening model for drug action on human rhinovirus HRV-1A. Proc Natl Acad Sci USA 1997;94:2466–2471.

Wang Q. Protease inhibitors as potential antiviral agents for the treatment of picornaviral infections. Prog Drug Res 1999;52:197–219.

von Herrath MG, Oldstone MB. Virus-induced autoimmune disease. Curr Opin Immunol 1996;8:878–885.

Zeitlin L, Cone R, Whaley K. Using monoclonal antibodies to prevent mucosal transmission of epidemic infectious diseases. Emerg Infect Dis 1999;5:54–64.

Antiviral activity of antiherpetic drugs in lymphoblast cells infected with human herpesvirus 6

L. Naesens*, L. De Bolle and E. De Clercq

Rega Institute for Medical Research, Katholieke Universiteit Leuven, B-3000 Leuven, Belgium

Introduction

Human herpesvirus 6 (HHV-6) is an ubiquitous virus for which the reported seroprevalence rates, depending on the age, vary between 40% and 100%. The virus is acquired early in childhood, and is the primary cause of exanthem subitum (roseola infantum) [1,2]. This febrile exanthematous syndrome has a benign course in the majority of infants, but is infrequently associated with febrile seizures or severe encephalitis [3]. In immunosuppressed adults, HHV-6 reactivation can result in generalized infections that are potentially life-threatening and manifested by pneumonia, hepatitis or encephalitis. There is increasing evidence that HHV-6 is associated with disease or graft failure after bone marrow or organ transplantation [4–7]. Finally, several investigators have recently proposed a link between HHV-6 and multiple sclerosis (MS), based on the laboratory observations that MS patients show elevated anti-HHV-6 IgM antibody titers, higher viral DNA load in serum, marked presence of HHV-6 proteins in active brain lesions or HHV-6 viremia in blood [8–12]. However, the inconsistency in the laboratory reports [13,14], and the perception that the reactivation of HHV-6 in these patients may be the consequence of immune dysfunction, rather than evidence for an active role of HHV-6 in MS pathogenesis, warrant against hasty conclusions. A clearer insight can be expected from larger multi-centered clinical studies using standardized and validated techniques to measure HHV-6, in combination with fundamental studies on the biological effects of HHV-6 in cultured lymphocytes or neural cells.

HHV-6 exists as two variants (HHV-6A and HHV-6B), which are clearly different in antigen reactivity to monoclonal antibodies, structure and function of selected genes and the efficiency by which they propagate in selected T-lymphoblast cell lines. It is believed that these distinct *in vitro* properties reflect a different cell tropism and, hence, clinical pathology for HHV-6A and HHV-6B. For instance, while HHV-6B is clearly the major cause of exanthem subitum, an epidemiological study on 668 CSF samples has indicated that HHV-6A may have higher potential to establish persistent infections of the brain [15].

HHV-6 belongs, together with human cytomegalovirus (CMV), to the betaherpesviruses. HHV-6 and CMV show some similarity with regard to morphology

---

*Correspondence address: Dr. L. Naesens, Rega Institute for Medical Research, Katholieke Universiteit Leuven, Minderbroedersstraat 10, B-3000 Leuven, Belgium. Tel.: +32-16-337345; Fax: +32-16-337340; E-mail: lieve.naesens@rega.kuleuven.ac.be

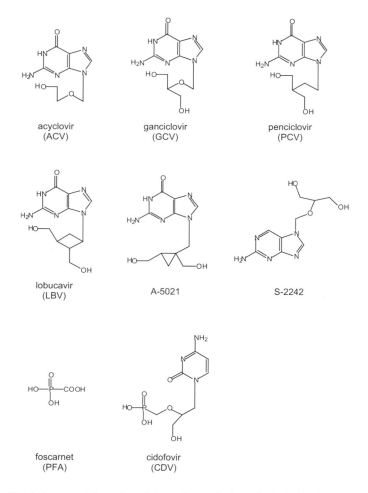

Fig. 1. Structural formulae of the antiherpetic drugs included in the study.

and genomic structure, and, hence, it has been presumed that these two viruses may have a comparable profile in their sensitivity to antiviral drugs. In the present report, we describe the anti-HHV-6 activity of a selection of antiviral compounds, mainly nucleoside analogues (Fig. 1). A number of these compounds, i.e., foscarnet, acyclovir, ganciclovir, penciclovir and cidofovir are commonly used in the therapy of infections with herpes simplex virus (HSV), varicella-zoster virus (VZV) or CMV. Our *in vitro* studies were performed in HHV-6-infected T-lymphoblast cells, and using a DNA hybridization assay to estimate viral replication. We also comment on the observed anti-HHV-6 efficacy and selectivity, by discussing the activation pathway of these nucleoside analogues in virus-infected cells and/or their inhibitory effect at the level of the target enzyme, i.e., the viral DNA polymerase.

Materials and methods

Cells

Two human immature T-lymphoblastoid cell lines were used: HSB-2, obtained from American Type Culture Collection (No. CCL 120.1), and MOLT-3, purchased from ABI Technologies (Columbia, MD, USA). The cells were propagated in RPMI 1640 medium, supplemented with 10% fetal calf serum, 2 mM L-glutamine, 0.075% sodium bicarbonate and gentamycin. The cultures were incubated at 37°C in a humidified and CO2-controlled incubator.

Virus stocks

The GS strain of HHV-6 variant A was kindly provided by Dr. R. Gallo (when at NCI, Bethesda, USA) [16], while the Z-29 strain of HHV-6 variant B was obtained from ABI Technologies (Columbia, MD, USA). For preparation of virus stocks, suspensions of HSB-2 cells infected with strain GS, or MOLT-3 cells infected with strain Z-29, were collected at 10–12 days post infection, when the cytopathic effect (CPE) was maximal and cells were ≥90% virus-positive (as determined by immunofluorescence detection of HHV-6 antigens). Since the titer of virus released in culture supernatant was insufficient, virus stocks were used as complete cell preparations, which were relatively stable when kept in aliquots at −80°C.

Antiviral drugs

The following antiherpetic drugs were used: foscarnet {PFA, Foscavir®; from Astra, Södertälje, Sweden}; acyclovir {ACV, Zovirax®; from Glaxo Wellcome, Research Triangle Park, NC, USA}; ganciclovir {GCV, Cymevene®; from Syntex, Palo Alto, CA, USA}; cidofovir {CDV, Vistide®; from Gilead Sciences, Foster City, CA, USA}, penciclovir {PCV; from Dr. I. Winkler, Hoechst AG, Frankfurt, Germany}, lobucavir {cyclobut-G, LBV, BMS-180194; from Bristol Myers Squibb, Princeton, NJ, USA}, S-2242 {2-amino-7-[(1,3-dihydroxy-2-propoxymethylpurine; from Hoechst AG, Frankfurt, Germany} and A-5021 {(1'S,2'R)-9-[[1',2'-bis(hydroxymethyl)-cycloprop-1'-yl]methyl]guanine; from Dr. T. Tsuji, Ajinomoto Co, Kawasaki, Japan}. Stock solutions of these compounds were prepared in dimethylsulfoxide or phosphate buffered saline (PBS). Working dilutions for the antiviral assays were prepared by appropriate dilution of the stock solutions in cell culture medium.

Antiviral assay

On day 0, the GS and Z-29 strains of HHV-6 were inoculated onto HSB-2 cells and MOLT-3 cells, respectively, at a multiplicity of infection of 0.001 CCID50 (50% cell culture infective dosis) per cell, and a final density of 5×10^6 cells per ml. After

2 hr virus adsorption at 37°C, the unadsorbed virus was removed by centrifugation, and the infected cells were resuspended and transferred to 48-well microtrays, containing the antiviral compounds in two- to five-fold dilutions. The final cell density was 0.8×10^6 cells per ml. The cells were incubated at 37°C and subcultivated on days 4 and 7, by two-fold dilution in medium containing fresh compound. On day 10–14, cells were examined by microscopy to score viral cytopathic effect (CPE, visible by the appearance of large ballooning cells) and drug toxicity. Then, total DNA was extracted from the infected cells with the QIAamp Blood kit (Qiagen, Germany), using the Manufacturer's instructions, and the DNA extracts were frozen at –20°C until further analysis.

Viral DNA detection by slot blot assay

A digoxigenin-labelled probe specific for strain GS or Z-29 was prepared by PCR reaction in the presence of digoxigenin-dUTP, on a 1-μl aliquot of a DNA extract prepared from infected cells showing manifest CPE. The 5'-3' primer sequence was GCTAGAACGTATTTGCTGCAGAACG and ATC-CGAAACAACTGTCTGACTGGCA, delimiting a 259 bp sequence within the U67 gene [17]. The amplified product was separated on agarose gel and extracted from the gel fragment with the QIAquick gel extraction kit (Qiagen, Germany), after which the purified probe was frozen at –20°C.

For quantitation of viral DNA in the infected and drug-treated cell cultures, appropriate aliquots of the total DNA extracts (containing 5 μg of total DNA, as determined spectrophotometrically at 260 nm) were boiled during 10 min, and blotted on a nylon membrane (Hybond-N from Amersham) using a Hoefer slot blot apparatus from Pharmacia (Sweden). After UV-cross linking, the membrane was prehybridized during 30 min at 42°C with DIG easy Hyb solution (Boehringer Mannheim, Germany). Then, the GS- or Z-29-specific digoxigenin-labelled probe was added and allowed to hybridize by overnight incubation at 42°C. After thorough washing (twice at room temperature in 2 x SSC, 0.1% SDS and twice at 65°C in 0.1 x SSC, 0.1% SDS), the membrane was treated with blocking reagent from Boehringer Mannheim. Then, the membrane was incubated during 1 hr with alkaline phosphatase-conjugated anti-digoxigenin antibody (from Boehringer Mannheim), and, after stringent washing, chemiluminescence detection was performed (using CSPD substrate from Clontech). After visualization on film, the intensity of the viral DNA bands was determined by densitometric scanning. The amounts of probe, anti-DIG antibody and CSPD substrate were standardized to assure that the DNA band intensity was linear to the amount of viral DNA loaded on the membrane. A typical example of a slot blot result is shown in Fig. 2.

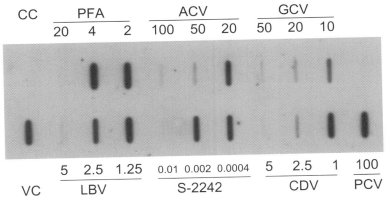

Fig. 2. DNA hybridization assay for evaluation of antiviral activity in HHV-6 (Z-29)-infected MOLT-3 cells. CC: uninfected cell control; VC: virus-infected control. The concentrations of the antiviral compounds are given in µg/ml.

Calculations

The antiviral IC_{50} was calculated by extrapolation, and defined as the compound concentration that produced 50% inhibition of virus replication, as determined by microscopic examination, or quantitation of viral DNA. Toxicity of the compounds was expressed as minimum cytotoxic concentration (MCC), or the lowest compound concentration that caused a microscopically visible alteration in cellular morphology.

Results and discussion

Previously published antiviral investigations for HHV-6 have mainly used, as a parameter for virus replication, the HHV-6-induced cytopathic effect, production of viral antigens (measured by immunofluorescence or dot-blot assay), or cell viability (using staining assays) [18–22]. Compared to these techniques, our newly developed DNA hybridization assay has the advantages of being both reliable, reproducible and easy-to-perform on a larger scale. In addition, the usage of a digoxigenin-labelled probe with chemiluminescence detection requires no special precautions (in contrast to ^{32}P-labelled probes). This technique can therefore be easily applied in screening procedures for new anti-HHV-6 compounds with favorable efficacy and selectivity. In the study presented here, we concentrated on a number of antiherpetic drugs that are either commonly used in the clinic or in preclinical development (Fig. 1).

As shown in Tables 1 and 2, the assays based on CPE and viral DNA detection yielded remarkably similar antiviral IC_{50} values. The pyrophosphate analogue foscarnet proved to be not only highly effective in suppressing HHV-6, but also relatively specific, its selectivity index (SI; ratio of cytotoxic over effective concentration) being 109 in HHV-6A (GS)-infected HSB-2 cells, and 60 in HHV-6B (Z-29)-

Table 1. Antiviral efficacy of selected antiherpetic drugs in HHV-6A (GS)-infected HSB-2 human T-lymphoblast cells.

	Antiviral activity[a]		Cytotoxicity[a]	
	IC_{50} by CPE[b] (μg/ml)	IC_{50} by DNA detection[c] (μg/ml)	MCC[d] (μg/ml)	SI[e]
foscarnet	2.0 ± 0.1	4.6 ± 2.6	500 ± 0	109
acyclovir	28 ± 9.5	45 ± 6	175 ± 50	3.9
ganciclovir	6.8 ± 0.4	7.2 ± 0.07	20 ± 0	2.8
penciclovir	71	72	100	1.4
lobucavir	2.8 ± 1.1	2.8 ± 1.1	10 ± 0	3.6
cidofovir	2.0 ± 0.9	2.9 ± 0.5	25 ± 15	8.6
S2242	0.005	0.004	0.75	188
A-5021	2.6 ± 1.3	3.5 ± 3.7	50 ± 0	15

[a]Data are mean values ± standard deviations from 2–5 individual experiments.
[b]IC_{50}: compound concentration that produces 50% inhibition of virus-induced cytopathic effect, as determined by microscopy.
[c]IC_{50}: compound concentration that produces 50% inhibition of viral replication, as determined by quantitation of viral DNA by hybridization assay.
[d]MCC: minimum cytotoxic concentration, or compound concentration that produces microscopically visible alteration in cell morphology.
[e]SI: selectivity index, or ratio of MCC over IC_{50} (by DNA detection).

infected MOLT-3 cells. Among the nucleoside analogues in our study, compound S-2242, an investigational compound with an N7-substituted purine base (Fig. 1), was found to be a remarkably potent anti-HHV-6 agent, with an IC_{50} <0.007 μg/ml and a SI of 33–188 (Tables 1 and 2) [23]. In contrast, the other commercially available antiherpetic drugs proved less potent and/or selective as inhibitors of HHV-6. Acyclovir and penciclovir (two potent inhibitors of HSV and VZV) suppressed HHV-6 replication only at high concentrations with low, if any, selectivity (SI: 3–4 and ≤1 for acyclovir and penciclovir, respectively). Cidofovir, a broad-spectrum antiherpetic nucleotide analogue, that has been licensed for the treatment of CMV retinitis in AIDS patients, displayed an appreciable selectivity for HHV-6 (SI: 6.5–8.6). A relatively high antiviral selectivity for HHV-6 (SI: 14) was noted for the cyclopropyl derivative A-5021, that is currently in clinical development for the treatment of HSV or VZV infections [24]. Little antiviral selectivity (SI: 1.7–3.6) was noted for lobucavir, a compound with good activity against HSV, VZV and CMV, that has been suspended following clinical evaluation.

The most striking observation in our study was the low efficacy of ganciclovir in the HHV-6-infected lymphoblast cells. In fact, both in the HSB-2 and in the MOLT-3 cells, ganciclovir was unable to suppress HHV-6 at non-toxic concentrations (≤20 μg/ml), resulting in a SI <3 (Tables 1 and 2). This result is in sharp contrast to the previously reported efficacy of ganciclovir in HHV-6-infected peripheral blood lymphocytes (PBL), with an IC_{50} value in the range of 0.3–2 μg/ml, and a SI >80 [19,20]. The most probable explanation for this discrepancy is that the efficiency by which ganciclovir is converted to its active triphosphate metabolite

Table 2. Antiviral efficacy of selected antiherpetic drugs in HHV-6B (Z-29)-infected MOLT-3 human T-lymphoblast cells.

	Antiviral activity[a]		Cytotoxicity[a]	
	IC$_{50}$ by CPE[b] (µg/ml)	IC$_{50}$ by DNA detection[c] (µg/ml)	MCC[d] (µg/ml)	SI[e]
foscarnet	6.8 ± 2.6	8.3 ± 1.8	500 ± 0	60
acyclovir	46 ± 21	44 ± 17	133 ± 58	3.0
ganciclovir	16 ± 4	15 ± 8	20 ± 0	1.3
penciclovir	>200	>100	≥100	≤1
lobucavir	2.4 ± 1.0	4.9 ± 1.7	8.3 ± 2.9	1.7
cidofovir	2.6 ± 1.4	3.1 ± 1.2	20 ± 9	6.5
S2242	0.0033 ± 0.0023	0.0066 ± 0.0039	0.20 ± 0.0	33
A-5021	3.4 ± 1.6	3.5 ± 1.1	50 ± 0	14

[a,b,c,d,e]See Table 1 for definitions.

is markedly cell type-dependent. It is likely that, compared to freshly isolated PBL, the tumor-derived cell lines used in our study have higher intracellular levels of deoxyguanosine (or deoxyguanosine 5'-triphosphate). This impedes the antiviral activity of ganciclovir due to a stronger competitive effect of the natural nucleosides/nucleotides at the level of the enzymes involved in its phosphorylation, or action target (i.e. viral DNA polymerase) [25]. On the other hand, the enzyme studies by Bapat et al. [26] have demonstrated that the triphosphate forms of ganciclovir and acyclovir are relatively poor inhibitors of HHV-6 DNA polymerase, with inhibitory concentrations that are markedly higher than the values for HSV-1 or CMV DNA polymerase, yet in the same range as the inhibitory concentrations for human DNA polymerases. Thus it appears that neither ganciclovir nor acyclovir are able of considerably discriminating between the HHV-6 and human DNA polymerases, thereby compromising their antiviral selectivity for HHV-6 at this enzymatic level.

Most antiherpetic drugs owe their antiviral selectivity to their specific activation by a virus-encoded kinase, i.e., HSV- or VZV-encoded thymidine kinase (that recognizes acyclovir, ganciclovir and penciclovir), or CMV UL97-encoded protein kinase (that recognizes ganciclovir) [25]. Whether HHV-6 encodes a functional enzyme that is able of phosphorylating nucleoside analogues, has not been thoroughly investigated. Since the discovery that the CMV UL97-encoded protein kinase is responsible for activation of ganciclovir in CMV-infected cells [27], it has been hypothesized that the HHV-6-encoded U69 protein may have a similar function, based on the sequence homology between the CMV UL97 and the HHV-6 U69 genes [28]. Indeed, Ansari and Emery [29] have recently shown that insertion of the HHV-6 U69 sequence into baculovirus renders the virus sensitive to inhibition by ganciclovir, as determined in infected insect cells. However, no data are as yet available on the phosphorylation of ganciclovir in HHV-6-infected human cells, whether fresh PBL or lymphoblast cells.

The low efficacy of acyclovir and penciclovir in HHV-6-infected cells, as observed in our studies, and in other investigations using fresh PBL, is in agreement with the general assumption that HHV-6 lacks a functional thymidine kinase, as encoded by HSV, VZV and EBV. In addition, it has been shown that HHV-6 infection is not associated with an increased activity in cellular thymidine kinase [30]. This explains our (unpublished) results on the metabolism of [8-^{3}H]-acyclovir in HHV-6-infected HSB-2 or MOLT-3 lymphoblast cells, showing that the phosphorylation of acyclovir is very similar in uninfected and infected cells, and relatively inefficient, since the levels of acyclovir triphosphate in cells with manifest CPE were only ≈20 pmol/million cells after 6 hours incubation with 100 μM (22 μg/ml) ACV. For a comparison, the intracellular concentration of acyclovir triphosphate in HSV-infected cells has been reported to be as high as ≈200 pmol per million cells, after 24-hour incubation with only 5 μM acyclovir [31].

In view of the clinical use of antiherpetic drugs for the treatment of HHV-6 infections, the following conclusions can be drawn from the currently available antiviral data for HHV-6. In a few case studies, foscarnet and ganciclovir (given alone or in combination, during maximally 7 weeks) have been found to suppress active HHV-6 infections (as occurring in bone marrow transplant recipients) [6,7]. Unfortunately, prolonged administration of these drugs appears to be excluded due to their severe side effects (i.e., nephrotoxicity for foscarnet, and myelosuppression for ganciclovir). On the other hand, since acyclovir suppresses HHV-6 replication only at higher concentrations (that would be difficult to achieve in patients), it is unlikely that this drug (or its oral prodrug valaciclovir) would show considerable efficacy against HHV-6 [4,7]. Recently, several proposals have been put forward to perform clinical trials with antiherpetic drugs in MS patients, in order to clarify the possible role of herpesviruses in MS pathogenesis. In fact, one clinical trial with oral acyclovir has been finished, and another study with valaciclovir is in progress [32]. In the first study, the acyclovir recipients showed 34% fewer relapses compared to the placebo group. This may implicate a role in MS for herpesviruses, other than HHV-6, that are more sensitive to the antiviral action of acyclovir than HHV-6 (i.e., HSV or VZV). On the other hand, if the purpose would be to establish the role of HHV-6 in MS, clinical studies should be conducted with antiviral drugs that are more efficacious against HHV-6 than acyclovir. These studies can be expected to yield the most conclusive answers in case of a careful selection and follow-up of the patients, not only on a neurological basis, but, particularly, in terms of viral load (of HHV-6 or other herpesviruses) in serum or CSF.

In conclusion, inspite of the available information on the anti-HHV-6 activity of common antiherpetic drugs, several questions remain to be solved regarding the virus-specific or cell type-dependent activation of nucleoside analogues in HHV-6-infected cells. Also, since the current repertory of drugs with favorable efficacy and selectivity for HHV-6 is clearly insufficient, systematic evaluation of newly developed antiherpetic drugs in HHV-6-infected cells seems warranted. To this purpose, the DNA hybridization assay described here should offer a highly valuable tool.

Acknowledgement

The authors wish to thank Dr. Johan Neyts for helpful advice. This investigation was supported in part by grants from the Flemish "Wetenschappelijk Onderzoek Multiple Sclerosis". Leen De Bolle is a research assistant from the "Fonds voor Wetenschappelijk Onderzoek (FWO) – Vlaanderen".

References

1. Yamanishi K, Okuno T, Shiraki K, Takahashi M, Kondo T, Asano Y, Kurata T. Identification of human herpesvirus-6 as a causal agent for exanthem subitum. Lancet 1988;1:1065–1067.
2. Pruksananonda P, Hall CB, Insel RA, McIntyre K, Pellett PE, Long CE, Schnabel KC, Pincus PH, Stamey FR, Dambaugh TR et al. Primary human herpesvirus 6 infection in young children. N Engl J Med 1992;326:1445–1450.
3. Kimberlin DW, Whitley RJ. Human herpesvirus-6: neurologic implications of a newly-described viral pathogen. J Neurovirol 1998;4:474–485.
4. Cone RW, Huang ML, Corey L, Zeh J, Ashley R, Bowden R. Human herpesvirus 6 infections after bone marrow transplantation: clinical and virologic manifestations. J Infect Dis 1999;179:311–318.
5. Imbert-Marcille BM, Tang XW, Lepelletier D, Besse B, Moreau P, Billaudel S, Milpied N. Human Herpesvirus 6 infection after autologous or allogeneic stem cell transplantation: a single-center prospective longitudinal study of 92 patients. Clin Infect Dis 2000;31:881–886.
6. Drobyski WR, Knox KK, Majewski D, Carrigan DR. Brief report: fatal encephalitis due to variant B human herpesvirus-6 infection in a bone marrow-transplant recipient. N Engl J Med 1994;330:1356–60.
7. Wang FZ, Linde A, Hagglund H, Testa M, Locasciulli A, Ljungman P. Human herpesvirus 6 DNA in cerebrospinal fluid specimens from allogeneic bone marrow transplant patients: does it have clinical significance? Clin Infect Dis 1999;28:562–568.
8. Soldan SS, Berti R, Salem N, Secchiero P, Flamand L, Calabresi PA, Brennan MB, Maloni HW, McFarland HF, Lin HC, Patnaik M, Jacobson S. Association of human herpes virus 6 (HHV-6) with multiple sclerosis: increased IgM response to HHV-6 early antigen and detection of serum HHV-6 DNA. Nat Med 1997;3:1394–1397.
9. Novoa LJ, Nagra RM, Nakawatase T, Edwards-Lee T, Tourtellotte WW, Cornford ME. Fulminant demyelinating encephalomyelitis associated with productive HHV-6 infection in an immunocompetent adult. J Med Virol 1997;52:301–8.
10. Challoner PB, Smith KT, Parker JD, MacLeod DL, Coulter SN, Rose TM, Schultz ER, Bennett JL, Garber RL, Chang M, et al. Plaque-associated expression of human herpesvirus 6 in multiple sclerosis. Proc Natl Acad Sci USA 1995;92:7440–7444.
11. Akhyani N, Berti R, Brennan MB, Soldan SS, Eaton JM, McFarland HF, Jacobson S. Tissue distribution and variant characterization of human herpesvirus (HHV)-6: increased prevalence of HHV-6A in patients with multiple sclerosis. J Infect Dis 2000;182:1321–1325.
12. Knox KK, Brewer JH, Henry JM, Harrington DJ, Carrigan DR. Human Herpesvirus 6 and multiple sclerosis: systemic active infections in patients with early disease. Clin Infect Dis 2000;31:894–903.
13. Martin C, Enbom M, Soderstrom M, Fredrikson S, Dahl H, Lycke J, Bergstrom T, Linde A. Absence of seven human herpesviruses, including HHV-6, by polymerase chain reaction in CSF and blood from patients with multiple sclerosis and optic neuritis. Acta Neurol Scand 1997;95:280–283.
14. Merelli E, Bedin R, Sola P, Barozzi P, Mancardi GL, Ficarra G, Franchini G. Human herpes virus 6 and human herpes virus 8 DNA sequences in brains of multiple sclerosis patients, normal adults and children. J Neurol 1997;244:450–454.

15. Hall CB, Caserta MT, Schnabel KC, Long C, Epstein LG, Insel RA, Dewhurst S. Persistence of human herpesvirus 6 according to site and variant: possible greater neurotropism of variant A. Clin Infect Dis 1998;26:132–137.
16. Salahuddin SZ, Ablashi DV, Markham PD, Josephs SF, Sturzenegger S, Kaplan M, Halligan G, Biberfeld P, Wong-Staal F, Kramarsky B, et al. Isolation of a new virus, HBLV, in patients with lymphoproliferative disorders. Science 1986;234:596–601.
17. Secchiero P, Carrigan DR, Asano Y, Benedetti L, Crowley RW, Komaroff AL, Gallo RC, Lusso P. Detection of human herpesvirus 6 in plasma of children with primary infection and immunosuppressed patients by polymerase chain reaction. J Infect Dis 1995;171:273–280.
18. Reymen D, Naesens L, Balzarini J, Holy A, Dvorakova H, De Clercq E. Antiviral activity of selected acyclic nucleoside analogues against humanherpesvirus 6. Antiviral Res 1995;28:343–357.
19. Yoshida M, Yamada M, Tsukazaki T, Chatterjee S, Lakeman FD, Nii S, Whitley RJ. Comparison of antiviral compounds against human herpesvirus 6 and 7. Antiviral Res 1998;40:73–84.
20. Takahashi K, Suzuki M, Iwata Y, Shigeta S, Yamanishi K, De Clercq, E. Selective activity of various nucleoside and nucleotide analogues against human herpesvirus 6 and 7. Antiviral Chem Chemother 1997;8:24–31.
21. Agut H, Aubin JT, Huraux JM. Homogeneous susceptibility of distinct human herpesvirus 6 strains to antivirals in vitro. J Infect Dis 1991;163:1382–1383.
22. Burns WH, Sandford GR. Susceptibility of human herpesvirus 6 to antivirals in vitro. J Infect Dis 1990;162:634–637.
23. Neyts J, Andrei G, Snoeck R, Jahne G, Winkler I, Helsberg M, Balzarini J, De Clercq E. The N-7-substituted acyclic nucleoside analog 2-amino-7-[(1,3-dihydroxy-2-propoxy)methyl]purine is a potent and selective inhibitor of herpesvirus replication. Antimicrob Agents Chemother 1994;38:2710–2716.
24. Iwayama S, Ono N, Ohmura Y, Suzuki K, Aoki M, Nakazawa H, Oikawa M, Kato T, Okunishi M, Nishiyama Y, Yamanishi K. Antiherpesvirus activities of (1'S,2'R)-9-[[1',2'-bis(hydroxymethyl)cycloprop-1'-yl]methyl]guanine (A-5021) in cell culture. AntimicrobAgents Chemother 1998;42:1666–1670.
25. Alrabiah FA, Sacks SL. New antiherpesvirus agents. Their targets and therapeutic potential. Drugs. 1996;52:17–32.
26. Bapat AR, Bodner AJ, Ting RC, Cheng YC. Identification and some properties of a unique DNA polymerase from cells infected with human B-lymphotropic virus. J Virol 1989;63:1400–1403.
27. Sullivan V, Talarico CL, Stanat SC, Davis M, Coen DM, Biron KK. A protein kinase homologue controls phosphorylation of ganciclovir in human cytomegalovirus-infected cells. Nature 1992;358:162–164.
28. Chee MS, Lawrence GL, Barrell BG. Alpha-, beta- and gammaherpesviruses encode a putative phosphotransferase. J Gen Virol 1989;70:1151–1160.
29. Ansari A, Emery VC. The U69 gene of human herpesvirus 6 encodes a protein kinase which can confer ganciclovir sensitivity to baculoviruses. J Virol 1999;73:3284–3291.
30. Di Luca D, Katsafanas G, Schirmer EC, Balachandran N, Frenkel N. The replication of viral and cellular DNA in human herpesvirus 6-infected cells. Virology 1990;175:199–210.
31. Smee DF, Boehme R, Chernow M, Binko BP, Matthews TR. Intracellular metabolism and enzymatic phosphorylation of 9-(1,3-dihydroxy-2-propoxymethyl)guanine and acyclovir in herpes simplex virus-infected and uninfected cells. Biochem Pharmacol 1985;34:1049–1056.
32. Bergström T. Herpesviruses — a rationale for antiviral treatment in multiple sclerosis. Antiviral Res 1999;41:1–19.

IL-1 receptor antagonist: a possible circulating and genetic marker of interferon-beta1b therapeutic effectiveness in MS?

F.L. Sciacca[1,*], L.M.E. Grimaldi[2], G. Comi[2], N. Canal[2]

[1]Neuromuscular disease department, National Neurological Institute "C. Besta" Milan; [2]Neuroscience Department, San Raffaele Scientific Institute, Milan, Italy

MS is an immune-mediated disease of the central nervous system (CNS) in which selective destruction of myelin and oligodendrocytes occur as a result of inflammation [1]. Complex cytokine and adhesion molecules profile favors and sustains inflammatory reactions in the CNS and this could lead to myelin damage [2]. The disease involves, in fact, the adhesion and transmigration of mononuclear cells across the blood brain barrier, which is regulated by an increase in the receptor-ligand interactions and expression of soluble mediators. Genetic and environmental factors (viral or bacterial infection, local hormonal / metabolic stress) may up-regulate the expression of adhesion molecules, proinflammatory cytokines and chemokines on endothelial cells and on circulating mononuclear cells. These factors may enhance the movement of mononuclear cells through the blood brain barrier, favoring their rolling, binding and diapedesis of into CNS. In this contest, pro-inflammatory cytokines and adhesion molecules represent potentially noxious molecules to the myelin sheet and/or oligodendrocyte. In acute MS, these molecules are expressed predominantly by microglial cells within and around the MS lesions [2].

On the other hand, immunomodulatory cytokines could play an important regulatory role controlling the inflammatory burst. The therapeutic effect shown by IFNβ could be explained, at least partially, by its capability to inhibit expression of pro-inflammatory cytokines, such as IL-1 and IL-12 [3–5], and to induce anti-inflammatory cytokines, such as IL-10, IL-4 [5–9] and IL-1 receptor antagonist (IL-1ra) [3,10].

Overall, the therapeutic efficacy of IFNβ still lacks a full understanding of its mechanism of action. Furthermore, clinical response of patients to the therapy is not homogeneous: in many cases improvement of disease status is not observable, despite the absence of adverse effect (such as anti-IFNβ neutralizing antibodies) that could compromise the IFNβ therapeutic activity. For all these reasons, marker of IFNβ pharmacological activity is been sought, by several authors, mainly among cytokine and adhesion molecules.

Among pro-inflammatory cytokines, IL-12 is been proposed as a candidate for

---

Correspondence address: Francesca Sciacca, Neuromuscular Disease Department, National Neurological Institute "C. Besta", via Celoria 11, 20133 Milano, Italy. Tel: (39)-02-2394371. Fax: (39)-02-70633874. E-mail: fsciacca@istituto-besta.it

the role of "peripheral marker" of IFNβ1b therapy responsiveness [11]. IL-12 is produced by antigen-presenting cells and has a central role in both innate and acquired immunity. IL-12 leads to development of T-helper1 (Th1) lymphocytes, induces IFNγ production in Th1 and NK cells, activates macrophage and B-cell production of complement-fixing antibodies. The importance of this crucial cytokine in autoimmune disease such as MS is been widely studied [12,13] and its exacerbating role in MS could be supposed on the basis of the observed increased levels of IL-12 preceding clinical relapses of the disease [14]. IFNβ has the ability of inhibit IL-12 production [5]. Patients that improved their disease status during the therapy had lower levels of IL-12 mRNA compared to "non responder" patients [11] and this observation sustains the role of biological marker for IL-12.

We have focused our attention to the anti-inflammatory cytokine IL-1ra, which is able to inhibit IL-1 action by binding to the IL-1 receptors without triggering intracellular signal [15]. Circulating levels of IL-1ra in patients with the relapsing remitting (RR) MS are normal during remission phases and significantly increase during exacerbation or in response to IFNβ treatment [16,17]. The latter might be due to the ability shown by IFNβ to directly induce IL-1ra mRNA expression and mature protein release in monocytic cell lines [10].

To assess the possible marker-role of IL-1ra in MS, we followed a group of patients with MS undergoing IFNβ1b treatment and monitored their clinical response with respect to their circulating levels of IL-1ra.

We studied 24 patients affected by definite RR MS. Sixteen of these patients [mean age 32.2 ± 7; mean expanded disability status scale (EDSS) score 2.4 ± 1.3] started treatment with systemic IFNβ1b (8×10^6 IU every one day) and 8 MS patients (mean age 29 ± 6.2; mean EDSS 2.3 ± 1.1) did not received any therapy. Plasma or serum was obtained from patients before starting the $IFNβ_{1b}$ treatment (time 0) and thereafter at 1, 3, 6, 9, 12 and 18 months. In untreated patients, plasma was obtained every 3 months for one year; the first samples have been considered as time 0 and the following as time 3, 6 and 12 months. We measured the circulating IL-1ra by a commercial double antibody ELISA kit (R. & D. System, Oxon, UK). The clinical response to the therapy was evaluated comparing the relapse rate and the EDSS progression measured in the 18 months of therapy with the same parameters in the 18 months immediately preceding the treatment. MS patients were grouped as follow:
1. Responder: patients showing a reduction of relapse rate and improvement or no progression of EDSS score during the 18 months of therapy compared to the 18 months immediately preceding the therapy;
2. Stable: patient having a relapse rate equal to that observed before treatment and no EDSS score progression;
3. Non responder: patient having a relapse rate worse than those observed before treatment and/or EDSS score progression.

In untreated patients, IL-1ra circulating levels fluctuated over a period of 12 months, showing a total mean increase of 25.9% compared to the arbitrary time 0. In MS patients treated with IFNβ1b, the IL-1ra circulating levels measured longi-

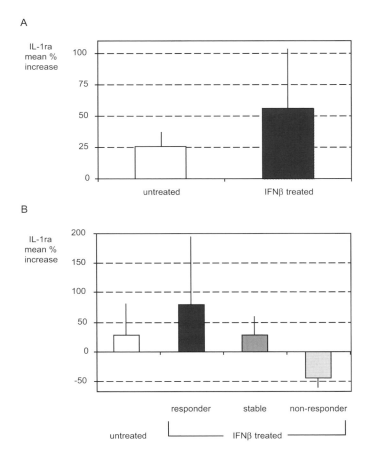

Fig. 1. Mean percentage of increase of IL-1ra circulating levels during 12 months in 8 untreated MS patients and during 18 months in 16 patients in therapy with IFNβ$_{1b}$ (Panel A). The panel B shows the variability of IL-1ra increase after the beginning of the IFNβ$_{1b}$ treatment in patients having different clinical response to the therapy.

tudinally during 18 months, showed a mean increase of 55.8% compared to time 0 (Fig. 1, panel A).

Patients that experience improvement of their clinical status during IFNβ1b treatment (responders, 9 patients), showed an elevated (79.95%) and rapidly growing percentage of increase of circulating IL-1ra levels; IL-1ra increase was milder and slower in 6 clinically stable patients (mean increase 27.8%) and, decreased (-44%) in a single non-responder patient (Fig. 1, panel B).

Based on our data and on other reports [18,19], we propose that the induction of IL-1ra could be one of the mechanisms by which IFNβ exerts its beneficial effects in patients with MS.

However, both basal and IFNβ-induced IL-1ra levels are extremely variable among healthy individuals and MS patients. We had previously observed that, in a group of healthy donors, this variability depended, at least partially, upon the IL-1ra intron 2 polymorphism: the carriage of the A2 allele was, in fact, associated with more elevated IL-1ra levels [20]. Although this association needs to be confirmed in a larger group of IFNβ treated MS patients, it is tempting to propose that the genetic background of MS patients (and the IL-1ra polymorphic status, in particular) may influence the response to IFNβ. If these results will be confirmed, only IFNβ1b responder patients, genetically predetermined, could favorable use this therapy, while non-responder would avoid inconvenience (expensive cost, parenteral administration) and adverse effect (flu-like symptoms) of an ineffective therapy.

References

1. Traugott U. Handbook of Multiple Sclerosis. In: Cook, S.D., editors. Marcel Dekker, New York, 1996;157–185.
2. Cannella B, Raine CS. The adhesion molecule and cytokine profile of multiple sclerosis lesions. Ann. Neurol. 1995;37;424–435.
3. Liu JSH, Amaral TD, Brosnan CF, Lee SC. IFNs are critical regulators of IL-1 receptor antagonist and IL-1 expression in human microglia. J Immunol 1998;161;1989–1996.
4. Karp C, Biron CA and Irani DN. Interferon β in multiple sclerosis: is IL-12 suppression the key? Immunol. Today 2000;21;24–28.
5. Wang X, Chen M, Wandinger KP, Williams G, Dhib-Jalbut S. IFN-beta-1b inhibits IL-12 production in peripheral blood mononuclear cells in an IL-10-dependent mechanism: relevance to IFN-beta-1b therapeutic effects in multiple sclerosis. J Immunol. 2000;165;548–557.
6. Rep MH, Hintzen RQ, Polman CH, van Lier RA. Recombinant interferon-beta blocks proliferation but enhances interleukin-10 secretion by activated human T-cell. J Neuroimmunol 1996;67;111–118.
7. Rudick RA, Ransohoff RM, Peppler R, VanderBrug Medendorp S, Lehmann P, Alam J. Interferon beta induces interleukin-10 expression: relevance to multiple sclerosis. Ann Neurol. 1996;40;618–627.
8. Rudick RA, Ransohoff RM, Lee JC, Peppler R, Yu M, Mathisen PM, Tuohy VK. In vivo effects of Interferon beta-1b on immunosuppressive cytokines in multiple sclerosis. Neurology 1998;50;1294–1230.
9. Rep MH, Schrijver HM, van Lopik T, Hintzen RQ, Roos MT, Ader HJ, Polman CH, van Lier RA. Interferon (IFN)-beta treatment enhance CD95 and interleukin 10 expression but reduces interferon-gamma producing T cells in MS patients. J Neuroimmunol. 1999;96;92–100.
10. Sciacca FL, Canal N, Grimaldi LME. Induction of IL-1 receptor antagonist by interferon beta: implication for the treatment of multiple sclerosis J Neurovirol. 2000;6 Suppl 2;S33–7.
11. van Boxel-Dezaire AHH, van Trigt-Hoff SCJ, Killestein J, Schrijver HM, van Houwelingen JC, Polman CH, Nagelkerken L. Contrasting responses to interferon beta-1b treatment in relapsing remitting multiple sclerosis: does baseline interleukin-12p35 messenger RNA predict the efficacy of the treatment? Ann. Neurol. 2000;48;313–322.
12. Becher B, Prat A, Antel JP. Brain-immune connection: immuno-regulatory properties of CNS-resident cells. Glia 2000;29:293–304.
13. Sinigaglia F, D'Ambrosio D, Rogge L. Type I interferons and the Th1/Th2 paradigm. Dev Comp Immunol 1999;23:657–63.
14. van Boxel-Dezaire AH, Hoff SC, van Oosten BW, Verweij CL, Drager AM, Ader HJ, van Houwelingen JC, Barkhof F, Polman CH, Nagelkerken L. Decreased interleukin-10 and increased interleukin-12p40 mRNA are associated with disease activity and characterize different disease stages in multiple sclerosis. Ann Neurol. 1999;45;695–703.

15. Dinarello CA. Biological basis for interleukin-1 in disease. Blood 1996;87;2095–2147.
16. Nicoletti F, Patti F, Di Marco R, Zaccone P, Nicoletti A, Meroni P, Reggio A. Circulating serum levels of IL-1ra in patients with relapsing remitting multiple sclerosis are normal during remission phases but significantly increase either during exacerbations or in response to IFN-beta treatment. Cytokine 1996;8;395–400.
17. Voltz R, Hartmann M, Spuler S, Scheller A, Mai N, Hohlfeld R, Yousry T. Longitudinal measurement of IL-1 receptor antagonist. J. Neurol. Neurosurg. Psychiatry 1997;62;200–201.
18. Perini P, Tiberio M, Sivieri S, Facchinetti A, Biasi G, Gallo P. Interleukin-1 receptor antagonist, soluble tumor necrosis factor-alpha receptor type I and II and soluble E-selectin serum levels in multiple sclerosis patients receiving weekly intramuscular injections of interferon-beta1a. Eur. Cytok. Netw. 2000;11;81–86.
19. Heesen C, Sieverding F, Buhmann C, Gbadamosi J. IL-1ra serum levels in disease stage of MS: a marker for progression? Acta Neurol. Scand. 2000;101;95–97.
20. Sciacca FL, Ferri C, Vandenbroeck K, Veglia F, Gobbi C, Martinelli F, Franciotta D, Zaffaroni M, Marrosu M, Martino G, Martinelli V, Comi G, Canal N, Grimaldi LME. Relevance of Interleukin-1 receptor antagonist intron 2 polymorphism in Italian multiple sclerosis patients. Neurology 1999;52;1896–1898.

Keyword index

16S rDNA, 222
16S rRna, 221,222,224,226,228
229 E, 210

acyclovir, 97,167
adenoviral vector, 73
adhesion molecules, 251
altered peptide ligands, 217
antibody, 166,210
antibody specific for the HLA-DR2-MBP 85–89 complex, 32
antigen-binding cleft, 26
antinflammatory cytokines, 72
antiviral, 241–247
antiviral action, 97
antiviral antibodies, 98
association, 10–15,18,19
astrocytes, 210
atrophy of the spinal cord, 84
autoantibodies, 2,3
autoimmune diseases, 39,64,65,67
autoimmune encephalomyelitis, 72,124
autoimmunity, 25,33,209
autoimmunogenic potential, 217
axon(s), 79,80,83,85,86
axonal damage, 79–80
axonal dysfunction, 84
axonal injury, 84–86
axonal transection 84

B-cell tolerance, 3
B-lymphoblastoid, 168
beta-interferon, 97
beta-interferon treated, 172
Biozzi AB/H mice, 73
blood brain barrier, 251
brain banks, 211
brain parenchyma, 215
brain viral flora, 215

bystander virus mediated inflammation, 216

candidate gene(s), 12–14
cationic antibody(ies), 221,223,226
CD4-positive T-helper lymphocytes, 26
CD8-positive cytotoxic T-cells, 26
Central Nervous System (CNS), 177, 215,217
cell body atrophy, 80
cerebrospinal fluid, 73,170,90
chemoattractant, 216
chemoattractant chemokines, 73
chemokines, 73,210
Chlamydia pneumoniae, 221,222,224, 226,228
chronic axonal dysfunction, 84
chronic demyelination, 83,84,86
chronic phase of disease, 79
clinical exacerbations, 97
clinical heterogeneity, 30
clinical phenotypes, 51
clinical trial 178
cluster studies, 174
CNS immunopathology, 217
CNS injury, 86
CNS persistence, 215
cognitive function, 98
common colds, 210
conformational homologies, 217
coronavirus(es) 169,209
course of MS, 93
cross-reactive T-cell clones, 216
cross-sectional study of active HHV-6 viremia, 186
CTLA-4, 38
cytokyne gene polymorphisms, 49
cytokines, 49,71–73,210,251
cytopathology, 209

Dementia Rating Scale, 99
demyelination, 57,58,71,79,80,83–86, 115–117,172,185,210
demyelination lesions, 105
disease progression, 52
DNA, 72, 97
dominant models, 29
DQB1*0602, 21
DRB5*0101, 21

ELISA, 99
encephalitits, 1,31,185
endogenous retrovirus(es), 169,195, 196,200,202,204
environmental factors, 209
ependymall cells, 73
epidemics of MS, 174
epidemiological studies, 209
Epstein-Barr virus, 171
etiologic fraction, 29
etiology, 97,163
exacerbation rate, 90
Expanded Disability Status Scale, 52, 102
Experimental Allergic Encephalitis, 31
experimental autoimmune encephalitis, 1
experimental autoimmune encephalomyelitis, 72

follow-up duration, 98

gadolinium enhancement, 98
GAMES, 19,20
gene amplification, 211
gene delivery strategies, 72
gene therapy, 72
gene-targeting, 3
genetic control, 57,58
genetic factors, 209
genetic susceptibility, 216
genetically-susceptible individuals, 217
genome, 11,14–16,18–20,43,44,102
genome screen (s), 11,14–16,18–20,

43,44
geographical clusters, 165
glial cells, 210
glial precursor cells, 74

hepatitis B, 115–118,121,126,127, 132–140
hepatitis B vaccination, 121,126, 136,137
hepatitis B vaccine(s), 135,138–140
hepatitis B virus, 115
herpes virus, 167,169,185,241
HERV-H, 195,196,199,201,203
histocompatibility antigens, 62,63
histocompatibility proteins, 61–64
HLA, 25,26,30,37,38,61
HLA Class I, 25,38
HLA Class II, 25,30
HLA Class III, 26
HLA complex, 25
HLA-B7, 27
HLA-DR1*1501, 27
HLA-DR2, 27,217
HLA-DR2+, 27,217
HLA-DRB, 38
(non Class II) HLA genes, 30
human cytomegalovirus, 97
Human Herpes Virus 6 (HHV-6), 153–159,169,185,186,190,241,243, 245–248
Human Herpes Virus (HHV-6) viremia, 186
humanized murine model of MS, 31
hypocretin (=orexin) 2 receptor, 30,32

IgG, 109,112,113,176
IgM detection, 188
immune system, 98,175
immunization, 143
immunogenic peptides, 216
immunohistochemical staining, 190
in situ hybridization, 211,213
inducible nitric oxide synthase (iNOS), 216

infection(s), 57,58,90–93,123–127, 147,149,150,153,154,158,231–236
infectious mononucleosis, 173
infectious pathogens, 209
inflammation, 79,83,86,216,251
influenza, 102,143,144,150,151
influenza syndrome, 102
influenza vaccine, 143,144
infulenza vaccine and relapse, 143,
influenza virus, 115
informed consensus, 98
interferon-β (IFN-β), 51,251
interferon-γ (IFN-γ), 50,211,213
interleukin-1 (IL-1), 50,211,13
interleukin-2 (IL-2), 212
interleukin-4 (IL-4), 50
interleukin-6 (IL-6), 213
interleukin-10 (IL-10), 50,51
interleukin-12 (IL-12), 251
interleukin-1RA, 50,51,252,254
interleukin-1RA intron 2, 254
intrathecal immunity, 109–113

linkage disequilibrium, 26
linkage tests, 41,43,44
logbook, 98
longitudinal study, 98,187
lymphocytes, 26,171
lymphotoxin, 50
magnetic resonance imaging (MRI), 98,144,178
matrix metalloproteinases (MMP), 210,211,214,216
MCP-1, 214
measles, 167,217
messenger RNA, 97
MHC proteins, 61,63–65,67,69
microglial cells, 210
migration studies in MS, 164
molecular analysis, 217
molecular genetics, 37
molecular mimicry, 123–125,216
motifs, 217
myelin antigen, 72

myelin basic protein (MBP), 31,38, 212,215
myelin epitopes, 216
Myelin Oligodendrocyte Glycoprotein (MOG), 3,73
narcolepsy, 30,32
narcoleptics, 33
natural killer cells, 68,68
neuroinvase human coronaviruses, 209,251,217
neuronal, 210
neuropsychological evaluation, 98
neurotropic viruses, 109,110,112,209
neurovirulence, 215
nitric oxide (NO), 210,214,216
nitric oxide synthase, 210
non-replicative viral vectors, 71
nucleocapsid (N) protein, 213
nucleoside analogues, 242,246–248

OC43, 210
occurrence of MS, 165
oligoclonal IgG bands 109,112,113
oligodendrocyte(s), 71
oligodendrocyte progenitor cells, 74
oligodendrocyte-myelin unit, 216
orexin, 30,32

pathogenesis of MS, 163
peripheral blood, 97,170
peripheral blood mononuclear cells, 170
persistent infection, 57,58
plasmid DNA, 72
plasmids, 72
point mutations, 215
Polymerase Chain Reaction (PCR), 97, 153,156,158,188
polymorphism, 254
prevalence of MS, 164
primary progressive (PP), 51
proinflammatory cytokines, 73
Proto Lipid Protein (PLP), 212,215

quantitative PCR, 158
quantitative realtime, 153,156

recessive models for MS, 29
relapse remitting (RR) course, 51
relapse remitting (RR) MS, 97
retrograde labelling experiments, 83
retroviral vectors, 72
retrovirus, 168,195–196
rhesus monkeys, 73
ribonuclases, 99
risk ratio, 173
RNA, 97,99,211
RNA extraction, 211

S protein, 215
safety, 136
seasonal variation in retrobulbar neutritis, 89,92
secondary progressive (SP), 51
semi-quantitative RT-PCR, 211
sensitivity, 213
seropositive for EBV, 172
serum, 98,170
socio-economic status, 166
southern hybridization, 213
specificity, 213
spinal cord, 79,80,84,85
spinal cord atrophy, 80,85
spinal cord demyelination, 80,85
subacute HHV-6 leukoencephalitis, 190
subclinical infections, 171
suppressor cells, 31
susceptibility genes, 41
susceptibility to MS, 25,30,164
synthetic peptides, 217

T-cell(s) 123–127,215,216
T-cell clones, 215,216

T-cell cross reactivity, 215
teeth extraction, 102
TGF-β, 50,51
Theiler's virus, 57,58
therapeutic strategy, 217
therapy, 72,231,251
transformation of lymphocytes, 173
transforming growth factor-α, 50
transmission, 175
treatment, 53,167
triggering factors, 97
trimolecular TCR-MHC-peptide complex, 217
tumor necrosis factor (TNF), 214
tumor necrosis factor-α (TNF-α), 214
tumor necrosis factor-β (TNF-β), 214
twin studies, 165

U 373MG, 211
upper respiratory tract, 215
upregulation of TNF-α mRNA expression, 216
urine, 98
vaccination, 121,123,126,136,137, 148–151
vaccinia virus, 72
valaciclovir, 178
viral persistence, 212
viron(s), 195–197,199,201,204
virulence, 175
virus, 57,58,72,79,83,85,115,167,169, 171,211,217
virus-like infections, 90–93
virus isolation, 177
(VV)-derived vectors, 72

Wechsler Memory Scale, 99

zymography, 211